20/1/06

03/ (29)

Pitman Research Notes in Mathematics Series

AF772744

Submission of proposals for consideration
Suggestions for publication, in the form of outlines and representative samples, are invited by the Editorial Board for assessment. Intending authors should approach one of the main editors or another member of the Editorial Board, citing the relevant AMS subject classifications. Alternatively, outlines may be sent directly to the publisher's offices. Refereeing is by members of the board and other mathematical authorities in the topic concerned, throughout the world.

Preparation of accepted manuscripts
On acceptance of a proposal, the publisher will supply full instructions for the preparation of manuscripts in a form suitable for direct photo-lithographic reproduction. Specially printed grid sheets can be provided and a contribution is offered by the publisher towards the cost of typing. Word processor output, subject to the publisher's approval, is also acceptable.

Illustrations should be prepared by the authors, ready for direct reproduction without furtner improvement. The use of hand-drawn symbols should be avoided wherever possible, in order to maintain maximum clarity of the text.

The publisher will be pleased to give any guidance necessary during the preparation of a typescript, and will be happy to answer any queries.

Important note
In order to avoid later retyping, intending authors are strongly urged not to begin final preparation of a typescript before receiving the publisher's guidelines. In this way it is hoped to preserve the uniform appearance of the series.

Longman Group Ltd
Longman House
Burnt Mill
Harlow, Essex, CM20 2JE
UK
(Telephone (0) 1279 426721)

Titles in this series. A full list is available from the publisher on request.

301 Generalized fractional calculus and applications
V Kiryakova
302 Nonlinear partial differential equations and their applications. Collège de France Seminar Volume XII
H Brezis and J L Lions
303 Numerical analysis 1993
D F Griffiths and G A Watson
304 Topics in abstract differential equations
S Zaidman
305 Complex analysis and its applications
C C Yang, G C Wen, K Y Li and Y M Chiang
306 Computational methods for fluid-structure interaction
J M Crolet and R Ohayon
307 Random geometrically graph directed self-similar multifractals
L Olsen
308 Progress in theoretical and computational fluid mechanics
G P Galdi, J Málek and J Necas
309 Variational methods in Lorentzian geometry
A Masiello
310 Stochastic analysis on infinite dimensional spaces
H Kunita and H-H Kuo
311 Representations of Lie groups and quantum groups
V Baldoni and M Picardello
312 Common zeros of polynomials in several variables and higher dimensional quadrature
Y Xu
313 Extending modules
N V Dung, D van Huynh, P F Smith and R Wisbauer
314 Progress in partial differential equations: the Metz surveys 3
M Chipot, J Saint Jean Paulin and I Shafrir
315 Refined large deviation limit theorems
V Vinogradov
316 Topological vector spaces, algebras and related areas
A Lau and I Tweddle
317 Integral methods in science and engineering
C Constanda
318 A method for computing unsteady flows in porous media
R Raghavan and E Ozkan
319 Asymptotic theories for plates and shells
R P Gilbert and K Hackl
320 Nonlinear variational problems and partial differential equations
A Marino and M K V Murthy
321 Topics in abstract differential equations II
S Zaidman
322 Diffraction by wedges
B Budaev
323 Free boundary problems: theory and applications
J I Diaz, M A Herrero, A Liñan and J L Vazquez
324 Recent developments in evolution equations
A C McBride and G F Roach
325 Elliptic and parabolic problems: Pont-à-Mousson 1994
C Bandle, J Bemelmans, M Chipot, J Saint Jean Paulin and I Shafrir
326 Calculus of variations, applications and computations: Pont-à-Mousson 1994
C Bandle, J Bemelmans, M Chipot, J Saint Jean Paulin and I Shafrir
327 Conjugate gradient type methods for ill-posed problems
M Hanke
328 A survey of preconditioned iterative methods
A M Bruaset
329 A generalized Taylor's formula for functions of several variables and certain of its applications
J-A Riestra
330 Semigroups of operators and spectral theory
S Kantorovitz
331 Boundary-field equation methods for a class of nonlinear problems
G N Gatica and G C Hsiao
332 Metrizable barrelled spaces
J C Ferrando, M López Pellicer and L M Sánchez Ruiz

J C Ferrando
M López Pellicer
L M Sánchez Ruiz
Universidad Politécnica de Valencia, Spain

Metrizable barrelled spaces

LONGMAN

Copublished in the United States with
John Wiley & Sons Inc., New York.

Longman Group Limited
Longman House, Burnt Mill, Harlow
Essex CM20 2JE, England
and Associated Companies throughout the world.

Copublished in the United States with
John Wiley & Sons Inc., 605 Third Avenue, New York, NY 10158

First published 1995

AMS Subject Classifications: (Main) 46A08, 46E30, 03E10
(Subsidiary) 46A30, 46E40, 46G10

ISSN 0269-3674

ISBN 0 582 28703 0

British Library Cataloguing in Publication Data

A catalogue record for this book is
available from the British Library

Library of Congress Cataloging-in-Publication Data

Ferrando, J. C.
Metrizable barrelled spaces / J.C. Ferrando, M. Lopez Pellicer, and L.M. Sanchez Ruiz.
p. cm. -- (Pitman research notes in mathematics series, ISSN 0269-3674 ; ???)
1. Barrelled spaces. 2. Vector-valued measures. I. Lopez Pellicer, M. II. Sanchez Ruiz, L. M. III. Title. IV. Series.
QA325.F47 1995
515'.73--dc20 95-30381
CIP

Printed and bound in Great Britain
by Biddles Ltd, Guildford and King's Lynn

A mis padres

Al Profesor D. Manuel Valdivia, mi maestro

A mis padres y hermanas

Contents

Preface

The purpose of these notes is to collect a number of recent results concerning barrelled locally convex — usually metrizable — spaces going from general facts involving cardinality and dimensionality to barrelledness conditions — even strong barrelledness properties — of some common vector-valued or scalar-valued normed spaces of functional analysis. We also present some consequences in measure theory deduced from the strong barrelledness properties that the space $\ell_0^\infty(\Sigma)$ of all scalar-valued Σ-simple functions defined on a set Ω, Σ being a σ-algebra of subsets of Ω, equipped with the supremum-norm, satisfies.

We feel these notes may be of interest to researchers and graduate students interested in locally convex spaces, normed spaces and measure theory.

In Chapter 1 we introduce barrelled spaces along with Baire-like, totally barrelled and unordered Baire-like spaces, their properties and importance in some basic results of functional analysis. In Chapter 2 we obtain some properties concerning dimensionality and cardinality in metrizable barrelled spaces, such as determining exactly the minimal size of a fundamental system of bounded subsets in a metrizable space, according as it is trivial, non-trivial and normable or non-normable, or stating precisely the minimal dimension of an infinite-dimensional metrizable barrelled space. In Chapter 3, we study the strictly decreasing classes of barrelled spaces of class n, giving distinguishing examples in the realm of metrizable locally convex spaces.

If a space E is barrelled of class n for every positive integer n, then E is said to be barrelled of class $\aleph_0$. Properties and distinguishing examples of this class of spaces are studied in Chapter 4. In Chapter 5 we study the space $\ell_0^\infty(\Sigma)$ by showing it is barrelled of class $\aleph_0$. However, in Chapter 6 we demonstrate that the space $\ell_0^\infty(\Sigma)$ is neither ultrabarrelled nor ultrabornological when Σ is an infinite σ-algebra and it may even fail to be barrelled for some classes of algebras. In Chapter 7 we apply the results of Chapter 5 to obtain some results in measure theory. In fact, we show how to obtain a generalization of Diestel and Faires' theorem that guarantees that some finitely additive vector measures are countably additive and have their range located, and how to strengthen the Nikodym-Grothendieck fundamental boundedness principle. In Chapter 8 we investigate strong barrelledness properties of the space of p-Bochner integrable X-valued functions defined on $L^p(\mu, X), 1 \leq p \leq \infty$, X being a normed space and (Ω, Σ, μ) a positive finite measure space. In Chapter 9 we study some properties of the space $\ell_0^\infty(\Sigma, E)$ of all Σ-simple functions on a set Ω, Σ being an infinite algebra of subsets of Ω, with values in a locally convex space E, as well as of the space $B(\Sigma, X)$ of all X-valued functions that are the uniform limit of a sequence of Σ-simple X-valued functions, both provided with the supremum-norm and X being a normed space. In Chapter 10 we make an extensive study of the Pettis integrable function space, with a special emphasis on its barrelledness properties. We finish by listing some open problems in Chapter 11. Proper credit is given at the end of each chapter, in the Notes and Remarks section.

Prof. L.M. Sánchez Ruiz wishes to thank DGICYT for support at the University of Florida where he has been able to work with Prof. S. Saxon on several occasions during 1992-95, as well as to UF and particularly to Prof. Murali Rao for their hospitality. Chapter 2 of this monograph is part of the fruit of these stays. A complement of this book collecting part of S. Saxon's work concerning barrelledness is M. Kunzinger's monograph [Ku], no. 298 in this series. In order to get a general view of the latest advances in strong barrelledness conditions, we also suggest the excellent monographs [V9] by M. Valdivia and [PB2] by P. Pérez Carreras and J. Bonet. Anyway, this book is self-contained with all proofs included and no references are given in the text. Readers with some knowledge on barrelled spaces may skip Chapter 1.

The authors are indebted to Profs M. Valdivia and S. Saxon who pioneered the research in this field and whose work has been the seed of a number of ideas developed throughout these pages. Finally, we are grateful to Dr. Saxon for his encouragement and helpful comments that have contributed to the final version of some parts of this manuscript.

Juan Carlos Ferrando
Manuel López Pellicer
Luis M. Sánchez Ruiz

1. Some facts about barrelled spaces

1.1 BARRELLED SPACES

In this chapter we will establish the general framework we are going to work in, and recall some of the principles that make barrelled spaces play a star role in some basic results of locally convex topological vector space theory, along with their main properties. First we recall some of the fundamentals.

A Hausdorff topological vector space E over the field $\mathbb{K}$ of the real or complex numbers is said to be a *locally convex space* if its topology τ is defined by a family of seminorms. This is equivalent to saying that τ has a base of (closed) absolutely convex neighbourhoods of the origin.

Definition 1.1.1 *A barrel of a locally convex space* E *is a subset of* E *which is closed, absolutely convex and absorbing.*

Therefore of some note is the fact that any locally convex space E has a base of neighbourhoods of the origin formed by barrels. But, in general, not every barrel needs to be a neighbourhood of the origin. When this happens we say that the space E is *barrelled*.

From now onwards, we shall use the word *space* to mean *locally convex space*. Let us recall that the Hahn-Banach theorem provides us with *many* linear continuous forms defined on a space E. In fact, the set of these linear continuous forms is the topological dual E' of E, and the pair (E,E') is a dual pair; this means that if $f(x) = 0$ for every $f \in E'$, then x must be 0, and that if $f(x) = 0$ for every $x \in E$, then f must be the null form. The latter is clear and the former follows directly from the Hahn-Banach theorem. Also the Hahn-Banach theorem provides the theorem of the bipolars, according to which if U is an absolutely convex subset of E, then the polar set of U,

$$U^{\circ} = \{f \in E' : |f(x)| \leq 1 \text{ for each } x \in U\},$$

verifies that $U^{\circ\circ} = U$.

Given a dual pair (E,E') we may consider several topologies on E. For the moment, we are just going to recall three of them:

- The weak topology, $\sigma(E,E')$, in which the polars of the finite subsets of E' are a base of neighbourhoods of the origin of E. This is the topology of the pointwise convergence over E'. The topology $\sigma(E,E')$ is weaker than or equal to the original topology of E, but the topological dual of $E(\sigma(E,E'))$ is E'. Sometimes $\sigma(E,E')$ is also called the topology of the uniform convergence on the finite subsets of E'.
- The strong topology, $\beta(E,E')$, in which a base of neighbourhoods of the origin is formed by taking the polars of all $\sigma(E',E)$-bounded subsets of E'. It is the topology of the uniform convergence on the $\sigma(E',E)$-bounded subsets of E'. This topology is finer than or equal to the original topology of E, and the topological dual of $E(\beta(E,E'))$ contains E'. The strong topology on E is also named the topology of the uniform convergence on the family of all $\sigma(E',E)$-bounded subsets of E'.
- The Mackey topology, $\mu(E,E')$, in which a base of neighbourhoods of the origin is formed by taking the polars of all absolutely convex $\sigma(E',E)$-compact subsets of E'. It is the topology of the uniform convergence on the absolutely convex $\sigma(E',E)$-compact subsets of E', together with their subsets. This topology is the finest locally convex topology on E compatible with the dual pair (E,E').

Since a subset V of a space E is a barrel if and only if V° is bounded in $\sigma(E',E)$ the following is clear.

Proposition 1.1.2 *For a space* $E(\tau)$ *the following assertions are equivalent:*

(i) $E(\tau)$ *is barrelled.*

(ii) $\tau = \beta(E,E')$.

(iii) Each $\sigma(E',E)$*-bounded subset of* E' *is equicontinuous.*

A subset of a space E is said to be *bornivorous* if it absorbs every bounded subset of E. A space E is said to be *quasibarrelled* if every bornivorous barrel of E is a neighbourhood of the origin. In a metrizable space every bornivorous barrel is a neighbourhood of the origin, thus every metrizable space is quasibarrelled, and clearly every barrelled space is quasibarrelled. A space E is said to have the Banach-Mackey property if every subset of E' that is pointwise bounded on E is uniformly bounded on every bounded subset of E, or equivalently if every $\sigma(E',E)$-bounded subset of E' is $\beta(E',E)$-bounded. Thus any quasibarrelled space with the Banach-Mackey property is barrelled. In fact, metrizable barrelled spaces are characterized by this property, that is, if E is a metrizable space then E is barrelled if and only if every $\sigma(E',E)$-bounded subset is $\beta(E',E)$-bounded.

In a barrelled space, every $\sigma(E',E)$-bounded subset of E', being equicontinuous, is weak* compact, and as each absolutely convex weak* compact subset of E' is $\sigma(E',E)$-bounded it follows that in each barrelled space its topology coincides with both the strong $\beta(E,E')$ and the Mackey $\mu(E,E')$ topologies.

Proposition 1.1.3 *Let* $\{E_n: n \in \mathbb{N}\}$ *be an increasing sequence of subspaces of* E. *If* E *is barrelled, then* E *is the inductive limit of the* $\{E_n: n \in \mathbb{N}\}$.

Proof. Let v be a continuous linear form defined on the inductive limit $E(\eta)$, and let v_n be the restriction of v to E_n. The Hahn-Banach theorem enables us to extend v_n to a continuous linear mapping $w_n \in E'$. Clearly $\{w_n(x): n \in \mathbb{N}\}$ converges to v(x) for each $x \in E$. The barrelledness of E implies that $\{w_n: n \in \mathbb{N}\}$ is a $\sigma(E',E)$-relatively compact subset of E'. Thus $v \in E'$ and hence $E' = E(\eta)'$. Then, as E and $E(\eta)$ are two Mackey spaces with the same dual, both topologies coincide, and consequently E is the inductive limit of the sequence $\{E_n: n \in \mathbb{N}\}$. ■

In section 1.2 we will see that every *Baire locally convex space*, i.e. each locally convex space which cannot be covered by a sequence of closed nowhere dense subsets, is barrelled. In particular every Fréchet space is barrelled, so every ℓ^p space, $1 \le p \le \infty$, is barrelled. However the subspace φ of the finite coordinate vectors $(\xi_1,\xi_2,\ldots,\xi_n,0,0,\ldots)$ is not barrelled since if that were the case, according to the previous result, its topology would be the finest locally convex topology.

The characterization of barrelled spaces by means of the equicontinuity of every subset H of E' such that $\{h(x): h \in H\}$ is bounded for every $x \in E$ admits the following generalization. Given two spaces E and F we denote by $\mathscr{L}(E,F)$ the linear space over $\mathbb{K}$ of all linear continuous mappings from E into F. A subset H of $\mathscr{L}(E,F)$ is said to be pointwise bounded if $\{h(x): h \in H\}$ is bounded for each $x \in E$.

Theorem 1.1.4 (Banach-Steinhaus) *The space* E *is barrelled if and only if for every space* F, *each pointwise bounded subset* H *of* $\mathscr{L}(E,F)$ *is equicontinuous.*

Proof. Let us suppose that H is a pointwise bounded subset of $\mathscr{L}(E,F)$. Given a neighbourhood of zero V in F, the closed absolutely convex set

$$U = \{x \in E: f(x) \in V, \text{ for each } f \in H\}$$

is a barrel, since the pointwise boundedness of H implies that if $y \in E$, then $\{f(y): f \in H\}$ is bounded, and therefore there exists a $\lambda > 0$ such that $\lambda\{f(y): f \in H\} \subseteq V$. Thus, $\lambda y \in U$ and U is absorbing.

Hence if E is barrelled, U is a neighbourhood of the origin in E, and H is equicontinuous.

The converse is clear. ■

Using now the original definition of barrelled spaces we are going to obtain a new characterization of them. Given the spaces E and F, the linear mapping T from E into F is said to be *nearly continuous* if for every neighbourhood V of the origin in F, $\overline{T^{-1}(V)}$ is a neighbourhood of the origin in E.

Proposition 1.1.5 *The space* E *is barrelled if, and only if, every linear mapping* T *from* E *into any space* F *is nearly continuous.*

Proof. If E is barrelled and V is a neighbourhood of the origin in F we have that the barrel $\overline{T^{-1}(V)}$ is a neighbourhood of the origin in E.

Therefore T is nearly continuous.

Conversely, if E were not barrelled there would be a barrel V in E which is not a neighbourhood of the origin. Then the identity i from E into $E(\beta(E,E'))$ is not nearly continuous since $i^{-1}(V) = V$ is not a neighbourhood of the origin in E. ■

Remark 1.1.6 A mapping f between the spaces E and F is said to have a *closed graph* if its graph, the set $\{(x,f(x)): x \in E\}$, is a closed subset in $E \times F$. Two well-known characterizations of when a linear mapping $T : E \longrightarrow F$ has a closed graph are: 1) T has a closed graph if, and only if, the space F has a locally convex topology τ weaker than the original on F such that $T : E \longrightarrow F(\tau)$ is continuous, and 2) T has a closed graph if, and only if, there exists a total subset H in $F'(\sigma(F',F))$, i.e. with a linear span $\sigma(F',F)$-dense in F', such that uT $\in E'$ for every $u \in H'$.

It is immediate that every linear continuous mapping $T : E \longrightarrow F$ has a closed graph, and from the Banach-Schauder theorem it follows that if E and F are Banach (or Fréchet) spaces, and T has a closed graph then T is continuous. When a situation like this holds, in which a linear mapping T with a closed graph is continuous, we say that T satisfies a *closed graph theorem*.

Remark 1.1.7 From Pták's homomorphism theorem, it follows that if E is a space, F is a Fréchet space (or more generally a Pták space) and $T : E \longrightarrow F$ is a nearly continuous linear mapping with a closed graph, then T is continuous.

Next we see that barrelled spaces comprise the class of locally convex spaces characterized by satisfying the closed graph theorem as domain class when Fréchet or Banach spaces are considered in the range.

Theorem 1.1.8 *Let* E *be a space. The following are equivalent:*
(i) E *is barrelled.*
(ii) Given any Fréchet space F, *if* $T : E \longrightarrow F$ *is a linear mapping with*

a closed graph in $E \times F$, *then* T *is continuous.*

(iii) Given any Banach space F, *if* $T : E \longrightarrow F$ *is a linear mapping with a closed graph in* $E \times F$, *then* T *is continuous.*

Proof. (i) $\Rightarrow$ (ii). The barrelledness of E implies that T is nearly continuous. Then Pták's homomorphism mapping theorem gives that T is continuous.

(ii) $\Rightarrow$ (iii). This is immediate.

(iii) $\Rightarrow$ (i). Let T be a barrel in E, $N := \bigcap\{n^{-1}T : n \in \mathbb{N}\}$, F the completion of E/N with the norm defined by the gauge of $k(T)$, and k the canonical mapping of E onto E/N. As T is closed, $k^{-1}(k(T)) = T + N = T$. Therefore if k, considered as a map from E into F, has a closed graph, T will be a neighbourhood of the origin in E and we will be done.

Let $G(k)$ denote the graph of k in $E \times F$, and choose $(x,y) \in \overline{G(k)}$. Then, denoting by B the completion of $k(T)$ in F, i.e. the closed unit ball of F, one has $[(x,y) + U \times \varepsilon B] \cap G(k) \neq \emptyset$ for each neighbourhood of the origin U in E and for all $\varepsilon > 0$. From this relation, it follows that $k(x) - y \in k(U) + \varepsilon B$. Now there is a $z_\varepsilon \in E$ so that $k(z_\varepsilon) \in y + \varepsilon B$ and consequently $k(x) - k(z_\varepsilon) \in (k(U) + 2\varepsilon B) \cap E/N \subseteq k(U) + 2\varepsilon \overline{k(T)}^{E/N} \subseteq k(U) + 3\varepsilon k(T)$. Hence, $x - z_\varepsilon \in U + 3\varepsilon T + N \subseteq U + 4\varepsilon T$ for all U, whereby $x - z_\varepsilon \in 4\varepsilon T$. So $k(x) - y \in k(x) - k(z_\varepsilon) + \varepsilon B \subseteq 4\varepsilon k(T) + \varepsilon B \subseteq 5\varepsilon B$. Thus as this is true for all $\varepsilon > 0$, then $y = k(x)$, i.e, $(x,y) \in G(k)$. ∎

In Exercise 1 of Chapter 5 we will consider Valdivia's Γ_r-spaces. For the moment, let us only mention that these spaces satisfy the condition that if E is a barrelled space, F a Γ_r-space and $T : E \longrightarrow F$ is a linear mapping with a closed graph in $E \times F$, then T is continuous, and if F is not a Γ_r-space, there is a barrelled space E and a non-continuous linear mapping $T : E \longrightarrow F$ with a closed graph. Therefore, Γ_r spaces form a maximal range class in the closed graph theorem when barrelled spaces are located in the domain.

Although Baire spaces do not have good permanence properties, the situation with barrelled spaces is quite different as we will have an opportunity to see in the later results.

Proposition 1.1.9 *Let* F *be a closed subspace of* E. *If* E *is barrelled, then* E/F *is barrelled.*

Proof. Let k the canonical mapping from E into E/F and let V be a barrel in E/F. Then $k^{-1}(V)$ is a barrel in E, and is a neighbourhood of the origin in E. Thus $k(k^{-1}(V))$ is a neighbourhood of zero in E/F. ■

Proposition 1.1.10 *Let* F *be a dense subspace of* E. *If* F *is barrelled, then* E *is barrelled.*

Proof. Let V be a barrel in E. Then the set $V \cap F$ is a neighbourhood of zero in F. Hence the closed set V is a neighbourhood of zero in E. ■

In order to prove that the product of a family of barrelled spaces is barrelled we need the following lemma.

Lemma 1.1.11 *Let* B *be a compact absolutely convex subset of the space* E. *If* V *is a barrel in* E, *then there is some* $\lambda > 0$ *such that* $\lambda B \subseteq V$.

Proof. Let E_B be the Banach space determined by the linear span of B, equipped with the gauge of B as norm. Then $V \cap E_B$ is a barrel in the barrelled space E_B. So there is a $\lambda > 0$ such that $\lambda B \subseteq V$. ■

Remark 1.1.12 As a consequence of the previous lemma, if V is a barrel in the product E of a family $\{E_i: i \in I\}$ of spaces, then there is a cofinite subset J of I such that the product of the family of spaces $\{E_i: i \in J\}$ is contained in V. As the algebraic direct sum $\oplus\{E_i: i \in J\}$ is dense in the topological product $\prod\{E_i : i \in J\}$ we need only show that each E_i, $i \in J$, is contained in V. If this were not so, then we would be able to find a sequence of spaces in $\{E_i: i \in I\}$, which we will denote by $\{E_n: n \in \mathbb{N}\}$, and a vector x_n in each space E_n such that, considering x_n as a vector of E, $x_n \notin nV$. Setting $K_n = \{\lambda x_n : |\lambda| \leq 1\}$ and $K = \prod\{K_n: n \in \mathbb{N}\}$, then we have that the barrel V does not absorb the compact set K, in contradiction with Lemma 1.1.11.

Proposition 1.1.13 *The topological product* E *of a family* $\{E_i: i \in I\}$ *of barrelled spaces is barrelled.*

Proof. Let V be a barrel of E. From the preceding remark we have that there is a cofinite subset J in I such that $\prod\{E_i: i \in I \setminus J\} \subseteq (1/2)V$. If the cardinality of $I \setminus J$ is p, then $(2p)^{-1}V$ contains a neighbourhood of the origin V_i in E_i for each $i \in I \setminus J$. Thus V is a neighbourhood of the origin in E since it contains $\prod\{F_i: i \in I\}$, where $F_i := E_i$ for $i \in J$, and $F_i := V_i$ for $i \in I \setminus J$. ■

In general, subspaces of barrelled spaces do not need to inherit barrelledness — for instance after 1.1.3 we have seen that the Hilbert space ℓ^2 contains the subspace φ which is not barrelled. However, we will prove next that countable-codimensional subspaces of a barrelled space certainly are barrelled. First we note the following

Lemma 1.1.14 *Let* F *be a hyperplane in a space* E. *If* U *is a barrel in* F, *then there exists some barrel* V *in* E *such that* $V \cap F = U$.

Proof. If U is a barrel in the hyperplane F, two cases may occur:

1. U is not closed in E. Then the closure V of U in E is a barrel in E such that $V \cap F = U$.
2. U is closed in E. Take $x \in E \setminus F$ and consider $K = \{\lambda x, |\lambda| \leq 1\}$. Then the set $V := U + K$ is a barrel in E such that $V \cap F = U$. ■

Proposition 1.1.15 *Let* F *be a countable-codimensional subspace of* E. *If* E *is barrelled, then* F *is barrelled.*

Proof. Let $\{x_n: n \in \mathbb{N}\}$ be a cobasis of F in E, and let F_n be the linear span of $F \cup \{x_1, x_2, \ldots, x_n\}$. Given a barrel U_0 in F, the preceding lemma enables us to determine a barrel U_1 in F_1 such that $U_1 \cap F = U_0$ and, for each positive integer $n > 1$, a barrel U_n in F_n such that $U_n \cap F_{n-1} = U_{n-1}$. We define $U := \bigcup\{U_n: n \in \mathbb{N}\}$; then $\bar{U}$ is a barrel in E. Now it is easily seen that if we show $\bar{U} \subseteq 2U$ then U will be a neighbourhood of the origin in E, so $U_0 = U \cap F$ will be a neighbourhood of the origin in F, and the proof would be over.

So to prove $\bar{U} \subseteq 2U$ is the issue at hand. Let us suppose that $x \notin 2U$. As $E = \bigcup\{F_n: n \in \mathbb{N}\}$, there is some $p \in \mathbb{N}$ such that $x \in F_m$ for $m \geq p$. The Hahn-Banach separation theorem applied to these F_m and the Hahn-Banach extension theorem enable us to find $v_m \in E'$ such that $v_m(x) = 2$

and $|v_m(y)| \leq 1$ for all $y \in U_n$. But as E is barrelled, the $\sigma(E',E)$-bounded set $\{v_m: m \in \mathbb{N}\}$ has a weakly adherent point $u \in E'$ such that $u(x) = 2$ and $|u(y)| \leq 1$ for every $y \in U$. So we have that $u \notin \bar{U}$. ■

1.2 BAIRE-LIKE SPACES

When E is a locally convex space, even a topological vector space, each scalar multiple of a nowhere-dense subset is nowhere-dense; this is also true of the translation of a nowhere-dense set. Therefore, if an open subset of E is covered by a countable family of nowhere-dense subsets, then E is covered by a countable (increasing) family of (closed) nowhere-dense subsets. Consequently, we have that a topological vector space E is a Baire space if and only if, given an increasing covering $\{A_n: n \in \mathbb{N}\}$ of E by closed sets, there is an A_p with non-empty interior. The condition $\bigcup\{A_n: n \in \mathbb{N}\} = E$ may be replaced by requiring $\bigcup\{A_n: n \in \mathbb{N}\}$ to be absorbing in E.

Closely related to Baire spaces are the so-called Baire-like spaces. A locally convex space E is said to be *Baire-like* if, given an increasing covering $\{A_n: n \in \mathbb{N}\}$ of E by closed absolutely convex subsets of E, there exists an A_p which is a neighbourhood of zero. We may replace $\bigcup\{A_n: n \in \mathbb{N}\} = E$ by $\bigcup\{A_n: n \in \mathbb{N}\}$ to be absorbing in E. Note that if A_p has non-void interior then it is a neighbourhood of the origin in E. Hence it is clear that every Baire space is Baire-like.

Every Baire-like space E is barrelled since if T is a barrel in E and we write $A_n = nT$, $n \in \mathbb{N}$, then some pT is a neighbourhood of zero; therefore T is a neighbourhood of zero and E is barrelled.

Proposition 1.2.1 *Let* E *be a barrelled space. If the completion of* E *is Baire, then* E *is Baire-like.*

Proof. To see why this is so, we only need to show that if $\{A_n: n \in \mathbb{N}\}$ is an increasing sequence of closed absolutely convex subsets of E covering E and $\bar{A}_n$ denotes the closure of A_n in the completion $\hat{E}$ of E, then $\hat{E} = \bigcup\{\bar{A}_n: n \in \mathbb{N}\}$. In fact then for $\hat{E}$ being a Baire space, some $\bar{A}_p$ must be a neighbourhood of the origin in $\hat{E}$, and therefore A_p will be a neighbourhood of the origin in E.

Suppose that there exists some point $x \in \hat{E} \setminus \bigcup\{\bar{A}_n : n \in \mathbb{N}\}$. Then for each n there exists some $u_n \in E'$ such that $u_n(A_n) = 0$ and $u_n(x) = 1$. The sequence $\{u_n : n \in \mathbb{N}\}$ converges in $\sigma(E',E)$ to the null linear form. Therefore, $\{u_n : n \in \mathbb{N}\}$ is an equicontinuous set in E and also in $\hat{E}$, which is not possible since if v is an adherent point of $\{u_n : n \in \mathbb{N}\}$ then $v(x) = 1$ and $v(y) = 0$ for every $y \in E$. The last equality yields $v(x) = 0$, a contradiction. ■

Next we will see a localization closed graph theorem. A locally convex space (F,τ) is said to be an (LB)-space it can be represented as the topological inductive limit of a properly increasing sequence

$$F_1(\tau_1) \subseteq F_2(\tau_2) \subseteq \ldots \subseteq F_n(\tau_n) \subseteq \ldots$$

of Banach spaces.

Theorem 1.2.2 *Let* E *be a Baire-like space and let* $F(\tau) := \text{ind}\{F_n(\tau_n) : n \in \mathbb{N}\}$ *be an (LB)-space. If the linear mapping* $T : E \longrightarrow F(\tau)$ *has a closed graph, then there exists a positive integer p such that* $T : E \longrightarrow F_p(\tau_p)$ *is continuous.*

Proof. There is an increasing sequence $\{B_n : n \in \mathbb{N}\}$ such that each B_n is a bounded absolutely convex neighbourhood of the origin in $F_n(\tau_n)$. In fact, let B_1 be the unit ball in $F_1(\tau_1)$. Assuming we have obtained B_i, $1 \leq i \leq n$ for some $n \in \mathbb{N}$, let B_{n+1} be the absolutely convex hull of the union of B_n and the unit ball of $F_{n+1}(\tau_{n+1})$. Then B_{n+1} is bounded in $F_{n+1}(\tau_{n+1})$ since, by recurrence, B_n is bounded in $F_n(\tau_n)$, which implies that B_n is bounded in $F_{n+1}(\tau_{n+1})$.

Let us define $E_n := T^{-1}(F_n)$, $C_n := T^{-1}(B_n)$ and let D_n be the closure of C_n in E. The set $\bigcup\{D_n : n \in \mathbb{N}\}$ is absorbing in the Baire-like space E; therefore there exists a D_p which is a neighbourhood of the origin in E with E_p being a dense subset of E. Therefore the restriction of the mapping T to E_p is nearly continuous, since D_p is a neighbourhood of the origin, and T has a closed graph in $E_p \times F_p$. Remark 1.1.7 shows that the restriction of T to E_p is a continuous mapping from E_p into $F_p(\tau_p)$. So, it has a continuous extension S from E into the Banach space $F_p(\tau_p)$. Taking into account now that T has a closed graph, there is a Hausdorff locally convex topology τ^* on F weaker than or equal to τ such that $T : E \longrightarrow F(\tau^*)$ is continuous, and as $S : E \longrightarrow F(\tau^*)$ is continuous, too, and T and S agree on the dense subspace E_p we obtain that T = S, and the conclusion follows. ∎

Next we study some of the properties of Baire-like spaces.

Proposition 1.2.3 *Let* F *be a dense subspace of* E. *If* F *is Baire-like, then* E *is Baire-like.*

Proof. If $\{A_n : n \in \mathbb{N}\}$ is an increasing covering of closed absolutely convex subsets of E, there is some A_p so that $A_p \cap F$ is a neighbourhood of the origin in F. Since F is dense in E, this implies that A_p is a neighbourhood of the origin in E. ∎

Proposition 1.2.4 *Let* F *be a closed subspace of* E. *If* E *is Baire-like then* E/F *is Baire-like.*

Proof. The proof is straightforward. Let k be the canonical mapping from E onto E/F, and suppose that E/F is covered by the increasing sequence $\{A_n : n \in \mathbb{N}\}$ of closed absolutely convex subsets of E/F. Then E is covered by the increasing sequence $\{k^{-1}(A_n) : n \in \mathbb{N}\}$, and there must be some $k^{-1}(A_p)$ which is a neighbourhood of the origin in E. Thus $k(k^{-1}(A_p)) = A_p$ is a neighbourhood of the origin in E/F. ∎

In order to prove that the product of a family of Baire-like spaces is Baire-like we need the following lemma, which is comparable to 1.1.11 in the case of barrelled spaces.

Lemma 1.2.5 *Let* B *be a compact absolutely convex subset of the space* E. *If* E *is covered by the increasing sequence* $\{A_n : n \in \mathbb{N}\}$ *of closed absolutely convex subsets, then there exists some* $p \in \mathbb{N}$ *and* $\lambda > 0$ *such that* $\lambda B \subseteq A_p$.

Proof. Let E_B be the Banach space generated by the linear span of B with the Minkowski functional of B as norm. Then there exists some $p \in$

$\mathbb{N}$ such that A_p contains a neighbourhood of the origin in the Baire-like space E_B. Therefore there exists a $\lambda > 0$ such that $\lambda B \subseteq A_p$. ■

Remark 1.2.6 As a consequence of the previous lemma, it happens that if E is the product of a family $\{E_i: i \in I\}$ of spaces which is covered by an increasing sequence $\{A_n: n \in \mathbb{N}\}$ of closed absolutely convex subsets of E, then reasoning as in 1.1.12, but using 1.2.5 instead of 1.1.11, the same argument given there shows that there is some $p \in \mathbb{N}$ and a cofinite subset J of I such that the product of all the spaces of the family $\{E_i: i \in J\}$ is contained in A_p.

Proposition 1.2.7 *Let* $\{E_i: i \in I\}$ *be a family of Baire-like spaces. Then* $E = \prod\{E_i: i \in I\}$ *is Baire-like.*

Proof. Suppose that E is covered by the increasing sequence $\{A_n: n \in \mathbb{N}\}$ of closed absolutely convex subsets of E. From the preceding remark it follows that there exists some $p \in \mathbb{N}$ and a cofinite subset J of I such that $\prod\{E_i: i \in J\}$ is contained in $(1/2)A_p$. Let us suppose that the index set $I \setminus J$ has p elements. Since every E_i is Baire-like, there exists some $q > p$ such that $(2p)^{-1}A_q$ contains a neighbourhood of the origin V_i in E_i for each $i \in I \setminus J$. Consequently A_q contains $F = \prod\{F_i: i \in I\}$, where $F_i = E_i$ for each $i \in J$, and $F_i = V_i$ for each $i \in I \setminus J$. As F is a neighbourhood of the origin in E, this shows that E is Baire-like . ■

Proposition 1.2.8 *Let* F *be a countable-codimensional subspace of* E. *If* E *is a Baire-like space, then* F *is Baire-like.*

Proof. Suppose that F is covered by the increasing sequence $\{A_n: n \in \mathbb{N}\}$ of closed absolutely convex subsets of F, and that $\{x_n: n \in \mathbb{N}\}$ is a cobasis of F in E. Let B_n be the absolutely convex hull of $A_n \cup \{x_1, x_2, \ldots, x_n\}$. Then $\{B_n: n \in \mathbb{N}\}$ is an increasing sequence of closed absolutely convex subsets of E and $\bigcup\{B_n: n \in \mathbb{N}\}$ is absorbing in E. Therefore, there exists some B_p which is a neighbourhood of the origin in E, and thus $A_p = B_p \cap F$ is a neighbourhood of the origin in F. ■

Let us consider a final result relating barrelled, Baire-like spaces and the space φ endowed with the finest locally convex topology. In order to prove this the following lemma will be helpful.

Lemma 1.2.9 *Let* $\{A_n : n \in \mathbb{N}\}$ *be an increasing sequence of closed absolutely convex subsets of the space* E *such that* $A_{n+1} \setminus \langle A_n \rangle \neq \emptyset$ *for every* $n \in \mathbb{N}$, *where* $\langle A_n \rangle$ *denotes the linear span of* A_n. *Let* $x_n \in A_{n+1} \setminus \langle A_n \rangle$, $S_n := \langle \{x_1, x_2, \ldots, x_n\} \rangle$, $S := \langle \{x_n : n \in \mathbb{N}\} \rangle$, *and* p *be a seminorm on* S. *Then there exists a sequence* $\{f_n : n \in \mathbb{N}\}$, *with* $f_n \in A_n^o$ *such that*

$$q_n(x) := \max \{|f_r(x)| : 1 \leq r \leq n\} \geq (1 + 2^{-n})p(x),$$

for each x *belonging to* S_n *and for each* $n \in \mathbb{N}$.

Proof. Replacing each x_k by a suitable scalar multiple if needed we may suppose that $p(x_k) \leq 1$ and thus $p(\sum_{1\leq i\leq n} a_i x_i) \leq \sum_{1\leq i\leq n}|a_i|$. Hence it is plain that it suffices to prove the lemma when $p(\sum_{1\leq i\leq n} a_i x_i) = \sum_{1\leq i\leq n}|a_i|$.

The Hahn-Banach theorem and $x_1 \notin (1 + 2^{-1})A_1$ imply that there is some $f_1 \in A_1^o$ such that $f_1(x_1) = 1 + 2^{-1}$. Hence

$$q_1(x) = |f_1(x)| \geq (1 + 2^{-1})p(x), \ x \in S_1.$$

Let us suppose that we have determined $f_i \in A_i^o$, with

$$q_i(x) \geq (1 + 2^{-i})p(x), \ x \in S_i, \ 1 \leq i \leq n.$$

Then, our task will be to show that there is some $f_{n+1} \in A_{n+1}^o$ such that

$$q_{n+1}(x) \geq (1 + 2^{-n-1})p(x), \ x \in S_{n+1},$$

and for this it clearly suffices to show that

$$q_{n+1}(x + x_{n+1}) \geq (1 + 2^{-n-1})p(x + x_{n+1}), \ x \in S_n.$$

Observe that

$$\lim_{\alpha\to\infty}[(1 + 2^{-n})\alpha - q_n(x_{n+1})]/(\alpha+1) = 1 + 2^{-n} > 1 + 2^{-n-1}.$$

Consequently, there exists some $M > 1$ such that for $\alpha \geq M$ we have

$$\frac{(1 + 2^{-n})\alpha - q_n(x_{n+1})}{\alpha + 1} > 1 + 2^{-n-1}.$$

Let $N = M + (1 + 2^{-n})\alpha - q_n(x_{n+1})$. The Hahn-Banach theorem, along with $x_{n+1} \notin NA_{n+1}$, yield the existence of $f_{n+1} \in A_{n+1}^o$ with $f_{n+1}(x_{n+1}) = N$.

Now we are going to show that, effectively, one has that

$$q_{n+1}(x + x_{n+1}) \geq (1 + 2^{-n-1})p(x + x_{n+1}), \ x \in S_n.$$

Given $x \in S_n$, there are two cases to be considered:

1. If $M \le p(x) = \alpha$ then, taking into account the recurrence, the fact that the seminorm q_n enjoys $q_{n+1}(z) \ge q_n(z)$, and that $q_{n+1}(x + x_{n+1}) = \max \{q_n(x + x_{n+1}), |f_{n+1}(x + x_{n+1})|\}$, we have that

$$q_{n+1}(x + x_{n+1}) \ge q_n(x + x_{n+1}) \ge q_n(x) - q_n(x_{n+1})$$
$$\ge (1 + 2^{-n})p(x) - q_n(x_{n+1}) = (1 + 2^{-n})\alpha - q_n(x_{n+1}).$$

As we have seen that

$$\frac{(1 + 2^{-n})\alpha - q_n(x_{n+1})}{\alpha+1} \ge 1 + 2^{-n-1}, \ \alpha \ge M,$$

then

$$q_{n+1}(x + x_{n+1}) \ge (1 + 2^{-n-1})(\alpha+1) = (1 + 2^{-n-1})(p(x) + 1)$$
$$\ge (1 + 2^{-n-1})p(x + x_{n+1}).$$

2. Now we consider the case $M > p(x) = \alpha$. From $x_i \in A_{n+1}$, for $i \le n$, and $f_{n+1} \in A^o_{n+1}$ it follows by a direct computation that, according to our assumption, $|f_{n+1}(x)| \le \sum_{1\le i\le n}|a_i| = p(x)$, $x \in S_n$ and then

$$q_{n+1}(x + x_{n+1}) \ge |f_{n+1}(x + x_{n+1})| \ge f_{n+1}(x_{n+1}) - |f_{n+1}(x)| \ge N - M.$$

But $N = M + (1 + 2^{-n})\alpha - q_n(x_{n+1})$, hence

$$q_{n+1}(x + x_{n+1}) \ge (1 + 2^{-n})\alpha - q_n(x_{n+1}).$$

But we have that $\dfrac{(1 + 2^{-n})M - q_n(x_{n+1})}{M + 1} \ge 1 + 2^{-n-1}$, hence

$$q_{n+1}(x + x_{n+1}) > (1 + 2^{-n-1})(M+1) > (1 + 2^{-n-1})(p(x) + 1)$$
$$\ge (1 +2^{-n-1})p(x + x_{n+1}). \blacksquare$$

Corollary 1.2.10 *Under the hypotheses of* 1.2.9 *if E is barrelled and* $\bigcup\{A_n : n \in \mathbb{N}\}$ *is absorbing, then S is isomorphic to* φ.

Proof. For $x \in E$ there exists $m, n \in \mathbb{N}$ such that $x \in mA_n$. Also $x \in mA_k$, for $k \ge n$. Let p be a seminorm on S and let $\{f_n : n \in \mathbb{N}\}$ be the elements of E' obtained in 1.2.9 with $f_n \in A^o_n$. From $x \in mA_k$ it follows that $|f_k(x)| \le m$, when $k \ge n$, and then, by 1.1.2, $\{f_n : n \in \mathbb{N}\}$ is equicontinuous. For $x \in S$ we have that $p(x) \le \sup \{|f_n(x)| : n \in \mathbb{N}\}$, hence p is continuous by equicontinuity. The countable dimensional space S has the finest locally convex topology and, therefore, is isomorphic to φ. ∎

Theorem 1.2.11 *Let* E *be a barrelled space. If* E *does not contain* φ*, then* E *is Baire-like.*

Proof. If the barrelled space E is not Baire-like then there exists an increasing covering $\{B_n: n \in \mathbb{N}\}$ in F such that every B_n is closed and it is not a neighbourhood of the origin in E. Then the linear span of B_n is a proper subspace of E, because if $\langle B_n \rangle = E$ then B_n would be a barrel in the barrelled space E, in contradiction with the fact that B_n is not a neighbourhood of the origin in E.

Let $n(1) = 1$ and let $x_1 \notin \langle B_1 \rangle$. Let $n(2) > n(1)$ be such that $x_1 \in B_{n(2)}$. Let us suppose that $n(1), n(2), \ldots, n(k)$ have been defined such that

$$x_i \in B_{n(i+1)} \setminus \langle B_{n(i)} \rangle \text{ for } 1 \leq i \leq k-1.$$

Let $x \notin \langle B_{n(k)} \rangle$ and let $n(k+1)$ be such that $x \in B_{n(k+1)}$; then if $x_k = x$ and $A_k = B_k$ we have obtained by recurrence the sequences $\{x_n: n \in \mathbb{N}\}$ and $\{A_n: n \in \mathbb{N}\}$ verifying the hypothesis of the preceding corollary. Hence E would contain a copy of φ. ■

1.3 UNORDERED BAIRE-LIKE SPACES

Leaving out the increasing condition in the definition of Baire-like spaces we obtain a particular subclass of spaces. It is said that the space E is *unordered Baire-like* if, given a covering $\{A_n: n \in \mathbb{N}\}$ of E formed by closed absolutely convex subsets, there exists an A_p such that it is a neighbourhood of the origin in E. Let us remark that every Baire space is unordered Baire-like.

In the next section and subsequent chapters we will consider other classes of barrelled spaces E defined in terms of barrelledness properties of sequences of subspaces covering E. We will use the following algebraic lemma in order to obtain a characterization of unordered Baire-like spaces by means of barrelledness properties of sequences of subspaces covering them. First we are going to give an algebraic lemma.

Lemma 1.3.1 *Let* $\mathcal{A} = \{A_n : n \in \mathbb{N}\}$ *and* $\mathcal{B} = \{B_n : n \in \mathbb{N}\}$ *be two countable families of subspaces of the vector space* E *whose union covers* E. *Then one of the two families,* $\mathcal{A}$ *or* $\mathcal{B}$, *covers* E.

Proof. If this were not the case, let us write $A = \bigcup\{A_n : n \in \mathbb{N}\}$ and $B = \bigcup\{B_n : n \in \mathbb{N}\}$. Then there would exist a vector $x \in A \setminus B$ and a vector $y \in B \setminus A$. But as $\{x + \lambda(y-x) : \lambda \in \mathbb{R}\}$ is an uncountable set of vectors we have that some A_p or B_q must contain two of these vectors. If A_p contains these two vectors then the vector space A_p also contains x and y, which is not possible. In the other case, i.e., if B_q contains two of these vectors, we obtain the same contradiction. ∎

Proposition 1.3.2. *For a space* E *the following are equivalent:*

(i) E *is unordered Baire-like.*

(ii) Given a sequence $\{E_n : n \in \mathbb{N}\}$ *of subspaces of* E *which covers* E, *there exists some* E_p *which is barrelled and dense in* E.

Proof. (i) ⇒ (ii). Suppose that (ii) fails and there is a sequence of subspaces $\{E_n : n \in \mathbb{N}\}$ of E which are not dense in E or not barrelled. If E_n is not dense then we write A_n equal to the closure of E_n in E. If E_n is not barrelled and B_n is a barrel in E_n which is not a neighbourhood of the origin then we take A_n equal to the closure of B_n in E. Then $E = \bigcup\{mA_n : m \in \mathbb{N}, n \in \mathbb{N}\}$ and none of the sets mA_n is barrelled. Hence E is not unordered Baire-like.

(ii) ⇒ (i). Let us suppose that $\{B_n : n \in \mathbb{N}\}$ is a sequence of closed absolutely convex subsets of E covering E. Then each B_n is a barrel in its linear span, say E_n. If E_p is barrelled and dense then we have that B_p is a neighbourhood of the origin in E_p. But E_p is dense in E and B_p is closed in E, hence B_p is a neighbourhood of the origin in E, and E is unordered Baire-like. ∎

Corollary 1.3.3 *A space* E *is unordered Baire-like if and only if, given a sequence* $\{E_n : n \in \mathbb{N}\}$ *of subspaces of* E *covering* E, *there exists some subcovering* $\{E_{n_p} : p \in \mathbb{N}\}$ *so that every* E_{n_p} *is barrelled and dense in* E.

Proof. It is a direct consequence of 1.3.1 and 1.3.2. ∎

Corollary 1.3.4 *A space* E *is unordered Baire-like if and only if, given a sequence* $\{E_n : n \in \mathbb{N}\}$ *of subspaces of* E *covering* E, *there exists some* E_p *which is unordered Baire-like.*

Proof. Suppose E is unordered Baire-like and there is a sequence $\{E_n : n \in \mathbb{N}\}$ of subspaces of E covering E such that no E_n is unordered Baire-like. Then Corollary 1.3.3 implies that every E_n has a covering $\{E_{nm} : m \in \mathbb{N}\}$ by subspaces such that every E_{nm} is non-barrelled or non-dense in E_n. Then E is not unordered Baire-like, because none of the subspaces E_{nm}, $n \in \mathbb{N}$ and $m \in \mathbb{N}$, is barrelled and dense in E despite the fact that $E = \bigcup\{E_{nm} : n \in \mathbb{N} \text{ and } m \in \mathbb{N}\}$.

Conversely, if E satisfies the condition of the theorem, applying it with every $E_n = E$, we conclude that E is unordered Baire-like. ■

We are now going to consider some permanence properties of unordered Baire-like spaces.

Proposition 1.3.5 *Let* F *be a dense subspace of* E. *If* F *is unordered Baire-like, then* E *is unordered Baire-like.*

Proof. Let $\{A_n : n \in \mathbb{N}\}$ be a covering of E formed by closed absolutely convex subsets of E. Then there exists some A_p such that $A_p \cap F$ is a neighbourhood of zero in F. Hence A_p is a neighbourhood of zero in E. ■

Proposition 1.3.6 *Let* F *be a closed subspace of a space* E. *If* E *is unordered Baire-like then* E/F *is unordered Baire-like.*

Proof. Let k be the canonical mapping from E onto E/F, and suppose that E/F is covered by the sequence $\{A_n : n \in \mathbb{N}\}$ of closed absolutely convex subsets of E/F. Then E is covered by the sequence $\{k^{-1}(A_n) : n \in \mathbb{N}\}$, and there is some $k^{-1}(A_p)$ which is a neighbourhood of the origin in E. Thus $k(k^{-1}(A_p)) = A_p$ is a neighbourhood of the origin in E/F. ■

Lemma 1.3.7 *Let* $\{E_m : m \in \mathbb{N}\}$ *be a sequence of spaces and* $\mathcal{B}$ *a countable family of closed absolutely convex subsets of* $E = \prod\{E_m : m \in \mathbb{N}\}$. *If* $\mathcal{B}$ *covers* E *and*

$$\mathcal{B}_1 = \{B \in \mathcal{B}: \prod\{E_m: m > p\} \subseteq B, \text{ for some } p \in \mathbb{N}\},$$

then $\mathcal{F}_1 = \{\langle B\rangle: B \in \mathcal{B}_1\}$ *covers* E.

Proof. Let $\mathcal{B}_2 = \mathcal{B} \setminus \mathcal{B}_1$, and let $\mathcal{F}_2 = \{\langle B\rangle: B \in \mathcal{B}_2\}$. Now $\mathcal{F}_1 \cup \mathcal{F}_2$ covers $\bigcup\mathcal{B}$. By 1.3.1 we need only show that $\mathcal{F}_2$ does not cover E, or equivalently, that $\{kB: B \in \mathcal{B}_2, k = 1, 2, \ldots\}$ does not cover E. If this latter family is empty, the result is trivial; otherwise let us index it as $\{B_n: n \in \mathbb{N}\}$, and note that $\prod\{E_m: m > p\}$ is not contained in any B_n, for n, $p \in \mathbb{N}$.

Since B_1 is a proper subset of E, there is some $x_1 \in E \setminus B_1$. Fixing $m_1 = 1$ we have that $x_1 \in \prod\{E_m: m \geq m_1\}$; if for each $i > m_1$ there is $y_i \in \prod\{E_m: m \geq i\}$ such that $x_1 + y_i \in B_1$, then $x_1 = \lim_i(x_1 + y_i)$ would belong to the closed set B_1, a contradiction. Hence there is $m_2 > m_1$ such that $(x_1 + \prod\{E_m: m \geq m_2\}) \cap B_1 = \emptyset$. If $x_1 + \prod\{E_m: m \geq m_2\} \subseteq B_2$, then $\prod\{E_m: m \geq m_2\} \subseteq B_2$ since B_2 is absolutely convex, a contradiction. Thus there is some $x_2 \in \prod\{E_m: m \geq m_2\}$ with $x_1 + x_2 \notin B_2$.

Now if, for each $i > m_2$, there is some $y_i \in \prod\{E_m: m \geq i\}$ such that $x_1 + x_2 + y_i \in B_2$, then $x_1 + x_2 = \lim_i(x_1 + x_2 + y_i)$ would belong to the closed set B_2, a contradiction. Hence there is $m_3 > m_2$ such that $(x_1 + x_2 + \prod\{E_m: m \geq m_3\}) \cap B_2 = \emptyset$. Going on in this way, by recurrence we obtain the coordinatewise convergent series $x = \sum_n x_n$ such that $x = \sum_{n<k} x_n + \sum_{n\geq k} x_n \notin B_k$. Hence $x \notin \bigcup\{B_n: n \in \mathbb{N}\} = \bigcup\mathcal{F}_2$, so $\mathcal{F}_2$ does not cover E. ■

Lemma 1.3.8 *Let* $\{E_m: m \in \mathbb{N}\}$ *be a sequence of linear spaces and* $\mathcal{F}$ *a countable family of proper linear subspaces of the product* $E = \prod\{E_m: m \in \mathbb{N}\}$. *Suppose each* F *in* $\mathcal{F}$ *contains* $\prod\{E_m: m \geq M\}$ *for some* $M \in \mathbb{N}$. *If* $\mathcal{F}$ *covers* E, *then for some* m, *the family*

$$\{F \in \mathcal{F}: F \text{ does not contain } E_m\}$$

covers E_m.

Proof. Let $\mathcal{F}_m = \{F \in \mathcal{F}: F$ *does not contain* $E_m\}$ and assume for each $m \in \mathbb{N}$ that $\mathcal{F}_m$ does not cover E_m. Then, on the one hand there exist $x_m \in E_m \setminus \bigcup\{F: F \in \mathcal{F}_m\}$ and on the other hand $\mathcal{F} = \bigcup\{\mathcal{F}_m: m \in \mathbb{N}\}$, for otherwise there would be an $F \in \mathcal{F}$ with $E_m \subseteq F$ for all $m \in \mathbb{N}$. But as there exists some $k \in \mathbb{N}$ such that $\prod\{E_m: m \geq k\} \subseteq F$, we would have $E \subseteq F$, which contradicts the assumption that F is a proper subspace of E.

Consider $\alpha_1 = 1$. Then $\alpha_1 x_1 \notin \bigcup\{F: F \in \mathcal{F}_1\}$. If two distinct points of the line determined by $\alpha_1 x_1$ and x_2 belong to one F of the family $\{F: F \in \mathcal{F}_1 \cup \mathcal{F}_2\}$, then this F contains $\alpha_1 x_1$ and x_2, in contradiction with $\alpha_1 x_1 \notin \bigcup\{F: F \in \mathcal{F}_1\}$ or $x_2 \notin \bigcup\{F: F \in \mathcal{F}_2\}$. Hence $\bigcup\{F: F \in \mathcal{F}_1 \cup \mathcal{F}_2\}$ covers only a countable subset of the uncountable set $\{\alpha_1 x_1 + \lambda x_2: \lambda \in \mathbb{R}\}$. Therefore, there is some α_2 such that $\alpha_1 x_1 + \alpha_2 x_2$ does not belong to $\bigcup\{F: F \in \mathcal{F}_1 \cup \mathcal{F}_2\}$, and so on. By an obvious recurrence we determine a sequence $\{\alpha_m: m \in \mathbb{N}\}$ of scalars such that $\sum_{i \leq m} \alpha_i x_i \notin \bigcup\{F: F \in \mathcal{F}_1 \cup \mathcal{F}_2 \cup \ldots \cup \mathcal{F}_m\}$.

Then given $F \in \mathcal{F}$ and $x = \sum_i \alpha_i x_i$, there exists some $p \in \mathbb{N}$ such that for $q > p$ we have that $\sum_{i>q} \alpha_i x_i \in \mathcal{F}$, and if $F \in \mathcal{F}_r$ we have, for every $m \geq r$, that $\sum_{i \leq m} \alpha_i x_i \notin F$. Hence $x \notin F$ for every $F \in \mathcal{F}$, a contradiction. ■

Proposition 1.3.9 *The countable product of unordered Baire-like spaces is Baire-like*

Proof. We are going to consider only the countable infinite case, since $\{0\}$ being an unordered Baire-like space, it follows that every finite product of unordered Baire-like spaces is isomorphic to a countable infinite product of such spaces.

Suppose that the product $E = \prod\{E_m: m \in \mathbb{N}\}$ of a sequence of unordered Baire-like spaces is covered by a countable family $\mathcal{B}$ of nowhere-dense closed absolutely convex subsets of E. By 1.1.13 E is barrelled, hence 1.1.11 implies that for each $B \in \mathcal{B}$ the linear span $\langle B \rangle$ is a proper subspace of E. Let

$$\mathcal{B}_1 = \{B \in \mathcal{B}: \prod\{E_m: m > k\} \subseteq B, \textit{ for some } k \in \mathbb{N}\}.$$

By 1.3.7, $\mathcal{F}_1 = \{\langle B \rangle: B \in \mathcal{B}_1\}$ covers E, and by 1.3.8 there exists some $m \in \mathbb{N}$ such that

$$\mathcal{F}_0 = \{F \in \mathcal{F}_1: F \textit{ does not contain } E_m\}$$

covers E_m. Obviously, there is a subfamily $\mathcal{B}_0 \subseteq \mathcal{B}_1$ such that

$$\mathcal{F}_0 = \{\langle B \rangle : B \in \mathcal{B}_0\}$$

and then $E_m \supseteq \{kB: k \in \mathbb{N}, B \in \mathcal{B}_0\}$. On the other hand every $(kB) \cap E_m$ is a closed absolutely convex subset of E_m without interior points, since in light of the definition of $\mathcal{F}_0$, it follows that its linear span is a proper subspace of E_m, a contradiction with the fact that E_m is a Baire-like space. ∎

Lemma 1.3.10 *Let* $\{B_n: n \in \mathbb{N}\}$ *be a sequence of closed subsets of an arbitrary product* $E = \prod\{E_i: i \in I\}$ *of topological vector spaces. If* $\bigcup\{B_n: n \in \mathbb{N}\} \supseteq \prod\{E_i: i \in J\}$ *for each countable subset* $J \subseteq I$, *then* $E = \bigcup\{B_n: n \in \mathbb{N}\}$.

Proof. Suppose that $x = (x_i)_{i\in I}$ is an element of E such that $x \notin B_n$ for every $n \in \mathbb{N}$. If for every finite subset $J_n \subseteq I$ there were some element $y \in \prod\{E_i: i \in I \setminus J_n\}$ such that $(x_i)_{i\in J_n} + y \in B_n$, then the fact that B_n is closed implies that $x \in B_n$. Therefore, for each $n \in \mathbb{N}$ there exists a finite subset J_n of I such that $\{x_i: i \in J_n\} + y \notin B_n$ for each $y \in \prod\{E_i: i \in I \setminus J_n\}$. Let $J = \bigcup\{J_n: n \in \mathbb{N}\}$. Then $(x_i)_{i\in J} = (x_i)_{i\in J_n} + (x_i)_{i\in J\setminus J_n} \notin B_n$ for each $n \in \mathbb{N}$, and J is countable, in contradiction with the hypothesis of the lemma. ∎

Theorem 1.3.11 *The topological product* E *of a family* $\{E_i: i \in I\}$ *of unordered Baire-like spaces is unordered Baire-like.*

Proof. Let us suppose that E is covered by a countable family $\mathcal{B}$ of nowhere-dense, closed and absolutely convex sets B. By 1.1.13 E is barrelled, hence no $B \in \mathcal{B}$ is absorbing in E, i.e., $E \neq \bigcup\{nB: n \in \mathbb{N}\}$. Thus applying 1.3.10 to the sequence $\{nB: n \in \mathbb{N}\}$ we obtain that there exists a countable index set $J_B \subseteq I$ such that B fails to absorb all points of $\prod\{E_i: i \in J_B\}$.

Writing $J := \bigcup\{J_B: B \in \mathcal{B}\}$ we have that the countable product $F := \prod\{E_i: i \in J\}$ is covered by the countable family $\{B \cap F: B \in \mathcal{B}\}$ formed by closed absolutely convex and non-absorbing subsets of F. Therefore every $B \cap F$, $B \in \mathcal{B}$, is nowhere-dense, in contradiction with 1.3.9, and the proof is over. ∎

Proposition 1.3.12 *Let* F *be a countable codimensional subspace of* E. *If* E *is unordered Baire-like, then* F *is unordered Baire-like.*

Proof. Let $\{A_n: n \in \mathbb{N}\}$ be a sequence of closed absolutely convex subsets of F which covers F, and let $\{b_n: n \in \mathbb{N}\}$ be a cobasis of F in E. Let B_n be the absolutely convex hull of $A_n \cup \{b_1, b_2, \ldots, b_n\}$. Then $\{mB_n: m \in \mathbb{N}, n \in \mathbb{N}\}$ is a sequence of closed absolutely convex subsets of E and $\bigcup\{mB_n: m \in \mathbb{N}, n \in \mathbb{N}\}$ is absorbing in E. Therefore, there exists some qB_p which is a neighbourhood of the origin in E, so $A_p = B_p \cap F$ is a neighbourhood of the origin in F and, consequently, F is unordered Baire-like. ∎

1.4 TOTALLY BARRELLED SPACES

As we commented in the previous section, unordered Baire-like spaces have arisen from removing the *increasing* condition in the definition of Baire-like spaces. Moreover, in 1.3.2 we have obtained that a space E *is unordered Baire-like if and only if given a sequence* $\{E_n: n \in \mathbb{N}\}$ *of subspaces of* E *which covers* E, *there exists some* E_p *which is barrelled and dense in* E. The weakening of the density condition produces the following class of barrelled spaces, located between Baire-like and unordered Baire-like spaces.

Definition 1.4.1 *The space* E *is totally barrelled if, given a sequence* $\{E_n: n \in \mathbb{N}\}$ *of subspaces of* E *which covers* E, *there is some* E_p *which is barrelled and its closure is finite-codimensional in* E.

Remark 1.4.2 Clearly every unordered Baire-like space is totally barrelled; and after a moment of reflection it is easy to give a straightforward proof showing that every totally barrelled space is Baire-like. In fact, suppose that E is a totally barrelled space and $\{A_n: n \in \mathbb{N}\}$ is an increasing covering of E formed by closed absolutely convex subsets. Let us denote by E_n the linear span of A_n. Then from the totally barrelledness it follows that there is some E_p which is barrelled and such that its closure $\overline{E_p}$ is finite-codimensional in E. On the other hand, as $\{A_n: n \in \mathbb{N}\}$ is an increasing covering of E there exists some A_q containing a cobasis of $\overline{E_p}$ in E. There is no problem in choosing such $q > p$. From 1.1.10 it follows that E_q is a dense barrelled subspace of E. Then A_q is a neighbourhood of the origin in E_q, so, by density, the closed set A_q is a neighbourhood of the origin in E. Hence E is Baire-like.

The following characterization of totally barrelled spaces brings to mind result 1.3.4 of unordered Baire-like spaces.

Proposition 1.4.3 *The space* E *is totally barrelled if and only if given any sequence of subspaces* $\{E_n: n \in \mathbb{N}\}$ *covering* E *there is some* E_p *which is totally barrelled and its closure is finite-codimensional in* E.

Proof. Considering $E_n = E$ for every $n \in \mathbb{N}$ gives us one half of the statement. For the other half, assume that E is totally barrelled and that $\{E_n: n \in \mathbb{N}\}$ is a covering of E by linear subspaces of E. Total barrelledness of E and Lemma 1.3.1 enable us to suppose that, taking a suitable subsequence of $\{E_n: n \in \mathbb{N}\}$ which will be also denoted by $\{E_n: n \in \mathbb{N}\}$, each $\overline{E_n}$ is finite codimensional in E. If no E_n were totally barrelled then for each E_n there would be a covering $\{F_{nm}: m \in \mathbb{N}\}$ by subspaces of E_n such that each F_{nm} is non-barrelled or its closure in E_n is infinite-codimensional in E_n. In the latter case the closure of F_{nm} in E is also infinite-codimensional in E since the topological product of this closure and one algebraic complement of $\overline{E_n}$ in E is obviously E-closed. Therefore we have that E may be covered by the countable family $\{F_{nm}: n \in \mathbb{N}, m \in \mathbb{N}\}$ of subspaces of E, such that each F_{nm} is non-barrelled or its closure is infinite-codimensional in E, a contradiction with the totally barrelledness of E. ∎

Proposition 1.4.4 *The space* E *is totally barrelled if and only if given any sequence of subspaces* $\{E_n: n \in \mathbb{N}\}$ *covering* E *there exists some* E_p *which is Baire-like.*

Proof. By 1.4.3 there exists some E_p which is totally barrelled, and therefore is Baire-like.

Conversely, let us suppose that the space E verifies that for any sequence of subspaces $\{E_n: n \in \mathbb{N}\}$ covering E there exists some E_p which is Baire-like. Then from Lemma 1.3.1 it follows that there exists a subsequence formed by Baire-like spaces covering E. Changing the notation if necessary we denote this subsequence also by $\{E_n: n \in \mathbb{N}\}$. If F_n is the closure of E_n in E we only have to show that there exists some F_q finite-codimensional in E. Let us suppose that every F_n is infinite-codimensional. Then for each $n \in \mathbb{N}$ we are able to find a sequence $\{x_{nm}: m \in \mathbb{N}\}$ of linearly independent vectors with respect to F_n, and denote by G_n and G_{nr} the linear span of $F_n \cup \bigcup\{x_{nm}: m \in \mathbb{N}\}$ and $F_n \cup \bigcup\{x_{nm}: m \leq r\}$ respectively. By the hypothesis some G_s is Baire-like, in contradiction with the existence of the increasing sequence $\{G_{sr}: s \in \mathbb{N}\}$ of closed proper subspaces of G_s.

Proposition 1.4.5 *Let* F *be a dense subspace of* E. *If* F *is totally barrelled, then* E *is totally barrelled.*

Proof. Let $\{E_n: n \in \mathbb{N}\}$ be a sequence of subspaces of E covering E. Setting $F_n = E_n \cap F$ for each $n \in \mathbb{N}$, there must exist a barrelled space F_p such that its closure in F is finite-codimensional in F. Therefore the closure of E_p in E is also finite-codimensional in E. By 1.1.10 E_p is barrelled and consequently E is totally barrelled. ∎

Proposition 1.4.6 *Let* F *be a closed subspace of* E. *If* E *is totally barrelled, then* E/F *is totally barrelled.*

Proof. Let k the canonical mapping between E and E/F. Given a sequence $\{G_n: n \in \mathbb{N}\}$ of subspaces of E/F covering E, by 1.4.4 there exists some $k^{-1}(G_p)$ which is Baire-like. By 1.2.4 $G_p = k(k^{-1}(G_p))$ is Baire-like, and the proposition follows using 1.4.4 again. ∎

Proposition 1.4.7 *The product* $E = \prod\{E_i: i \in I\}$ *of totally barrelled spaces is totally barrelled.*

Proof. By 1.1.13, E is barrelled. If E were not totally barrelled then there would be a covering $\{F_n: n \in \mathbb{N}\}$ of E formed by subspaces in such a way that for each F_n, its closure is infinite-codimensional in E or there is a barrel U_n in F_n which is not a neighbourhood of the origin in F_n. The latter would imply that the linear span of the closure of U_n in E is not barrelled since otherwise $\overline{U_n}$ would be a neighbourhood of the origin in its linear span, and thus U_n would be a neighbourhood of the origin in F_n which is not the case. Then 1.1.15 implies that $\langle\overline{U_n}\rangle$ is infinite-codimensional in E. Clearly $\{\overline{E_n}: n \in \mathbb{N}\}$ or $\{m\overline{U_n}: m \in \mathbb{N}, n \in \mathbb{N}\}$ covers E. Therefore, if E were not totally barrelled, there would be a covering $\{W_n: n \in \mathbb{N}\}$ of E formed by closed absolutely convex subsets such that each linear span $\langle W_n\rangle$ is infinite codimensional in E. Now, from Lemma 1.3.7 and the forthcoming claim 1.4.8, it follows that there is a subsequence of $\{W_n: n \in \mathbb{N}\}$, which we will also denote by $\{W_n: n \in \mathbb{N}\}$ enjoying the following two properties:

(i) $\{\langle W_n\rangle: n \in \mathbb{N}\}$ is a covering of E.

(ii) For each $n \in \mathbb{N}$, there exists a finite subset J_n such that

$$W_n \supseteq \prod\{E_i: i \in I \setminus J_n\}.$$

Let $\mathcal{F}_i = \{\langle W_n\rangle: n \in \mathbb{N},\ \langle W_n\rangle \cap E_i \textit{ is infinite-codimensional in } E_i\}$. Clearly, $\bigcup\{(\bigcup\mathcal{F}_i): i \in I\} = \{\langle W_n\rangle: n \in \mathbb{N}\}$, since if $\langle W_n\rangle \cap \prod\{E_i: i \in J_n\}$ is infinite-codimensional in $\prod\{E_i: i \in J_n\}$, there is some $i \in J_n$ such that $\langle W_n\rangle \cap E_i$ is infinite-codimensional in E_i. But there exists some index j such that $\mathcal{F}_j$ covers E_j since otherwise, for each $i \in I$ there would be some vector $x_i \in E_i$ such that $x_i \notin \bigcup\{\langle W_n\rangle: \langle W_n\rangle \in \mathcal{F}_i\}$. Let $L_i = \langle x_i\rangle$ and $L := \prod\{L_i: i \in I\}$. Then L is a Baire space and is covered by $\{mW_n: m, n \in \mathbb{N}\}$. Thus some W_n must contain a neighbourhood of the origin of L. So $L \subseteq W_n$, and if $W_n \in \mathcal{F}_i$ we have $x_i \in \bigcup\{\langle W_n\rangle: \langle W_n\rangle \in \mathcal{F}_i\}$, a contradiction.

Then $\mathcal{F}_j = \{\langle W_n\rangle: n \in \mathbb{N},\ \langle W_n\rangle \cap E_j \textit{ is infinite-codimensional in } E_j\}$ is a covering of the totally barrelled space E_j. Hence there is some index s such that $\langle W_s\rangle \cap E_j$ is barrelled and its closure is finite-codimensional in E_j, but $\langle W_s\rangle \cap E_j$ is infinite-codimensional in E_j. We have reached a contradiction between the two aforementioned codimensions since, reasoning as in 1.2.1, we obtain that the closure of the barrelled space $\langle W_s\rangle \cap E_j$ in E_j is the union of the closures of $\langle nW_s\rangle \cap E_j$, $n \in \mathbb{N}$, but every $\langle nW_s\rangle \cap E_j$ is closed, hence the closure of $\langle W_s\rangle \cap E_j$ in E_j is $\langle W_s\rangle \cap E_j$. ■

Claim 1.4.8 *Let* $\{E_i: i \in I\}$ *be a family of spaces and let* $\mathcal{B} = \{B_n: n \in \mathbb{N}\}$ *be a countable family of closed absolutely convex sets of* $E = \prod\{E_i: i \in I\}$. *If* $\mathcal{B}$ *covers* E *and*
$\mathcal{B}_1 = \{B \in \mathcal{B}: B \supseteq \prod\{E_i: i \in I\backslash J(B)\}$, J(B) being some finite subset of I$\}$, *then* $\mathcal{F}_1 = \{\langle B\rangle: B \in \mathcal{B}_1\}$ *covers* E.

Proof. Assume that $\mathcal{B}_1$ is void. If a closed convex set B_n contains E_i for every $i \in I$ then $\prod\{E_i: i \in I\} \subseteq B_n$. Therefore for every n there must exist an index i(1,n) such that $E_{i(1,n)}$ is not contained in B_n for each $n \in \mathbb{N}$. But $\prod\{E_i: i \in I \setminus \{i(1,n)\}\}$ is not contained in B_n for each $n \in \mathbb{N}$; this implies that for each n there exists an index i(2,n) such that B_n does not contain $E_{i(2,n)}$ for each $n \in \mathbb{N}$, and so on. In this way we determine for each $n \in \mathbb{N}$ an infinite countable set of indexes $I_n = \{(1,n), (2,n), \ldots.\}$ such that B_n does not contain E_j for each $j \in I_n$. Then if $M = \bigcup\{I_n : n \in \mathbb{N}\}$, the preceding property of the B_n is in contradiction with 1.3.7 in the product space $\prod\{E_i: i \in M\}$.

If $\mathcal{F}_1$ does not cover E then, by 1.3.1, the family $\mathcal{F}_2 = \{\langle B\rangle : B \notin \mathcal{B}_1\}$ covers E. E is also covered by the countable family $\mathcal{B}_2 = \{mB: \langle B\rangle \in \mathcal{F}_2, m \in \mathbb{N}\}$, but according to what we have established above there would be some $mB \in \mathcal{B}_2$ — which implies that $B \notin \mathcal{B}_1$ — such that $mB \supseteq \prod\{E_i: i \in I \setminus J(B)\}$ for some finite subset $J(B) \subseteq I$. Also $B \supseteq \prod\{E_i : i \in I \setminus J(B)\}$, and from this follows $B \in \mathcal{B}_1$, a contradiction ■

Proposition 1.4.9 *Let* F *be a countable-codimensional subspace of* E. *If* E *is a totally barrelled space, then* F *is totally barrelled.*

Proof. Let G be a countable dimensional complement of F in E and let $\{F_n: n \in \mathbb{N}\}$ be a covering of F, each F_n being a subspace of F. By 1.4.4 there is some $p \in \mathbb{N}$ such that $F_p + G$ is Baire-like. Now, 1.2.8 shows that F_p is Baire-like. Hence, the new use of 1.4.4 yields that F is totally barrelled. ■

1.5 ULTRABORNOLOGICAL SPACES

It is easy to check that the inductive limit of a family of barrelled spaces is barrelled. A particular situation arises when every space of the family is a Banach space. In this case we say that the space defined by means of the inductive limit is an *ultrabornological space*. Clearly every ultrabornological space is barrelled and bornological, i.e., an inductive limit of normed spaces. On the other hand, every sequentially complete bornological space is ultrabornological, and this includes Fréchet spaces and barrelled spaces which are strong duals of metrizable spaces. Separated quotients, locally convex direct sums and finite-codimensional subspaces of ultrabornological spaces are ultrabornological. In this section, we just want to present two particular features concerning ultrabornological spaces, namely inheritance under topological products and a closed graph theorem concerning them.

The question of whether products of ultrabornological spaces are ultrabornological is a deep one, related to a problem of set theory, namely to the existence of Ulam-measurable cardinals. A cardinal number $\mathfrak{m}$ is said to be *measurable* if given a set M of cardinality $\mathfrak{m}$ there is a *Ulam measure* defined on 2^M, that is, a non-trivial countably additive $\{0,1\}$-valued measure μ defined on the power set 2^M such that $\mu(\{\omega\}) = 0$ for every $\omega \in M$. A cardinal $\mathfrak{i}$ is called *strongly inaccessible* if it satisfies the following three conditions: $\mathfrak{i} > \aleph_0$; $2^\kappa < \mathfrak{i}$ for every cardinal number $\kappa < \mathfrak{i}$; and $\sum_{\alpha\in A}\kappa_\alpha < \mathfrak{i}$ for every family $\{\kappa_\alpha : \alpha \in A\}$ of size $< \mathfrak{i}$ of cardinals $\kappa_\alpha < \mathfrak{i}$. If there exists a set M with measurable cardinal $\mathfrak{m}$, then $\mathfrak{m}$ must be strongly inaccesible.

The *Mackey-Ulam theorem* asserts that the product space $\mathbb{R}^I$ is (ultra)bornological if and only if 2^I does not admit a Ulam measure. Hence if κ is smaller than the first strongly inaccessible cardinal, then $\mathbb{R}^\kappa$ is (ultra)bornological. This machinery helps us to answer the problem of the permanence of the property of being ultrabornological under arbitrary products since a product of κ ultrabornological spaces is ultrabornological if and only if $\mathbb{R}^\kappa$ is ultrabornological.

Next we provide a closed graph theorem with ultrabornological spaces in the domain and the so-called webbed spaces in the range. A *web* in a space E is a family of absolutely convex subsets $\{C_{n_1n_2\ldots n_i} : n_1, n_2, \ldots, n_i, i \in \mathbb{N}\}$ of E such that $\bigcup\{C_{n_1} : n_1 \in \mathbb{N}\}$ is absorbing in E and for each i, $n_1, \ldots, n_i \in \mathbb{N}$, the family $\{C_{n_1\ldots n_{i+1}} : n_{i+1} \in \mathbb{N}\}$ is contained in $(1/2)C_{n_1n_2\ldots n_i}$ and its union absorbs every element of $C_{n_1n_2\ldots n_i}$. The web is said to be *compatible* with the topology if for each increasing sequence of positive integers $\{n_i : i \in \mathbb{N}\}$ and each neighbourhood of the origin U in E there exists some $p \in \mathbb{N}$ such that $C_{n_1n_2\ldots n_p} \subseteq U$. We say that a space E is a *webbed space* if it contains a compatible web such as the aforementioned such that given any increasing sequence of positive integers $\{n_i : i \in \mathbb{N}\}$ and any $x_i \in C_{n_1n_2\ldots n_i}$, then $\sum_i x_i$ converges in E. Note that every Fréchet space is a webbed space. In order to check this it is enough to consider every $C_{n_1n_2\ldots n_i}$ equal to the i-th element of a nested base $\{U_i : i \in \mathbb{N}\}$ of absolutely convex neighbourhoods of the origin.

Theorem 1.5.1 *Let* T *be a linear mapping between the spaces* E *and* F *with (closed) sequential graph. If* E *is ultrabornological and* F *a webbed space, then* T *is continuous.*

Proof. Let $\{E_\alpha : \alpha \in I\}$ be a family of Banach spaces and $\{A_\alpha : \alpha \in I\}$ a family of mappings such that E is the inductive limit of $\{E_\alpha : \alpha \in I\}$ by means of $\{A_\alpha : \alpha \in I\}$. We have to show that $T_\alpha := T \circ A_\alpha$ is continuous for each $\alpha \in I$.

First of all it is easily seen that each T_α has a (sequentially) closed graph since if $\{x_n : n \in \mathbb{N}\}$ is a sequence that converges to some x in E_α such that $\{T_\alpha(x_n) : n \in \mathbb{N}\}$ does converge to some $y \in F$, then the continuity of A_α implies that $\{A_\alpha(x_n) : n \in \mathbb{N}\}$ converges to $A_\alpha(x)$ and so $y = (T \circ A_\alpha)(x) = T_\alpha(x)$.

Consider now the case that $\{C_{n_1n_2\ldots n_i} : n_1, n_2, \ldots, n_i, i \in \mathbb{N}\}$ is a web in F that makes F a webbed space and consider any $\alpha \in I$. Then the family $\{T_\alpha^{-1}(C_{n_1 n_2 \ldots n_i}): n_1, n_2, \ldots, n_i, i \in \mathbb{N}\}$ is a web in E and the Baire category theorem enables us to find an increasing sequence of positive integers $\{n_i: i \in \mathbb{N}\}$ such that each $W_{\alpha,i} := \overline{T_\alpha^{-1}(C_{n_1 n_2 \ldots n_i})}$ is a neighbourhood of the origin in E. Therefore for this $\alpha \in I$, there is a base $\{U_i: i \in \mathbb{N}\}$ of neighbourhoods of the origin in E with $U_i \subseteq W_{\alpha,i}$ and, therefore, given any neighbourhood of the origin U in E, $U_i \subseteq T_\alpha^{-1}(C_{n_1 n_2 \ldots n_i}) + U$. Hence taking U_{i+1} as U, it follows that

$$T_\alpha(U_i) \subseteq C_{n_1 n_2 \ldots n_i} + T_\alpha(U_{i+1}). \tag{1}$$

In order to show that T_α is continuous, we will prove that given any closed absolutely convex neighbourhood of the origin V in F there is some $p \in \mathbb{N}$ such that $T_\alpha(U_p) \subseteq V$. Since the web considered is compatible with the topology we may assume that there is some $p \in \mathbb{N}$ such that $C_{n_1 n_2 \ldots n_{p-1}} \subseteq V$. Consider now any $x_0 \in U_p$; (1) implies that there must be some $x_1 \in U_{p+1}$ such that $T_\alpha(x_0) - T_\alpha(x_1) \in C_{n_1 n_2 \ldots n_p}$. Now again (1) provides some $x_2 \in U_{p+2}$ such that $T_\alpha(x_1) - T_\alpha(x_2) \in C_{n_1 n_2 \ldots n_p n_{p+1}}$. Going on in this way we obtain a sequence of elements $x_j \in U_{p+j}$, and consequently converging to 0, in such a way that $T_\alpha(x_j) - T_\alpha(x_{j+1}) \in C_{n_1 n_2 \ldots n_{p+j}}$. Bearing in mind that the considered web makes F a webbed space, $\sum_j (T_\alpha(x_j) - T_\alpha(x_{j+1}))$ converges. From this it follows that $\{T_\alpha(x_j): j \in \mathbb{N}\}$ converges to some $z \in F$. Since T_α has a (sequentially) closed graph, z must be 0 and hence $\{T_\alpha(x_j): j \in \mathbb{N}\}$ converges to 0. Finally note that

$$T_\alpha(x_0) - T_\alpha(x_j) = \sum_{0 \leq m \leq j-1} (T_\alpha(x_m) - T_\alpha(x_{m+1})) \in C_{n_1 n_2 \ldots n_{p-1}},$$

and therefore $T_\alpha(x_0) \in \overline{C_{n_1 n_2 \ldots n_{p-1}}} \subseteq \overline{V} = V$. ■

1.6 ELEMENTARY FACTS CONCERNING TENSOR PRODUCTS

Let E and F be two linear spaces over the field $\mathbb{K}$. Denote by $B(E,F)$ the linear space of all bilinear forms defined in $E \times F$ and let $B(E,F)^*$ be the algebraic dual of $B(E,F)$. Consider the so-called tensorialization map $\otimes: E \times F \longrightarrow B(E,F)^*$ defined by $\otimes(x,y)(B) = B(x,y)$ for each $(x,y) \in E \times F$ and $B \in B(E,F)$. It is plain that $\otimes$ is a $B(E,F)^*$-valued bilinear mapping defined in $E \times F$. The tensor product $E \otimes F$ is defined as the linear span of the range of $\otimes$ in $B(E,F)^*$, i.e, $E \otimes F = \langle\otimes(E \times F)\rangle$. As usual we will write $x \otimes y$ instead of $\otimes(x,y)$. Note that according to the bilinearity of $\otimes$, each $z \in E \otimes F$ can be written as $z = \sum_{1\le i\le n} x_i \otimes y_i$, with $x_i \in E$, $y_i \in F$ for $1 \le i \le n$, although this representation is by no means unique.

Assuming that E, F and G are linear spaces over $\mathbb{K}$ let us denote by $B(E,F;G)$ the linear space of all bilinear mappings from $E \times F$ into G, and by $L(E \otimes F, G)$ the linear space of all linear mappings from $E \otimes F$ into G. As it is easily proven, the mapping $\phi: B(E,F;G) \longrightarrow L(E \otimes F, G)$ defined by $\phi(B)(x \otimes y) = B(x,y)$ for each $(x,y) \in E \times F$ and $B \in B(E,F;G)$ is an isomorphism from $B(E,F;G)$ onto $L(E \otimes F,G)$, named linearization. In particular, $B(E,F)$ is isomorphic to the algebraic dual $(E \otimes F)^*$ of $E \otimes F$ and we may identify $B(E,F)$ and $(E \otimes F)^*$.

Suppose E and F are two (Hausdorff locally convex) spaces, and let $\mathcal{U}(E)$ and $\mathcal{U}(F)$ be two bases of absolutely convex neighbourhoods of the origin in E and F, respectively. Writing $U \otimes V$ instead of $\otimes(U,V)$, where $U \in \mathcal{U}(E)$ and $V \in \mathcal{U}(F)$, it is easy to notice that $U \otimes V$ is an absorbent subset of $E \otimes F$. Hence the family $\{\Gamma(U \otimes V): U \in \mathcal{U}(E),\ V \in \mathcal{U}(F)\}$ is a base of absolutely convex neighbourhoods of the origin of some locally convex topology τ_π on $E \otimes F$. This topology τ_π is precisely the finest locally convex topology on $E \otimes F$ under which $\otimes$ is a continuous map, and it is called the *projective topology* on $E \otimes F$. In what follows we will write $E \otimes_\pi F$ whenever $E \otimes F$ is provided with such a topology.

If ϕ is the isomorphism between $B(E,F)$ and $(E \otimes F)^*$ defined above, given any $B \in \mathcal{B}(E,F)$, $\mathcal{B}(E,F)$ being the linear subspace of $B(E,F)$ of all continuous bilinear forms, it is plain that $\phi(B)$ is a continuous linear form on $E \otimes F$. So the restriction of ϕ to $\mathcal{B}(E,F)$ is an isomorphism from $\mathcal{B}(E,F)$ onto $(E \otimes F)'$. Thus $\mathcal{B}(E,F) \cong (E \otimes F)'$ algebraically. Now the map $\Theta: E' \times F' \longrightarrow \mathcal{B}(E,F)$ defined by $\Theta(u,v)(x,y) = \langle x,u\rangle_E \langle v,y\rangle_F$, with $\langle\cdot,\cdot\rangle_E$

and $\langle\cdot,\cdot\rangle_F$ standing for the bilinear forms corresponding to the dual pairs (E,E') and (F,F') respectively, is a one-to-one linear operator of E' × F' into $\mathcal{B}$(E,F). This shows that E' × F' may be identified with a subspace of (E ⊗ F)'. So $E \otimes_\pi F$ is a Hausdorff locally convex space.

Letting Π be the Minkowski functional of Γ(U⊗V), p be the gauge of $U \in \mathcal{U}(E)$ and q be the gauge of $V \in \mathcal{U}(F)$, it is easy to check that

$$\Pi(z) = \inf \{\textstyle\sum_{1\le i\le n} p(x_i)\cdot q(y_i): z = \sum_{1\le i\le n} x_i \otimes y_i\}$$

where the infimum extends over all possible representations $\sum_{1\le i\le n} x_i \otimes y_i$ of z in E ⊗ F. Thus, $\Pi(x \otimes y) = p(x)\cdot q(y)$ for $x \in E$ and $y \in F$. Besides, if E and F are normed spaces, it follows that

$$\|z\|_\pi = \inf \{\textstyle\sum_{1\le i\le n} \|x_i\|\cdot\|y_i\|: z = \sum_{1\le i\le n} x_i \otimes y_i\}$$

We will denote by $E \hat{\otimes}_\pi F$ the completion of $E \otimes_\pi F$.

Proposition 1.6.1 *Assuming* E *and* F *are both Fréchet or normed spaces, then* (E ⊗ F)' *is isomorphic to* $\mathcal{L}$(E,F'(β(F',F))), *the linear space of all continuous linear mapping from* E *to* F'(β(F',F)).

Proof. We will prove only the Fréchet case, since the other is similar but easier than this.

Define Ψ: $\mathcal{B}$(E,F) ⟶ $\mathcal{L}$(E,F'(β(F',F))) by ⟨y,Ψ(B)x⟩ = B(x,y), where ⟨•,•⟩ denotes the bilinear form corresponding to the dual pair (F,F'). Then Ψ is an isomorphism onto its range and thus (E ⊗ F)' is isomorphic to a subspace of $\mathcal{L}$(E,F'(β(F',F))). Let $T \in \mathcal{L}$(E,F'(β(F',F))) and denote by $\{V_n, n \in \mathbb{N}\}$ a decreasing base of absolutely convex neighbourhoods of the origin of the Fréchet space F. Since each polar V_n° is a Banach disk in F'(β(F',F)), if L_n denotes the Banach space of the linear span of V_n° equipped with the norm of the gauge of V_n°, $\{L_n, n \in \mathbb{N}\}$ is an increasing sequence of Banach spaces whose union covers F'. So the inductive limit F'(η) of that sequence is an (LB)-space.

Since E is Baire-like and T: E ⟶ F'(η) has a closed graph because η ≥ β(F',F), according to Theorem 1.2.2, T is continuous and there is some $m \in \mathbb{N}$ such that $T(E) \subseteq L_m$. Consequently, $T: E \longrightarrow L_m$ is continuous and $U = T^{-1}(V_m^\circ)$ is a neighbourhood of the origin in E. If B ∈ B(E,F) is such that B(x,y):= ⟨y,Tx⟩, clearly Ψ(B) = T. Besides, B is continuous, since if $x \in U$ and $y \in V_m$ then $Tx \in V_m^\circ$ and so $|<y,Tx>| \leq 1$. Therefore, $|B(x,y)| \leq 1$ for each $(x,y) \in U \times V_m^\circ$. This implies that Ψ is surjective and we are done. ■

For the next result, E and F will be two normed spaces. We suppose $\mathcal{B}(E,F)$ is provided with the norm $\|B\|_{\mathcal{B}} = \sup\{|B(x,y)|, \|x\| \leq 1, \|y\| \leq 1\}$. $(E \otimes F)'$ is endowed with the dual norm, and $\mathcal{L}(E,F'(\beta(F',F)))$ is equipped with the norm $\|T\|_{\mathcal{L}} = \sup\{\|Tx\|, \|x\| \leq 1\}$ of the uniform convergence on the unit ball.

Proposition 1.6.2 *If* E *and* F *are normed spaces, then* $(E \otimes_\pi F)'$, $\mathcal{B}(E,F)$ *and* $\mathcal{L}(E,F'(\beta(F',F)))$ *are linearly isometric.*

Proof. If $B \in \mathcal{B}(E,F)$, and Ψ represents the algebraic isomorphism of the previous proposition, then

$$\|B\|_{\mathcal{B}} = \sup\{|B(x,y)|, \|x\| \leq 1, \|y\| \leq 1\} =$$

$\sup\{|<y,\Psi(B)x>|, \|x\| \leq 1, \|y\| \leq 1\} = \sup\{\|\Psi(B)\|, \|x\| \leq 1\} = \|\Psi(B)\|_{\mathcal{L}}$, establishing that $\mathcal{B}(E,F) \equiv \mathcal{L}(E,F'(\beta(F',F)))$, where here $\equiv$ means (linear) isometry. On the other hand, regarding B also as an element of $(E \otimes F)'$, it is obvious that, if $|\cdot|$ denotes the norm of the strong topology of F', $|B| = \sup\{|<z,B>|, \|z\|_\pi \leq 1\} \geq \sup\{|<x \otimes y,B>|, \|x\| \leq 1, \|y\| \leq 1\} = \sup\{|B(x,y)|, \|x\| \leq 1, \|y\| \leq 1\} = \|B\|_{\mathcal{B}}$. But given $\varepsilon > 0$ there exists $z \in E \otimes F$ with $\|z\| \leq 1$ such that $|B| < |<z,B>| + \varepsilon$. Hence, assuming that $z = \sum_{1\leq i\leq n} x_i \otimes y_i$, then $|<z,B>| \leq \sum_{1\leq i\leq n}|B(x_i,y_i)| \leq \|B\|_{\mathcal{B}}\sum_{1\leq i\leq n}\|x_i\|\|y_i\| \leq \|B\|_{\mathcal{B}}\|z\|_\pi \leq \|B\|_{\mathcal{B}}$. This shows that $|B| = \|B\|_{\mathcal{B}}$. So, $(E \otimes F)' \equiv \mathcal{B}(E,F)$. ■

As $(E \otimes F, E' \otimes F')$ is a dual pair, the locally convex topology on $E \otimes F$ of the uniform convergence of the family $\{U^\circ \otimes V^\circ : U \in \mathcal{U}(E), V \in \mathcal{U}(F)\}$ of subsets of $E' \otimes F'$, where the polars have been taken in E' and F' respectively, is a Hausdorff topology weaker than the projective one, since $U \otimes V \subseteq (U^\circ \otimes V^\circ)^\circ$. This locally convex topology is called the *injective topology* of $E \otimes F$. We will write $E \otimes_\varepsilon F$ when $E \otimes F$ is endowed with such a topology. It can be shown that the injective topology may be defined by the seminorms

$$\varepsilon_{UV}(z) = \sup \{ |\sum_{1\le i\le n} \langle u,x_i\rangle\langle v,y_i\rangle|,\ u \in U^\circ,\ v \in V^\circ \},$$

for each $U \in \mathcal{U}(E)$ and each $V \in \mathcal{U}(F)$, ε_{UV} being independent of the representation of z being considered.

On the other hand, notice that each vector $\sum_{1\le i\le n} x_i \otimes a_i \in E \otimes \mathbb{K}$, with $x_i \in E$ and $a_i \in \mathbb{K}$, $1 \le i \le n$, can be written as $x \otimes 1$, with $z = \sum_{1\le i\le n} a_i x_i$. Hence, the map $J\colon E \otimes_\varepsilon \mathbb{K} \longrightarrow E$ defined by $J(z \otimes 1) = z$ is a topological isomorphism. In fact, if p is the gauge of $U \in \mathcal{U}(E)$ and V is the unit ball of $\mathbb{K}$, $\varepsilon_{UV}(z \otimes 1) = \sup \{ |\langle u,z\rangle\langle v,1\rangle|,\ u \in U^\circ,\ v \in V^\circ \}$ $= \sup \{ |\langle u,z\rangle|,\ u \in U^\circ \} = p(z)$. This establishes that $E \otimes_\varepsilon \mathbb{K} \cong E$, where here $\cong$ means topological isomorphism. Now, if $U \in \mathcal{U}(E)$, $V \in \mathcal{U}(V)$, $x \in E$ $y \in F$, p denotes the gauge of U and q the gauge of V, then $\varepsilon_{UV}(x \otimes y) = p(x)\cdot q(y) = \Pi(x \otimes y)$. According to what we have seen above, this implies that $E \otimes_\varepsilon \mathbb{K} \cong \mathbb{K} \cong E \otimes_\pi \mathbb{K}$. In the following $E \hat{\otimes}_\varepsilon F$ will denote the completion of $E \otimes_\varepsilon F$, though this is also commonly written $E \check{\otimes}_\varepsilon F$. A space E is said to be nuclear if $E \otimes_\varepsilon F = E \otimes_\pi F$ for each space F.

Theorem 1.6.3 *Assume* E *and* F *are Baire-like spaces. If* E *is metrizable, then* $E \otimes_\pi F$ *is Baire-like.*

Proof. Let $\{W_n,\ n \in \mathbb{N}\}$ be an increasing sequence of closed absolutely convex subsets of $E \otimes_\pi F$ covering $E \otimes F$ and let $\{U_n,\ n \in \mathbb{N}\}$ denote some decreasing base of closed absolutely convex neighbourhoods of the origin in E. Setting $V_n = \{y \in F\colon x \otimes y \in W_n,\ \forall x \in U_n\}$ for each $n \in \mathbb{N}$, then $\{V_n,\ n \in \mathbb{N}\}$ is an increasing sequence of closed absolutely convex subsets of F. Let us show that $\{V_n,\ n \in \mathbb{N}\}$ covers F.

If $z \in F$, define $Z_n = \{x \in E\colon x \otimes z \in W_n\}$ for each $n \in \mathbb{N}$. Again, $\{Z_n,\ n \in \mathbb{N}\}$ is an increasing sequence of closed absolutely convex sets of E covering E. As E is Baire-like there is some $m \in \mathbb{N}$ such that Z_m is a neighbourhood of the origin in E, and hence there is some $r \ge m$ such that $U_r \subseteq Z_m$. This implies that $x \otimes z \in W_m$ for each $x \in U_r$. Thus $z \in V_m$ and we are done.

Since F is Baire-like, there is some $k \in \mathbb{N}$ such that V_k contains a neighbourhood of the origin V of F. From $V \subseteq V_k$, it follows that $U_k \otimes V \subseteq W_k$. Consequently, $\Gamma(U_k \otimes V) \subseteq W_k$ and hence $E \otimes_\pi F$ is Baire-like. ∎

Corollary 1.6.4 *If* E *and* F *are metrizable barrelled spaces, then* $E \otimes_\pi F$ *is barrelled.*

Theorem 1.6.5 *Let* E *and* F *be unordered Baire-like. If* E *is metrizable, then* $E \otimes_\pi F$ *is unordered Baire-like.*

Proof. Let $\{W_n, n \in \mathbb{N}\}$ be an increasing sequence of closed absolutely convex subsets of $E \otimes_\pi F$ covering $E \otimes F$ and let $\{U_n, n \in \mathbb{N}\}$ denote some decreasing base of closed absolutely convex neighbourhoods of the origin in E. Setting $V_{nm} = \{y \in F: x \otimes y \in W_n, \forall x \in U_m\}$ for $n, m \in \mathbb{N}$, then $\{V_{nm}: n, m \in \mathbb{N}\}$ is a sequence of closed absolutely convex sets of F. Let us show that $\{V_{nm}: n, m \in \mathbb{N}\}$ covers F.

If $z \in F$, define $Z_n = \{x \in E: x \otimes z \in W_n\}$ for each $n \in \mathbb{N}$. Again, $\{Z_n, n \in \mathbb{N}\}$ is a sequence of closed absolutely convex sets of E covering E. As E is unordered Baire-like there are $m, r \in \mathbb{N}$ with $r \geq m$, such that $U_r \subseteq Z_m$. This implies that $x \otimes z \in W_m$ for each $x \in U_r$. Hence, $z \in V_{mr}$. Then, there are $i, j \in \mathbb{N}$ such that V_{ij} is a neighbourhood of the origin in F. Since F is unordered Baire-like, there are $p, q \in \mathbb{N}$ such that V_{pq} contains a neighbourhood of the origin V of F. From $V \subseteq V_{pq}$, it follows that $U_q \otimes V \subseteq W_p$.

Hence $\Gamma(U_q \otimes V) \subseteq W_p$, and so $E \otimes_\pi F$ is unordered Baire-like. ∎

Theorem 1.6.6 *Let* E *and* F *be barrelled spaces. If* E *is normed, then the product* $E \otimes_\pi F$ *is barrelled.*

Proof. Let T be a barrel in $E \otimes_\pi F$ and denote by U the closed unit ball of E. Then define $V = \{y \in F: x \otimes y \in T\ \forall x \in U\}$. Clearly V is a closed absolutely convex subset of F. We are going to show that V is absorbing in F. In fact, given $z \in F$ define $Z = \{x \in E: x \otimes z \in T\}$ which happens to be a barrel in E, and thus a neighbourhood of the origin in E. Hence there is some $m \in \mathbb{N}$ such that $U \subseteq mZ$, and consequently, $x \otimes z \in mT$ for each $x \in U$. This implies that $z \in mV$ and so V is absorbing. Since F is a barrelled space, V is a neighbourhood of the origin in F. So, we reach the conclusion as above. ∎

1.7 EXERCISES

1. *Some necessary and sufficient conditions for a metrizable space to be barrelled.* Let E be a metrizable space. The following are equivalent

(i) E is barrelled.

(ii) $E'(\sigma(E',E))$ is sequentially complete.

(iii) $E'(\sigma(E',E))$ is locally complete.

(iv) Each $\sigma(E',E)$-bounded sequence in E' is equicontinuous.

(v) Each $\sigma(E',E)$-null sequence in E' is equicontinuous.

2. *Distinguishing between barrelled and Baire-like spaces.* Assume E is a barrelled space.

(i) If E is metrizable, then E is a Baire-like space.

(ii) If E has a complemented copy of φ, the locally convex direct sum of $\aleph_0$ copies of $\mathbb{K}$, then E is not Baire-like.

3. *Distinguishing between Baire-like and totally barrelled spaces.* Each non-complete metrizable (LF)-space is a Baire-like space which is not totally barrelled.

4. *An ultrabornological space.* A subset $A \subseteq \mathbb{N}$ is said to be of density zero if $\lim_n |\{m \in A,\ m \le n\}|/n = 0$, where $|B|$ denotes the cardinality of the set B. Let $\mathcal{Z}$ denote the ring of all subsets of $\mathbb{N}$ of density zero, and denote by $\mathcal{Z}(c_0)$ the linear subspace of c_0 formed by all those vectors $\xi \in c_0$ whose support, i.e. the set $\{n \in \mathbb{N},\ \xi_n \neq 0\}$, is a member of $\mathcal{Z}$, equipped with the relative topology.

(i) Let $\xi \in \omega$. If $\sum_{n\in A}|\xi_n| < \infty$ for each $A \in \mathcal{Z}$, then $\xi \in \ell^1$.

(ii) If $A \subseteq \mathbb{N}$ denote by $c_0(A)$ the sectional subspace of c_0 of all those $\xi \in c_0$ whose support is contained in A. Then each $c_0(A)$ is a copy of c_0 or it is finite-dimensional and $\mathcal{Z}(c_0)$, equipped with the hull topology τ of the family $\{c_0(A),\ A \in \mathcal{Z}\}$, is an ultrabornological space.

(iii) If u is a continuous linear form defined on $\mathcal{Z}(c_0)$ and $\xi_n := \langle u, e_n \rangle$ for each $n \in \mathbb{N}$, where (e_n) denotes the unit vector bases of c_0, then we have that $(\xi_n) \in \ell^1$.

(iv) The topological dual of $(\mathcal{Z}(c_0),\tau)$ is isometric to ℓ^1, and hence τ coincides with the original topology of $\mathcal{Z}(c_0)$.

5. *The subspace of ℓ^1 of all those absolutely summable sequences whose supports have density zero.* Denote by $\mathcal{Z}(\ell^1)$ the linear subspace of the classic Banach space ℓ^1 formed by all $\xi = (\xi_n) \in \ell^1$ whose support is a set of density zero.

(i) $\mathcal{Z}(\ell^1)$ is not unordered Baire-like.

(ii) If $\{E_n, n \in \mathbb{N}\}$ is a sequence of subspaces of $\mathcal{Z}(\ell^1)$ covering $\mathcal{Z}(\ell^1)$ and L denotes the linear span of the unit vector bases of ℓ^1, there is $p \in \mathbb{N}$ such that every barrel in $E_p + L$ absorbs the unit vectors of ℓ^1.

(iii) $\mathcal{Z}(\ell^1)$ is a totally barrelled dense subspace of ℓ^1.

(iv) $\mathcal{Z}(\ell^1)$ is ultrabornological.

6. *An unordered Baire-like space which is not Baire.* Let p be a fixed real number with $0 < p < 1$. We denote by E the linear subspace of ℓ^1 of all those sequences $\{x_n, n \in \mathbb{N}\}$ in $\mathbb{K}$ such that $\sum_i |x_i|^p < \infty$. Now, define $d(x,y) = \sum_i |x_i - y_i|^p$ for each x, $y \in E$ and for each $n \in \mathbb{N}$ denote by U_n the set $\{x \in E : \sum_i |x_i|^p \le 1/n\}$.

(i) (E, d) is a complete metrizable topological vector space which is not locally convex.

(ii) The metric d on E is invariant under translations, and for each x in E, $\{x + U_n, n \in \mathbb{N}\}$ is a base of neighbourhoods of x in (E, d).

(iii) The topology on E induced by the metric d is finer than the one induced by the norm of ℓ^1.

(iv) E, with the topology induced by ℓ^1, is unordered Baire-like. *[Hint: If $\{A_n, n \in \mathbb{N}\}$ is a sequence of closed absolutely convex sets of* E *covering* E *there is some* $(r,s) \in \mathbb{N}^2$ *with* $U_s \subseteq 2A_r$. *Prove that* $(1/s)^p B \cap E \subseteq 2A_r$, *where* B *denotes the closed unit ball of* ℓ^1.*]*

(v) E, equipped with the relative topology of ℓ^1, is not Baire. *[Hint: Show that* $(1/s)^p(U_n+U_n) \subseteq U_n$ *for each* $n \in \mathbb{N}$.*]*

7. *On injective tensor products.* Let E and F be two (as usual Hausdorff locally convex) spaces. Let $\mathcal{U}(E)$ and $\mathcal{U}(F)$ be two bases of absolutely convex neighbourhoods of the origin in E and F respectively, and assume that $U \in \mathcal{U}(E)$ and $V \in \mathcal{U}(F)$.

(i) Show that $\varepsilon_{UV}(z)$ does not depend of any particular representation of z in $E \otimes F$.

(ii) If E, F and G are three spaces over $\mathbb{K}$, we have that

$$E \otimes_\varepsilon (F \times G) \cong (E \otimes_\varepsilon F) \times (E \otimes_\varepsilon F)$$

where here $\cong$ denotes topological isomorphism.

(iii) If F is finite-dimensional $E \otimes_\varepsilon F = E \otimes_\pi F \cong E^{\dim F}$. Thus if F is finite-dimensional, then F is nuclear.

(iv) Let A and B be compact Hausdorff topological spaces, and denote by C(A) and C(B) the Banach spaces of all continuous real-valued functions defined in A and B, respectively, provided with the supremum norm. Then we have that $C(A) \hat{\otimes}_\varepsilon C(B) \cong C(A \times B)$.

(v) If X is a Banach space and Ω is a compact Hausdorff topological space, denote by $C(\Omega)$ the Banach space of all continuous real valued functions defined in Ω and let $C(\Omega,X)$ denote the Banach space of all continuous X-valued functions defined in Ω, both equipped with the supremum norm. Then $C(\Omega) \hat{\otimes}_\varepsilon X \cong C(\Omega,X)$.

1.8 NOTES AND REMARKS

Our treatment of barrelled spaces is standard (see for instance [H], [J] or [Sh]). Some characterizations of linear Baire spaces are given in [S2]. Our characterization (1) of Remark 1.1.6 is a classic result due to Komura, while characterization (2) may be found in [RR]. Pták's homomorphism theorem can be found in [Pt]. Part (iii) of Theorem 1.1.8 is a old result of Mahowald which states that if E is not a barrelled space, then there is a Banach space F and a non-continuous linear mapping φ from E into F which has a closed graph. Our sections 1.2 and 1.3 are based upon the original papers of Saxon [S1], Todd and Saxon [TS], and in Pérez Carreras and Bonet's book [PB2]. The definition of Baire-like spaces was originally motivated by a well-known result of Amemiya and Komura [AK] concerning metrizable barrelled spaces. Our section 1.4 is adapted from Valdivia and Pérez Carreras [VP1]. Besides the results given in that section concerning totally barrelled spaces, it should be pointed out that this class of spaces satisfy the following closed graph theorem [PB2, 9.1.46].

Theorem 1.8.1 *Let* T *be a linear mapping between the spaces* E *and* F *with closed graph. If* E *is totally barrelled and* F *is a webbed space, then there exists a subspace* H *of* F *dominated by a Fréchet space* $H(\tau)$, *i.e., with* τ *finer than the original topology of* H, *such that* $T : E \longrightarrow H(\tau)$ *is continuous.*

To our knowledge, the problem of whether there exists a measurable set is still unsolved. A proof of the Mackey-Ulam measure may be found in Jarchow [J, §13.5.4] and Pérez Carreras and Bonet [PB2, 6.2.23]. Theorem 1.5.1 is a version of De Wilde's closed graph theorem for which we have basically followed Robertson and Robertson [RR]. Our introduction to tensor products is standard. Theorems 1.6.3 and 1.6.5 may be found in [V10], while Theorem 1.6.6 may be seen in [PB2]. Exercises 5 and 6 are based on Examples 2 and 3, respectively, of pp. 277-287 of Valdivia's book [V9], even though the subspace of ℓ^1 considered in Exercise 6 was really introduced by Dierolf et al. [DDD]. Exercise 7 collects a number of elementary properties of injective tensor products. One can have a look at Jarchow's book [J].

Let us also mention some other topics related to barrelledness for which there has been no room in the exposition we have developed in this chapter. Recently, S. Dierolf and J. Kakol [DK] have introduced some spaces with strong barrelledness properties, namely s-barrelled spaces. These spaces have been defined as domain spaces for which the sequentially closed graph theorem holds when Fréchet spaces are considered in the range. Clearly, replacing *sequentially closed* by *closed* in the previous definition one obtains the class of barrelled spaces. Hence, every s-barrelled space is barrelled. These authors show $\mathbb{R}^{\mathbb{R}}$ contains a dense linear subspace E which is Baire and bornological but not s-barrelled, although $\mathbb{R}^{\mathbb{R}}$ is itself a non-metrizable s-barrelled Baire space; moreover E is not the inductive limit of metrizable barrelled spaces. This example answers in the negative a question of Roelcke. On the other hand they also show that there exist non-bornological s-barrelled spaces. Replacing *Fréchet* by *Banach* in the above definition, one obtains the CS-barrelled spaces [KSW], and although every s-barrelled space is CS-barrelled, it is unknown if the converse holds. Let us remark that, under metrizability, these spaces coincide with barrelled spaces.

Though it lies outside our scope, let us provide a guide to the weak barrelledness world for which (until 1990) there is a good insight in Pérez Carreras and Bonet [PB2], Kunzinger [Ku] and Jarchow [J]. A space is said to be countably barrelled, or $\aleph_0$-barrelled [Hus] if every $\aleph_0$-barrel, i.e. every barrel which is the intersection of a sequence of closed absolutely convex neighbourhoods of the origin, is itself a neighbourhood of the origin. A space E is c_0-barrelled [W] if every $\sigma(E',E)$-null sequence of E' is equicontinuous. A space E is called C-barrelled [PB2, p.237] if every $\aleph_0$-barrel $U = \bigcap\{U_n : n \in \mathbb{N}\}$ in E, such that for every $x \in E$ there is a positive integer p with $x \in U_n$ for each $n \geq p$, is a neighbourhood of the origin in E. A space $E(\tau)$ is said to have property (L) [Ru] if for every increasing sequence of absolutely convex sets $\{A_n : n \in \mathbb{N}\}$ in E such that each $x \in E$ is absorbed by some A_n, τ is the finest locally convex topology on E agreeing with τ on the set A_n for every $n \in \mathbb{N}$. A space E is ω-barrelled [LS] (σ-barrelled in [DWH], ℓ^∞-barrelled in [PB2]) if every countable $\sigma(E',E)$-bounded subset of E' is equicontinuous. A space E has property (C) [LS] if every $\sigma(E',E)$-bounded subset of E' is relatively $\sigma(E',E)$-countably compact. A space E has property (S) [LS] if E' is $\sigma(E',E)$-sequentially complete. A space E is called dual locally complete [V4] if $E'(\sigma(E',E))$ is locally complete, that is, if every $\sigma(E',E)$-bounded closed absolutely convex subset of E' is a Banach disk. A space E is said to be weakly barrelled [FS1] if given an increasing sequence of (closed) subspaces $\{E_n : n \in \mathbb{N}\}$ of E covering E, E happens to be the strict inductive limit of $\{E_n : n \in \mathbb{N}\}$. The relationships between these classes are

$$\text{barrelled} \Rightarrow \aleph_0\text{-barrelled} \Rightarrow \omega\text{-barrelled} \Rightarrow \text{property (C)} \Rightarrow$$

$$\text{property (S)} \Rightarrow \text{dual locally complete},$$

$$\aleph_0\text{-barrelled} \Rightarrow \text{C-barrelled} \Rightarrow \text{property (L)} \Rightarrow \text{weakly barrelled}$$

$$\omega\text{-barrelled} \Rightarrow c_0\text{-barrelled} \Rightarrow \text{dual locally complete},$$

$$\text{C-barrelled} \Rightarrow c_0\text{-barrelled},$$

and in the category of Mackey spaces one has

$$c_0\text{-barrelled} \Leftrightarrow \text{property (L)} \Leftrightarrow \text{dual locally complete}.$$

All the one-directional arrows cannot be reversed (these and other examples may be found in [DWH], [LS], [NS], [PB2], [FL4], [FS1]). Under metrizability all the above classes of spaces, apart from the weakly barrelled spaces, coincide with the barrelled spaces. There exist c_0-barrelled spaces which are not weakly barrelled [FL4]. Notice that each non-normable Fréchet space contains a dense barrelled space which is weakly barrelled but not barrelled [FS1]. These weak barrelledness conditions are currently being studied by Saxon and Sánchez Ruiz, [SS5] and [SS6].

Another way of weakening the conditions usually required in topological vector spaces (t.v.s) theory is to get rid of working with local convex conditions. Robertson and Iyahen pioneered this line of research. Adasch, Ernst and Keim's monograph [AEK] is a good introduction to this world. The definitions of Baire-like space and unordered Baire-like space were extended to t.v.s. by Kakol [Ka1-2] and Kakol and Roelcke [KR1]. Among the most recent results, we have that Sánchez Ruiz [Sa3] has introduced and characterized subbarrelled spaces as the class of topological vector spaces without local convexity conditions that satisfies the closed graph theorem as domain spaces when Fréchet or Banach spaces are considered in the range. Other contributions in this field may be found in [Sa2-7] and [MS].

In what concerns barrelled dual spaces, let us recall that a space E is said to be distinguished if its strong dual is barrelled. If E is metrizable and distinguished then $E'(\beta(E',E))$ is even ultrabornological (see for instance [RR, p.161]). Each proper (LB)-space and each strict inductive limit of reflexive Fréchet spaces is distinguished. The class of distinguished Fréchet spaces is also very wide, since for example every reflexive Fréchet space and every quasinormable Fréchet space is distinguished; there also exist non-distinguished Fréchet spaces ([G2] and [K1]). Let us remark that the strong dual of a non-distinguished Fréchet space is a non-barrelled $\aleph_0$-barrelled space.

An important line of research during recent years has been to characterize classes of distinguished Fréchet spaces as wide as possible (cf. Bierstedt and Meise [BM]). Another fundamental question in the theory of distinguished spaces is the study of its stability properties. Groethendieck studied in [G3] the projective tensor product of two distinguished Fréchet spaces. S. Dierolf [De] has shown that in general $\ell^1 \hat{\otimes}_\pi F$ is not distinguished even if F is a reflexive Fréchet space. D. Bierstedt and J. Bonet [BB], J.C. Díaz and M.A. Miñarro [DM], have studied the stability of distinguished spaces under projective tensor products. The case of injective tensor products of distinguished Fréchet spaces has been investigated by J.C. Díaz, J.A. López Molina and M.J. Rivera [DLR], and stability under other tensor products (see [DF] for the definitions) has been studied in [LR], obtaining positive answers under smoother conditions than in the projective case.

2. Dimensionality and cardinality

2.1 INTRODUCTION

We start this chapter with two easy remarks that will put us on the track of the kind of things that we are going to study next. Given any set A, we will write $|A|$ to denote its cardinality.

Proposition 2.1.1 *Let* E *be a metrizable barrelled space. Then either* $\dim(E) < \aleph_0$ *or* $\dim(E) \geq \aleph_1$.

Proof. Assume $\dim(E) = \aleph_0$ and let $\{e_n : n \in \mathbb{N}\}$ be a basis of E. If we write $E_n := \langle\{e_1, \ldots, e_n\}\rangle$ for each $n \in \mathbb{N}$, then $E = \bigcup\{E_n : n \in \mathbb{N}\}$ and, since E is metrizable and barrelled, and therefore Baire-like, some E_n must coincide with E. Contradiction. ∎

Therefore an infinite-dimensional metrizable barrelled space E has $\dim(E) \geq \aleph_1$. But, is $\aleph_1$ really the least possible dimension of a space of this kind?

On the other hand, if $\{U_n : n \in \mathbb{N}\}$ is a base of neighbourhoods of the origin in a metrizable locally convex space, given a bounded subset B of E, for each $n \in \mathbb{N}$, there exists some $b_n \in \mathbb{K}$ such that $B \subseteq b_n U_n$. Thus the collection $\mathfrak{B}$ formed by all the $\bigcap\{b_n U_n : n \in \mathbb{N}\}$, $(b_n) \in \mathbb{K}^{\mathbb{N}}$, is a fundamental system of bounded sets in E, in the sense that every $A \in \mathfrak{B}$ is bounded, and for each bounded subset B of E, there is some $A \in \mathfrak{B}$ such that $B \subseteq A$. Thus,

Proposition 2.1.2 *Every metrizable locally convex space* E *has a fundamental system of bounded sets* $\mathfrak{B}$ *in* E *of size* $|\mathfrak{B}| \leq c$.

But, what is the smallest size that such a fundamental system has?

In the forthcoming pages we will answer these and other basic questions concerning cardinality and dimensionality in metrizable (barrelled) spaces.

2.2 THE BOUNDING AND DOMINATING CARDINALS

We assume that the reader is familiar with the cardinals $\aleph_0$, $\aleph_1$ and the continuum $\mathfrak{c}$. In this section we are going to introduce two cardinals, the bounding and dominating cardinals, which play a predominant role in the study of cardinality/dimensionality in metrizable barrelled spaces. To give their simple definitions we will consider $\mathbb{N}^{\mathbb{N}}$ endowed with the relation, $\leq^*$, defined by

$$f \leq^* g \text{ if } \{n \in \mathbb{N}: f(n) > g(n)\} \text{ is finite.}$$

It is easy to prove that this relation is indeed reflexive and transitive, i.e. a quasi-order, but not symmetric. For the latter consider for example $f = \{1, 1, 1, 1, \dots\}$ and $g = \{2, 1, 1, 1, \dots\}$, note that $f \leq^* g$ and $g \leq^* f$ hold, but $f \neq g$.

Definition 2.2.1 *A subset* $A \subseteq \mathbb{N}^{\mathbb{N}}$ *is said to be unbounded in* $\mathbb{N}^{\mathbb{N}}$ *if* A *is unbounded in* $(\mathbb{N}^{\mathbb{N}},\leq^*)$, *i.e. if there is no* $f \in \mathbb{N}^{\mathbb{N}}$ *such that for every* $g \in A$, $g \leq^* f$.

Definition 2.2.2 *A subset* $A \subseteq \mathbb{N}^{\mathbb{N}}$ *is said to be dominating in* $\mathbb{N}^{\mathbb{N}}$ *if* A *is cofinal in* $(\mathbb{N}^{\mathbb{N}},\leq^*)$, *that is if for each* $f \in \mathbb{N}^{\mathbb{N}}$ *there is some* $g \in A$ *such that* $f \leq^* g$.

Definition 2.2.3 *The bounding cardinal* $\mathfrak{b}$ *and the dominating cardinal* $\mathfrak{d}$ *are defined as the minimum size that a subset of* $\mathbb{N}^{\mathbb{N}}$ *must have to be unbounded and cofinal respectively, i.e.*

$$\mathfrak{b} = \min\{|B|: B \textit{ is unbounded in } \mathbb{N}^{\mathbb{N}}\},$$
$$\mathfrak{d} = \min\{|B|: B \textit{ is cofinal in } \mathbb{N}^{\mathbb{N}}\}.$$

In general,

$$\aleph_1 \leq \mathfrak{b} \leq \mathfrak{d} \leq \mathfrak{c},$$

and equality or strict < between any of the above numbers is consistent with ZFC. The well-known Continuum Hypothesis (CH) is $\aleph_1 = c$ and Martin's Axiom (MA) implies $\mathfrak{b} = \mathfrak{d} = c$. The following result will later be useful.

Lemma 2.2.4 *There is a set* $\mathbb{D}$ *in* $\mathbb{N}^{\mathbb{N}}$ *of size* $\mathfrak{d}$ *such that for each* $f \in \mathbb{N}^{\mathbb{N}}$ *there is some* $g \in \mathbb{D}$ *such that* $f(n) \le g(n)$ *for all* $n \in \mathbb{N}$.

Proof. Let B be a cofinal subset of $\mathbb{N}^{\mathbb{N}}$ of size $\mathfrak{d}$, then $\mathbb{D} := \{mg\colon m \in \mathbb{N}, g \in B\}$ fulfils the lemma. ∎

2.3 CARDINALITY OF RELEVANT FAMILIES OF BOUNDED SETS IN METRIZABLE SPACES

Definition 2.3.1 *Given a metrizable space* E, *we define*

$$\mathfrak{d}_b(E) := \min\{|\mathfrak{B}|\colon \mathfrak{B} \textit{ is a fundamental system of bounded sets in } E\}$$

From Proposition 2.1.2, we know that $\mathfrak{d}_b(E) \le c$ for any metrizable space. The next result settles its value with accuracy.

Theorem 2.3.2 *For* E *a metrizable space,*

$$\mathfrak{d}_b(E) = \begin{cases} 1 & \textit{if } E = \{0\}. \\ \aleph_0 & \textit{if } E \textit{ is normable, } E \neq \{0\}. \\ \mathfrak{d} & \textit{if } E \textit{ is non-normable.} \end{cases}$$

Proof. The first case is trivial and for the second one, note that if B_E denotes the unit ball of E, then $\mathfrak{B} = \{nB_E\colon n \in \mathbb{N}\}$ is a fundamental system of bounded sets in E, and no such system may be finite.

Assume E is non-normable and $\{U_n\colon n \in \mathbb{N}\}$ is a nested base of balanced neighbourhoods of the origin. Let us first show that if $\mathfrak{B}$ is a family of bounded sets in E with size $|\mathfrak{B}| < \mathfrak{d}$, then $\mathfrak{B}$ cannot be a fundamental system of bounded sets, that is there is some bounded set in E which is not contained in any $B \in \mathfrak{B}$. Since E is non-normable, no neighbourhood is bounded and there is a sequence $\{f_n\colon n \in \mathbb{N}\}$ in E' such that $\{f_n(U_n)\colon n \in \mathbb{N}\}$ is unbounded. For each $B \in \mathfrak{B}$ let $g_B \in \mathbb{N}^{\mathbb{N}}$ be the element defined in such a way that $g_B(n)$ is the smallest positive integer satisfying

$$g_B(n) > \sup\{|f_n(x)|: x \in B\}.$$

As $\{g_B: B \in \mathfrak{B}\}$ has size less than $\mathfrak{d}$, there exists some $f \in \mathbb{N}^{\mathbb{N}}$ such that for each $B \in \mathfrak{B}$, $f(n) > g_B(n)$ for infinitely many $n \in \mathbb{N}$, and since each $f_n(U_n)$ is unbounded, we may choose $x_n \in U_n$ with $|f_n(x_n)| > f(n)$. The set $A := \{x_n: n \in \mathbb{N}\}$, being a sequence that converges to 0, is bounded. However A is not contained in any $B \in \mathfrak{B}$ since, if given $B \in \mathfrak{B}$ we take $n \in \mathbb{N}$ such that $f(n) > g_B(n)$, then

$$\sup\{|f_n(x)|: x \in A\} \geq |f_n(x_n)| > f(n) > g_B(n) > \sup\{|f_n(x)|: x \in B\}.$$

Next let us build a fundamental system $\mathfrak{B}$ of bounded sets in E of size $\leq \mathfrak{d}$. Let $\mathbb{D}$ be the set provided by Lemma 2.2.4 and consider as $\mathfrak{B}$ the family given by all the $\bigcap\{h(n)U_n: n \in \mathbb{N}\}$, $h \in \mathbb{D}$. Clearly every such subset of E is bounded and if B is any bounded subset of E, then for each $n \in \mathbb{N}$ there exists some $g(n) \in \mathbb{N}$ such that $B \subseteq g(n)U_n$. Let $h \in \mathbb{D}$ such that $g(n) \leq h(n)$ for all $n \in \mathbb{N}$. Then $B \subseteq \bigcap\{g(n)U_n: n \in \mathbb{N}\} \subseteq \bigcap\{h(n)U_n: n \in \mathbb{N}\} \in \mathfrak{B}$. Hence $\mathfrak{B}$ is a fundamental system of bounded sets in E of size $|\mathfrak{B}| \leq |\mathbb{D}| = \mathfrak{d}$.

Putting together the above two results, we obtain $\mathfrak{d}_b(E) = \mathfrak{d}$. ■

Corollary 2.3.3 *If* E *is a metrizable space and has a fundamental sequence of bounded sets then* E *is normable.*

Corollary 2.3.4 *The minimal size for a base of neighbourhoods of the origin for the strong dual of a non-normable metrizable space is* $\mathfrak{d}$.

On the other hand it is clear that if $\mathfrak{B}$ is a fundamental family of bounded sets in a metrizable space E, then the union of its elements spans E. Thus if we consider $\mathfrak{sp}_b(E)$ to be the minimum cardinality of a family of bounded sets in E whose union spans E, we have

$$\mathfrak{sp}_b(E) \leq \mathfrak{d}_b(E) \leq \mathfrak{d},$$

and for any normed space E,

$$\mathfrak{sp}_b(E) = 1.$$

This gives one half of the following,

Theorem 2.3.5 *Let* E *be a barrelled space. Then,*

$$E \text{ is normable} \Leftrightarrow \mathfrak{sp}_b(E) = 1.$$

Proof. If B is a bounded subset of E and E = $\langle B \rangle$, then the closed absolutely convex hull of B is a bounded barrel, hence a bounded neighbourhood and E is normable. ∎

The next result will help us to provide further information on $\mathfrak{sp}_b(E)$.

Lemma 2.3.6 *Let* $\mathfrak{B}$ *be a family of bounded sets in a metrizable space* E. *If* $|\mathfrak{B}| < \mathfrak{b}$, *there exist positive integers* ε_B *such that* $A := \bigcup\{\varepsilon_B B: B \in \mathfrak{B}\}$ *is bounded.*

Proof. Let $\{U_n: n \in \mathbb{N}\}$ be a base of balanced neighbourhoods of the origin. For each $B \in \mathfrak{B}$ and $n \in \mathbb{N}$, there is some $f_B(n) \in \mathbb{N}$ with $B \subseteq f_B(n)U_n$. Since $\mathfrak{B}$ has size less than $\mathfrak{b}$, there is some $f \in \mathbb{N}^{\mathbb{N}}$ such that $f_B \leq^* f$ for all $B \in \mathfrak{B}$. Therefore, for each $B \in \mathfrak{B}$, $B \subseteq f(n)U_n$ for almost all $n \in \mathbb{N}$. Hence given any $B \in \mathfrak{B}$, there is some $\varepsilon_B \in \,]0, 1]$ such that $\varepsilon_B B \subseteq \bigcap\{f(n)U_n: n \in \mathbb{N}\}$. Therefore $A := \bigcup\{\varepsilon_B B: B \in \mathfrak{B}\}$ is a bounded set. ∎

Since each element of a space gives rise to singleton bounded subsets, the previous lemma yields:

Theorem 2.3.7 *In any metrizable space* E *of dimension* $< \mathfrak{b}$,

$$\mathfrak{sp}_b(E) = 1.$$

Theorem 2.3.8 *For* E *a metrizable space either* $\mathfrak{sp}_b(E) = 1$ *or*

$$\mathfrak{b} \leq \mathfrak{sp}_b(E) \leq \mathfrak{d},$$

$\mathfrak{b}$ *and* $\mathfrak{d}$ *being the largest and smallest possible cardinals in these inequalities.*

Proof. By Theorem 2.3.2, the former statement about $\mathfrak{d}$ is clear. Let us see that there is indeed some metrizable space E with $\mathfrak{sp}_b(E) = \mathfrak{d}$. Consider $\mathbb{K}^{\mathbb{N}}$ as E and let us show that if $\mathfrak{B}$ is a family of bounded sets in $\mathbb{K}^{\mathbb{N}}$ with $|\mathfrak{B}| < \mathfrak{d}$, then its union cannot span $\mathbb{K}^{\mathbb{N}}$. Let $\mathfrak{B}'$ be the set of all positive integral multiples of absolutely convex hulls of unions of finite subsets of $\mathfrak{B}$. Then $\bigcup\mathfrak{B}' = \langle\bigcup\mathfrak{B}\rangle$. Now for each $n \in \mathbb{N}$ and $B \in \mathfrak{B}'$, let $g_B(n)$ be the smallest positive integer such that

$$g_B(n) > \sup\{|u_n(x)|: x \in B\},$$

where u_n is the n-th coordinate functional defined on $\mathbb{K}^{\mathbb{N}}$. Since the set $\{g_B: B \in \mathfrak{B}'\}$ has size less than $\mathfrak{d}$, there is some $f \in \mathbb{N}^{\mathbb{N}} \subset \mathbb{K}^{\mathbb{N}}$ such that, given any $B \in \mathfrak{B}'$, $u_n(f) = f(n) > g_B(n)$ for infinitely many n. Therefore, $f \notin B$ and $\bigcup\mathfrak{B}'=\langle\bigcup\mathfrak{B}\rangle \neq \mathbb{K}^{\mathbb{N}}$.

On the other hand, if $\mathfrak{B}$ is a family of bounded sets in E with $|\mathfrak{B}| < \mathfrak{b}$, then the previous lemma provides a bounded subset A in E whose linear span coincides with the linear span of the union of the elements of $\mathfrak{B}$. Hence either $\mathfrak{sp}_b(E) = 1$ or $\mathfrak{sp}_b(E) \geq \mathfrak{b}$. In section 2.5 we show that there exist metrizable spaces F for which $\mathfrak{sp}(F) = \mathfrak{b}$. ■

2.4 DIMENSIONALITY IN METRIZABLE BARRELLED SPACES

From Proposition 2.1.1, we know that any infinite-dimensional metrizable barrelled space has uncountable dimension. In order to fix the least possible infinite dimensionality $\mathfrak{b}_\infty$ for these kind of spaces, let us first prove,

Lemma 2.4.1 *Let* A *be a set of the space* E *with size* $|A| < \mathfrak{b}$. *If* $\{f_n: n \in \mathbb{N}\}$ *is a sequence in* E' *such that* $\{f_n(x): n \in \mathbb{N}\}$ *converges to* 0 *for each* $x \in A$, *then there is a sequence of positive integers* $\{a_i: i \in \mathbb{N}\}$ *diverging to* $+\infty$, *with* $1 \leq a_i \leq i + 1$, *such that* $\{a_i f_i(x): i \in \mathbb{N}\}$ *is bounded for each* $x \in A$.

Proof. For each $x \in A$, let us consider $f_x \in \mathbb{N}^{\mathbb{N}}$ defined by

$$f_x(n) = \min\{k \in \mathbb{N}: |f_i(x)| \leq 1/(n+1) \text{ for all } i \geq k\}.$$

Thus for any $n \in \mathbb{N}$ and $i \geq f_x(n)$, $|f_i(x)| \leq 1/(n+1)$. Now since $|A| < \mathfrak{b}$, there exists some $h \in \mathbb{N}^{\mathbb{N}}$ such that $f_x \leq^* h$ for all $x \in A$, and we may assume h is strictly increasing and $h(1) = 1$. Therefore for almost all $n \in \mathbb{N}$, if $i \geq h(n)$, then

$$|f_i(x)| \leq \frac{1}{n+1}.$$

In particular, for almost all $n \in \mathbb{N}$, if $h(n+1) > i \geq h(n)$ then

$$(n+1)|f_i(x)| \leq 1.$$

Hence $\{(n+1)f_i(x)\colon h(n+1) > i \geq h(n),\ n \in \mathbb{N}\}$ is bounded for each $x \in A$, and the proof is complete if we take, e.g., $a_i = n + 1$ for $h(n+1) > i \geq h(n)$, $n \in \mathbb{N}$. ■

Theorem 2.4.2 $\mathfrak{b}$ *is the least dimension* $\mathfrak{b}_\infty$ *for any infinite-dimensional metrizable barrelled space.*

Proof. In section 2.5 we will build a metrizable barrelled space which has dimension $\mathfrak{b}$, so $\mathfrak{b}_\infty \leq \mathfrak{b}$.

To show that $\mathfrak{b}_\infty \geq \mathfrak{b}$, note first that if E is non normable, Theorems 2.3.5 and 2.3.8 imply $\mathfrak{sp}_b(E) \geq \mathfrak{b}$, and hence Theorem 2.3.7 yields $\dim(E) \geq \mathfrak{b}$.

If E is an infinite-dimensional normable space, choose a Josefson-Nyssenzweig sequence $\{f_n\colon n \in \mathbb{N}\}$ in $E' = \hat{E}'$, i.e. a normalized sequence of E' which $\sigma(E',E)$-converges to 0. Now, if $\dim(E) < \mathfrak{b}$, by the previous lemma it is quickly seen that there is a sequence of positive integers $\{a_i\colon i \in \mathbb{N}\}$ with $\{a_i f_i\colon i \in \mathbb{N}\}$ $\sigma(E',E)$-bounded but not norm bounded, so that E is not barrelled, which is not possible. Hence $\dim(E) \geq \mathfrak{b}$. ■

Corollary 2.4.3 *Let* E *be a metrizable barrelled space. If* F *is a subspace of* E *of codimension* $< \mathfrak{b}$, *then* $\overline{F}$ *finite-codimensional,* $\mathfrak{b}$ *being the largest cardinal for which this property holds in general.*

Proof. Since the metrizable barrelled quotient $E/\overline{F}$ has dimension less than $\mathfrak{b}$, the previous theorem implies that the closure $\overline{F}$ of F is finite codimensional.

$\mathfrak{b}$ is the largest cardinal for which this poperty is true since $\{0\}$ is a $\mathfrak{b}$-codimensional subspace of $\psi_\mathfrak{b}$ for which this property fails. ■

Theorem 2.4.4 *Let* E *be a metrizable barrelled space. If* F *is a subspace of* E *of codimension* $< \mathfrak{b}$, *then* F *is barrelled,* $\mathfrak{b}$ *being the largest cardinal for which this property holds in general.*

Proof. The previous corollary implies that the closure $\overline{F}$ of F is finite codimensional, and so $\overline{F}$ is barrelled. Hence there is no loss of generality if we assume $E = \overline{F}$ and identify E' with F'. Let C be a cobasis for F in E and suppose that F is not barrelled, so that there is a balanced non-equicontinuous subset A of E' bounded at points of F. Two cases may occur:

(a) Given any neighbourhood U of the origin in E, there is some f $\in$ A such that f(U) is unbounded. Then we can take a basic sequence $\{U_n: n \in \mathbb{N}\}$ of neighbourhoods of the origin in E and a sequence $\{f_n: n \in \mathbb{N}\}$ in A such that every $f_n(U_n)$ is unbounded.

On the other hand, since $|C| < \mathfrak{b}$, there is some $g \in \mathbb{N}^{\mathbb{N}}$ such that, for each $x \in C$,

$$|f_n(x)| \le g(n) \quad \text{for almost all} \quad n \in \mathbb{N}.$$

Thus $\{\frac{1}{g(n)} f_n: n \in \mathbb{N}\}$ is pointwise bounded on C as well as on F, therefore is $\sigma((E',E)$-bounded. However there is no neighbourhood of the origin in E on which this set is uniformly bounded, contradicting the barrelledness of E.

(b) There is some neighbourhood U of the origin in E, such that, for all $f \in A$, f(U) is unbounded. Let $\{U_n: n \in \mathbb{N}\}$ be a sequence of basic neighbourhoods of the origin in E contained in U. Since A is not uniformly bounded on any of the U_n, for each $n \in \mathbb{N}$, there is a sequence $\{f_k: k \in \mathbb{N}\}$ in A such that, for each $k \in \mathbb{N}$,

$$k + 1 < \sup\{|f_k(x)|: x \in U_n\} = M_k < \infty.$$

Thus, $\{\frac{1}{M_k} f_k: k \in \mathbb{N}\}$ is equicontinuous, and since $\{f_k: k \in \mathbb{N}\}$ is $\sigma(E',F)$-bounded and $\{\frac{1}{M_k}: k \in \mathbb{N}\}$ converges to 0,

$$\lim_{k\to\infty} \frac{1}{M_k} f_k(x) = 0$$

for each $x \in F$, as well as for each $x \in C$ since otherwise the Alaougli-Bourbaki theorem would yield a non-zero $\sigma(E',E)$-limit point in E' vanishing on F, contradicting the density of F in E. Now Lemma 2.4.1 provides an unbounded set $\{a_k: k \in \mathbb{N}\}$ of positive integers, with $1 \le a_k \le k + 1$, such that

$$A_n := \{a_k \frac{1}{M_k} f_k: \quad k \in \mathbb{N}\}$$

is pointwise bounded on C. The A_n so defined are pointwise bounded on F, and therefore are $\sigma(E',E)$-bounded, but no A_n is uniformly bounded on

U_n. Applying now the condition that $|C| < \mathfrak{b}$, there must be some $g \in \mathbb{N}^{\mathbb{N}}$ such that, for each $x \in C$,

$$\sup\{|f(x)|: f \in A_n\} \leq g(n) \text{ for almost all } n \in \mathbb{N}.$$

Therefore $\{|f(x)|: f \in \frac{1}{g(n)} A_n\}$ is finite for all $n \in \mathbb{N}$ and, moreover is ≤ 1 for almost all n. Hence $B := \bigcup\{\frac{1}{g(n)} A_n : n \in \mathbb{N}\}$ is a $\sigma(E',E)$-bounded subset of the balanced A which is not uniformly bounded on any U_n. Consequently B is not equicontinuous, again contradicting the barrelledness of E.

$\mathfrak{b}$ is the largest cardinal for which this property holds since, by Proposition 2.1.1, any $\aleph_0$-dimensional subspace of any $\mathfrak{b}$-dimensional metrizable barrelled space — in the next section we will see that such spaces do exist — is non-barrelled and $\mathfrak{b}$-codimensional. ■

2.5 THE SPACE $\psi_{\mathfrak{b}}$

As any cardinal κ is the well ordered set of all ordinals β of smaller size, that is satisfying $|\beta| < \kappa = |\kappa|$, we will index sets of size κ by κ itself. Thus, for ordinals α and β, one has $\alpha < \beta$ if and only if $\alpha \in \beta$. Given $i \in \mathbb{N}$, throughout this section u_i will denote the i-th coordinate functional defined on $\mathbb{K}^{\mathbb{N}}$.

Lemma 2.5.1 *There is a family $\{g_\beta: \beta \in \mathfrak{b}\}$ in $\mathbb{N}^{\mathbb{N}}$ such that if $\{n_k: k \in \mathbb{N}\}$ is a strictly increasing sequence of positive integers and for each $k \in \mathbb{N}$, $v_k = \sum_{i=1}^{n_k} \lambda_{ki} u_i$, $\lambda_{kn_k} \neq 0$, then there exists some $\alpha_0 \in \mathfrak{b}$ such that $\{v_k(x): k \in \mathbb{N}\}$ is unbounded whenever $x \in \langle\{g_\beta: \beta \in \mathfrak{b}\}\rangle \subset \mathbb{K}^{\mathbb{N}}$ has a component beyond α_0.*

Proof. Let us define a family $\{g_\beta: \beta \in \mathfrak{b}\}$ in $\mathbb{N}^{\mathbb{N}}$ by induction. Let $\{f_\beta: \beta \in \mathfrak{b}\}$ be an unbounded subset of $\mathbb{N}^{\mathbb{N}}$ formed by increasing functions and suppose that for some $\beta \in \mathfrak{b}$ we have already chosen $\{g_\alpha: \alpha < \mathfrak{b}\}$. Since the size of this set is less than $\mathfrak{b}$, there is some $f \in \mathbb{N}^{\mathbb{N}}$, which we may assume strictly increasing, upper bound in $\mathbb{N}^{\mathbb{N}}$ for

$$\{f_\gamma + \sum_{\alpha\in\sigma} p_\alpha g_\alpha: \gamma \in \beta,\ \sigma \subseteq \beta \text{ finite},\ p_\alpha \in \mathbb{N}\}.$$

Now define g_β by

$$g_\beta(1) = f(1),\ g_\beta(n+1) = f(n+1) + 2g_\beta(n)f(n+1),\ n \in \mathbb{N}.$$

Let us check that the set $\{g_\beta\colon \beta \in \mathfrak{b}\}$ has the required property. Suppose $\{v_k\colon k \in \mathbb{N}\}$ is as in the hypothesis. Write $n_0 = 0$ and take $h \in \mathbb{N}^{\mathbb{N}}$ such that for each $k \in \mathbb{N}$ and $n_{k-1} < j \le n_k$,

$$h(j)|\lambda_{kn_k}| - \sum_{i=1}^{n_k-1}|\lambda_{ki}| > j.$$

Now fix $\alpha_0 \in \mathfrak{b}$ such that $f_{\alpha_0}(n) > h(n)$ for infinitely many $n \in \mathbb{N}$ and, for simplicity, we will assume that x is of the form

$$x = g_\beta + \sum_{\alpha\in\sigma} a_\alpha g_\alpha \text{ for some finite subset } \sigma \text{ of } \beta,\ \beta > \alpha_0.$$

Let f be as above corresponding to β and choose $p_\alpha \in \mathbb{N}$ with $p_\alpha \ge |a_\alpha|$, $\alpha \in \sigma$. Let $M \in \mathbb{N}$ such that for $n \ge M$,

$$g_\beta(n) \ge f(n) \ge \left(f_{\alpha_0} + \sum_{\alpha\in\sigma} p_\alpha g_\alpha\right)(n). \tag{1}$$

Then for every $n \ge M$,

$$\begin{aligned} |x(n)| &= |g_\beta(n) + \sum_{\alpha\in\sigma} a_\alpha g_\alpha(n)| \le g_\beta(n) + \sum_{\alpha\in\sigma} p_\alpha g_\alpha(n) \\ &\le g_\beta(n) + g_\beta(n) = 2g_\beta(n). \end{aligned} \tag{2}$$

$$\begin{aligned} |x(n+1)| &= |g_\beta(n+1) + \sum_{\alpha\in\sigma} a_\alpha g_\alpha(n+1)| \\ &\ge g_\beta(n+1) - \sum_{\alpha\in\sigma} p_\alpha g_\alpha(n+1) \ge g_\beta(n+1) - f(n+1) \\ &= 2g_\beta(n)f(n+1), \end{aligned}$$

and consequently,

$$\left|\frac{x(n+1)}{x(n)}\right| \ge \frac{2g_\beta(n)f(n+1)}{2g_\beta(n)} = f(n+1).$$

Hence, since $f(n+1) \ge n+1$ for all $n \in \mathbb{N}$, there is some positive integer $N > M$ such that for each $n \ge N$,

$$|x(n)| \ge \max\{|x(i)|\colon 1 \le i < n\}. \tag{3}$$

On the other hand, as $f_{\alpha_0} \le^* f$ but $f_{\alpha_0}(n) > h(n)$ for infinitely many $n \in \mathbb{N}$, there are arbitrarily large $j > N$ with

$$f(j) \ge f_{\alpha_0}(j) > h(j),$$

and for some $k \in \mathbb{N}$, $n_{k-1} < j \le n_k$. Let us now consider

$$|v_k(x)| = \left|\sum_{i=1}^{n_k}\lambda_{ki}x(i)\right| \ge |\lambda_{kn_k}||x(n_k)| - \sum_{i=1}^{n_k-1}|\lambda_{ki}x(i)|$$

$$= |\lambda_{kn_k}| \left| \left(g_\beta + \sum_{\alpha\in\sigma} a_\alpha g_\alpha \right)(n_k) \right| - \sum_{i=1}^{n_k-1} |\lambda_{ki} x(i)|$$

$$\geq |\lambda_{kn_k}| \left(g_\beta - \sum_{\alpha\in\sigma} p_\alpha g_\alpha \right)(n_k) - \sum_{i=1}^{n_k-1} |\lambda_{ki} x(i)|$$

(by (3)) $\geq |\lambda_{kn_k}| \left(g_\beta - \sum_{\alpha\in\sigma} p_\alpha g_\alpha \right)(n_k) - |x(n_k-1)| \sum_{i=1}^{n_k-1} |\lambda_{ki}|$

(by the definition of g_β and (2))

$$\geq |\lambda_{kn_k}| \left[f(n_k)[1 + 2g_\beta(n_k-1)] - \sum_{\alpha\in\sigma} p_\alpha g_\alpha(n_k) \right] - 2g_\beta(n_k-1) \sum_{i=1}^{n_k-1} |\lambda_{ki}|$$

$$= |\lambda_{kn_k}| \left(f - \sum_{\alpha\in\sigma} p_\alpha g_\alpha \right)(n_k) + 2g_\beta(n_k-1) \left[|\lambda_{kn_k}| f(n_k) - \sum_{i=1}^{n_k-1} |\lambda_{ki}| \right]$$

$$\geq |\lambda_{kn_k}| \left(f-f_{\alpha_0} - \sum_{\alpha\in\sigma} p_\alpha g_\alpha \right)(n_k) + 2g_\beta(n_k-1) \left[|\lambda_{kn_k}| f(n_k) - \sum_{i=1}^{n_k-1} |\lambda_{ki}| \right]$$

(taking into account (1) and that $g_\beta(n_k-1) \geq 1$)

$$\geq 2\left[|\lambda_{kn_k}| f(n_k) - \sum_{i=1}^{n_k-1} |\lambda_{ki}| \right]$$

(bearing in mind that $f(n_k) \geq f(j) > h(j)$)

$$> 2\left[|\lambda_{kn_k}| h(j) - \sum_{i=1}^{n_k-1} |\lambda_{ki}| \right]$$

(by the choice of h)

$$> 2j.$$

Consequently, $\{v_k(x): k \in \mathbb{N}\}$ is unbounded. ∎

Theorem 2.5.2 *There is a dense barrelled subspace* $\psi_{\mathfrak{b}}$ *of* $\mathbb{K}^{\mathbb{N}}$ *of dimension* $\mathfrak{b}$ *such that:*

(i) Given any subspace F *of* $\psi_{\mathfrak{b}}$,

F *is barrelled* $\Leftrightarrow$ dim(F) $< \aleph_0$ *or* dim(F) $= \mathfrak{b}$.

(ii) Given any $\mathfrak{b}$*-dimensional subspace* F *of* $\psi_{\mathfrak{b}}$, *its closure* $\overline{F}$ *has finite codimension in* $\psi_{\mathfrak{b}}$.

(iii) Given any bounded subset B *of* $\psi_{\mathfrak{b}}$, dim(⟨B⟩) $< \mathfrak{b}$.

(iv) Given any $\mathfrak{b}$*-dimensional subspace* F *of* $\psi_{\mathfrak{b}}$, $\mathfrak{sp}_{\mathfrak{b}}(F) = \mathfrak{b}$.

Proof. Take $\{g_\beta: \beta \in \mathfrak{b}\}$ in $\mathbb{N}^{\mathbb{N}}$ as in the previous lemma and denote by $\{e_i: i \in \mathbb{N}\}$ the unit vectors. Let us write

$$\psi_{\mathfrak{b}} := \langle \{g_\beta: \beta \in \mathfrak{b}\} \cup \{e_i: i \in \mathbb{N}\} \rangle,$$

and see that this space complies with the promised properties.

(i) ($\Rightarrow$) Suppose F is an infinite-dimensional barrelled subspace of $\psi_{\mathfrak{b}}$.

Putting together Theorem 2.4.2 and the obvious fact that $\dim(F) \leq \dim(\psi_{\mathfrak{b}}) = \mathfrak{b}$, we deduce that $\dim(F) = \mathfrak{b}$.

($\Leftarrow$) If F is a $\mathfrak{b}$-dimensional subspace of $\psi_{\mathfrak{b}}$, then

$$E := F + \langle\{e_i : i \in \mathbb{N}\}\rangle$$

is dense in $\mathbb{K}^{\mathbb{N}}$ and has dual $E' = \langle\{u_i : i \in \mathbb{N}\}\rangle$. Now given any infinite-dimensional set $\{v_k : k \in \mathbb{N}\}$ in E', the previous lemma yields $\alpha_0 \in \mathfrak{b}$ such that whenever $x \in \langle\{g_\beta : \beta \in \mathfrak{b}\}\rangle$ has a component beyond α_0, the scalar set $\{v_k(x) : k \in \mathbb{N}\}$ is unbounded. Therefore any $\sigma(E',E)$-bounded subset of E' is finite-dimensional, and so equicontinuous. Therefore E is barrelled. Corollary 2.4.3 implies F is barrelled.

(ii) Suppose $\overline{F}$ is infinite codimensional in $\psi_{\mathfrak{b}}$. Let G be a subspace of $\psi_{\mathfrak{b}}$ such that $\overline{F}$ is $\aleph_0$-codimensional in G. Then $\dim(G) = \mathfrak{b}$ and, by (i) G is barrelled. Hence the $\aleph_0$-dimensional metrizable space $G/\overline{F}$ is barrelled, contradicting Proposition 2.1.1.

(iii) If for some bounded subset B of $\psi_{\mathfrak{b}}$, $\dim(\langle B\rangle) = \mathfrak{b}$, then $F = \langle B\rangle$ would be a barrelled space with $\mathfrak{sp}_b(F) = 1$, thus normable by Theorem 2.3.5, a contradiction.

(iv) This is an immediate consequence of (iii). ■

2.6 EXERCISES

1. *The bounding cardinal* $\mathfrak{b}$.

(i) Let $\mathfrak{b}_{\langle B\rangle}$ be defined by

$$\mathfrak{b}_{\langle B\rangle} := \sup\{\kappa : E \textit{ is metrizable} \text{ and } \dim(E) < \kappa \Rightarrow \mathfrak{sp}_b(E) = 1\}.$$

$\mathfrak{b}_{\langle B\rangle}$ coincides with $\mathfrak{b}$.

(ii) The cardinal $\mathfrak{b}$ changes if we replace $\leq^*$, of its definition in $\mathbb{N}^{\mathbb{N}}$, by $\leq^{**}$ where $f \leq^{**} g$ if $f(n) \leq g(n)$ for all $n \in \mathbb{N}$.

2. *The dominating cardinal* $\mathfrak{d}$.

(i) The cardinal $\mathfrak{d}$ does not change if we replace $\leq^*$ of its definition in $\mathbb{N}^{\mathbb{N}}$, by the relation $\leq^{**}$ of the previous exercise.

(ii) $\mathfrak{d}$ is the optimal cardinal in order to get the least cardinal κ for which E^{κ} contains ϕ for each space E not endowed with its weak topology.

3. *The space* ψ_δ.

(i) There exists a dense barrelled subspace ψ_δ of $\mathbb{K}^{\mathbb{N}}$ of dimension $\mathfrak{d}$ such that:

(ii) Given any bounded subset B of ψ_δ, $\dim(\langle B\rangle) < \mathfrak{d}$.

(iii) Given any $\mathfrak{d}$-dimensional subspace F of ψ_δ, $\mathrm{sp}_b(F) = \mathfrak{d}$.

4. *The separable quotient problem* consists in finding out whether every infinite-dimensional Banach space has a properly, i.e. a Hausdorff infinite-dimensional, separable quotient. The density character of an infinite-dimensional topological vector space is the smallest dimension of its dense subspaces.

(i) A Banach space has a properly separable quotient if and only if it has a dense non-barrelled subspace.

(ii) By transitivity of quotients the separable quotient problem reduces to Banach spaces of density character $\le \mathfrak{c}$.

(iii) Using only ZFC theory the separable quotient problem is reduced to Banach spaces E with

$$\mathfrak{b} \le \text{density character of } E \le \mathfrak{c}.$$

2.7 NOTES AND REMARKS

All the results of this chapter involving $\mathfrak{b}$, $\mathfrak{d}$ and metrizable (barrelled) spaces have been taken from Saxon and Sánchez Ruiz [SS1] and [SS2]. These two papers along with [SS3] settle a fruitful marriage between the bounding cardinal of Rothberger [Rt] and Katekov [Ka], which date from 1939 and 1960, respectively, and metrizable (barrelled) spaces. This line of research has subsequently been followed by Burke and Todorcevic [BT]. These two cardinals really do happen to behave as *optimal cardinals* in studying these spaces. An excellent exposition of these two and another four cardinals may be found in van Dowen [vD], where the content of section 2.2 may be found apart from Lemma 2.2.4 which comes from [SS1].

One of the most relevant uses of these two cardinals has been to solve the Barrelled Countable Enlargement (BCE) problem for infinite-dimensional metrizable barrelled spaces. The BCE problem, a term used by Robertson, Tweddle and Yeomans [RTY] for the first time (see also Pérez Carreras and Bonet [PB2, 4.5] and Tweddle, Saxon and Sánchez Ruiz [TSS] for a review on this topic), deals with finding out whether for every barrelled space E with $E' \neq E^*$, there is some infinite countable-dimensional subspace M of E^* transverse to E' such that the Mackey topology $\mu(E,E'+M)$, called a countable enlargement of E, is barrelled. Then E is said to have a BCE. To date this problem is unsolved, though partial answers were obtained by Robertson, Tweddle and Yeomans [RTY] who showed that spaces with dense barrelled c-, or greater, codimensional subspaces do have BCE's; Tweddle and Yeomans [TY] where Proposition 2.1.2 of this chapter may be found, enabling them to solve the BCE problem for infinite-dimensional normed spaces assuming CH and for metrizable spaces with dimension > c; and Bonet and Pérez Carreras[BP2] who showed that countable-codimensional subspaces of spaces with a BCE do have a BCE. These themes are focused by Saxon and Sánchez Ruiz, [SS2] and [SS3], improving the aforementioned results, the latter optimally, with the aid of $\mathfrak{b}$ and $\mathfrak{d}$. Moreover, let us remark that in [SS2] the BCE problem was finally solved for any infinite-dimensional normed space using only ZFC theory, and in [SS3] for any metrizable barrelled space of dimension ≥ d, thus covering the c-dimensional case, and solving the BCE problem for infinite-dimensional metrizable barrelled spaces assuming that scales do exist, that is $\mathfrak{b} = \mathfrak{d}$, a condition weaker than MA. Very recently Saxon [S7] has finally solved this problem just using ZFC theory. Tweddle and Catalán [TC] have discussed the preservation of some weak barrelledness conditions under countable enlargements showing that some of the results on existence of BCE's do carry over to the classes of spaces considered by them but there is a negative answer for arbitrary countably barrelled and ω-barrelled spaces.

Grothendieck [G2] showed that every separable metrizable space E is quasidistinguished, i.e. every bounded subset of the completion of E is contained in the closure of a bounded subset of E, and asked whether separability was redundant in the above. Assuming CH, Dieudonné [Di] showed it was not and Amemiya [A] also answered in the negative only using ZFC theory. Clearly normed spaces have this property and Tsirulnikov [Ts] questioned whether the metrizable spaces originated by countable enlargements of normed spaces would also keep it. Assuming CH, Catalán and Tweddle [CT] answered negatively to this question and Saxon and Sánchez Ruiz [SS4], by using $\mathfrak{b}$ and $\mathfrak{d}$, have obtained a negative answer with no extra-ZFC assumptions.

Let us also add that these two venerable cardinals have been used by Saxon and Tweddle [ST] while studying Robertson, Saxon and Robertson's splitting theorem [RSR], which says that any given Hamel basis of a barrelled space E determines two topologically complementary subspaces, one of them endowed with its finest locally convex topology.

The contents of sections 2.3, 2.4 and 2.5 are taken from [SS1] and [SS2]. Proposition 2.1.1 states that the least infinite dimensionality for a metrizable barrelled space is greater than or equal to $\aleph_1$ and is due to Bourbaki (cf. [SL, Corollary C]). In Theorem 2.4.2 ([SS2, Theorem 1]) $\aleph_1$ is optimally replaced by $\mathfrak{b}$ in Bourbaki's result, and this recalls Mazur's result according to which $\mathfrak{c}$ is the smallest infinite dimensionality for a Fréchet space (see e.g. [PB2, p.76] and [PT]). A proof of the existence of a Josefson-Nyssenzweig sequence, that is, a weak* null sequence of norm-one functionals in an infinite-dimensional dual Banach space, used in Theorem 2.4.2, may be seen in [D, Ch. XII]. Corollary 2.4.3 and Theorem 2.4.4 are included in [SS2, Theorem 2], which shows that $\mathfrak{b}$ also plays in metrizable barrelled spaces the same role as $\mathfrak{c}$ does in Fréchet spaces in Saxon and Levin's statement that every subspace of codimension $< \mathfrak{c}$ in a Fréchet space is barrelled [SL, p.95] (cf. [V11]). This recalls Saxon and Levin [SL] and Valdivia [V1] result according to which every subspace of codimension $< \aleph_1$ of a barrelled space is barrelled. Lemma 2.4.1 is [SS2, Lemma]. Corollary 2.3.3 is [SS1, Corollary 1] and gives a proof of [K1, §29.1(2)]. Theorem 2.3.8 ([SS1, Theorem 3]) is not as neat as Theorem 2.3.2 ([SS1, Theorem 1] since as the Remark to [SS1, Theorem 3] points out there are non-normable spaces E with $\mathfrak{sp}_b(E) = 1$: consider ℓ^p with the topology induced by $\mathbb{K}^{\mathbb{N}}$. Theorem 2.3.5 is included in that Remark, too. In a note supplied by M. Burke, he considers the minimum cardinality $cov(\mathfrak{B}_E)$ of a family of bounded sets which covers E, showing that $cov(\mathfrak{B}_E) = \aleph_0 \cdot \mathfrak{sp}_b(E)$. Thus Theorem 2.3.8 provides $cov(\mathfrak{B}_E) \in \{\aleph_0\} \cup [\mathfrak{b}, \mathfrak{d}]$. Burke notes that if $\mathfrak{d} \le \mathfrak{b}^+$, then obviously $cov(\mathfrak{B}_E) \in \{\aleph_0, \mathfrak{b}, \mathfrak{d}\}$, and using ideas of [BT] points out that whether $\mathfrak{b} < cov(\mathfrak{B}_E) < \mathfrak{d}$ may hold is independent of "ZFC + $\mathfrak{b}^+$" $< \mathfrak{d}$, by showing that if there is a Luzin set of cardinality $\mathfrak{d}$ in $\mathbb{R}^{\mathbb{N}}$, then for any $\lambda \in [\mathfrak{b}, \mathfrak{d}]$, there is a subspace E of $\mathbb{R}^{\mathbb{N}}$ such that $cov(\mathfrak{B}_E) = \lambda$, and that if there is a cofinal subset of $(\mathbb{N}^{\mathbb{N}}, \le^*)$ isomorphic to $\omega_1 \times \omega_3$ with the product ordering, then for every metrizable space E, $cov(\mathfrak{B}_E) \in \{\aleph_0, \mathfrak{b}, \mathfrak{d}\}$.

Lemma 2.5.1 is a *simplified* version of Saxon and Sánchez Ruiz [SS1, Lemma], whose additional strength was used by the authors to study the BCE problem. Theorem 2.5.2 appears in [SS1, Theorem 6], and is a skilful CH-free reconstruction of Tweddle's space ψ [Tw], crucial in the locally convex characterizations of $\mathfrak{b}$. Exercises 1, 2 and 3 are taken from [SS1]. In particular, Exercise 2(ii) [SS1, Theorem 2] shows that $\mathfrak{d}$ optimally replaces $\mathfrak{c}$ in Saxon [S1, Theorem 1.4].

Seemingly considered since 1932, explicit references/study of the separable quotient problem may be found in Lacey [L], Rosenthal [Rs1], Saxon and Narayanaswami [SN2], and Saxon and Wilansky [SW]. Properly separable quotients have been studied by Robertson [R2]. Part (i) of Exercise 4 comes from [SW], the other parts are found in [SS2]. Related to density characters is the study of the largest codimension of dense subspaces studied by Saxon [S6].

3. Barrelled spaces of class n

3.1 INTRODUCTION

As we have seen in Chapter 1, a space E is said to be respectively

- *totally barrelled*, if given a sequence of subspaces of E covering E, there is one that is barrelled and its closure is of finite codimension in E;
- *(unordered) Baire-like*, if given any (arbitrary) increasing sequence of closed absolutely convex subsets of E covering E, there is one that is a neighbourhood of the origin;
- *Baire*, if the intersection of any sequence of open dense subsets of E is dense.

The relationship among all these classes of barrelled spaces is

Barrelled ⊇ Baire-like ⊇ Totally barrelled ⊇
Unordered Baire-like ⊇ Baire.

Actually, if E is Baire-like, given any barrel B in E one has that $E = \bigcup\{nB,\ n \in \mathbb{N}\}$, and so B needs to be a neighbourhood of the origin, hence E is a barrelled space. If E is totally barrelled and $\{E_n,\ n \in \mathbb{N}\}$ is any increasing sequence of subspaces covering E, there is a $p \in \mathbb{N}$ such that E_p is barrelled and $\overline{E_p}$ is finite-codimensional, hence for some $q \geq p$, E_q is barrelled and dense in E. Assume E is an unordered Baire-like space and there is a sequence of subspaces $\{E_n,\ n \in \mathbb{N}\}$ of E covering E such that either $\overline{E_n}$ is infinite-codimensional or E_n is not barrelled for each $n \in \mathbb{N}$. In the first case, take $B_n := \overline{E_n}$, in the second take B_n to be the closure in E of a barrel T_n of E_n that is not a neighbourhood of the origin in E_n. Now $\{mB_n,\ m,\ n \in \mathbb{N}\}$ is a sequence of closed absolutely convex sets of E covering E. Hence, there is a $p \in \mathbb{N}$ such that B_p is a neighbourhood of zero in E, obtaining in both cases a contradiction. On the other hand, we know that barrelled, Baire-like, totally barrelled and unordered Baire-like spaces are stable for separated quotients, completions, countable-codimensional subspaces, topological products, and all of them satisfy the three-space problem, i.e., if F is a closed subspace of E such that F and E/F belong to one of these classes, then E belongs to the same class. Now we are going to demonstrate that a Baire-like space cannot contain any non-Baire-like barrelled dense subspace.

Lemma 3.1.1 *Let* F *be a barrelled dense subspace of* E. *If* $\{A_n, n \in \mathbb{N}\}$ *is an increasing sequence of absolutely convex subsets of* F *covering* F, *then* $E = \bigcup\{\overline{A_n}, n \in \mathbb{N}\}$.

Proof. Suppose there is some $x \in E \setminus \bigcup\{\overline{A_n}, n \in \mathbb{N}\}$. Since $x \notin \overline{A_n}$, by the Hahn-Banach theorem there is a $u_n \in E'$ such that $u_n x > 1$ and $|u_n y| \le 1$ for all $y \in A_n$. The set $\{u_n, n \in \mathbb{N}\}$ is weakly* bounded in E'. In fact, if $z \in F$ there exists some $p \in \mathbb{N}$ with $z \in A_p$ and hence $|u_n z| \le 1$ for each $n \ge p$, since $u_n \in A_n^o \subseteq A_p^o$ for each $n \ge p$. As F is barrelled, $\{u_n, n \in \mathbb{N}\}$ is F-equicontinuous and, F being dense in E, E-equicontinuous. Then by the Alaoglu-Bourbaki theorem, the sequence $\{u_n, n \in \mathbb{N}\}$ has a weakly* adherent point u in E'. But $u \in A_n^o$ for each $n \in \mathbb{N}$ since $u_m \in A_n^o$ for each $m \ge n$. Now, $\bigcap\{A_n^o, n \in \mathbb{N}\} = \{0\}$ and hence $u = 0$. Hence, $ux = 0$. But, since $u_n x > 1$ for all $n \in \mathbb{N}$, it follows that $ux \ge 1$, which is a contradiction. ■

Proposition 3.1.2 *Let* F *be a dense barrelled subspace of* E. *Assuming* E *is Baire-like, then* F *is Baire-like.*

Proof. Let $\{A_n, n \in \mathbb{N}\}$ be an increasing sequence of closed absolutely convex subsets of F covering F. By the previous lemma $E = \bigcup\{\overline{A_n}, n \in \mathbb{N}\}$, the closure of the A_n being taken in E. Therefore, there must be some positive integer p such that $\overline{A_p}$ is a neighbourhood of the origin in E and, consequently, A_p is a neighbourhood of the origin in F. ■

Corollary 3.1.3 *A metrizable space is Baire-like if and only if it is barrelled.*

Proof. Notice that the closure of a metrizable space is Baire. ■

3.2 BARRELLED SPACES OF CLASS s

We are going to introduce a decreasing sequence of classes of barrelled spaces with strong barrelledness properties, namely barrelled spaces of class $s \in \mathbb{N}$, located between Baire-like and totally barrelled spaces.

Definition 3.2.1 *Given a positive integer* s, *a countable family of subspaces* $W = \{L_{m_1 m_2 \ldots m_p}, m_r \in \mathbb{N}, 1 \le r \le p \le s\}$ *of a linear space* L *is an* s*-net in* L *if the sequence* $\{L_m, m \in \mathbb{N}\}$ *is increasing, covers* L *and, for each* $p \in \{2, \ldots, s\}$, $\{L_{m_1 \ldots m_{p-1} m}, m \in \mathbb{N}\}$ *is increasing and covers* $L_{m_1 \ldots m_{p-1}}$. *By* W_s *we denote the family* $\{L_{m_1 \ldots m_s}, m_i \in \mathbb{N}, 1 \le i \le s\}$.

Definition 3.2.2 *Given a positive integer* s, *a space* E *is said to be barrelled of class* s, *or briefly,* $E \in \mathscr{C}_s$, *if given any* s*-net* W *in* E, *there is some* $F \in W_s$ *that is barrelled and dense in* E.

A space $E \in \mathscr{C}_1$ is called *suprabarrelled* or *db*. Notice that if E is suprabarrelled and $\{A_n, n \in \mathbb{N}\}$ is an increasing sequence of closed absolutely convex subsets of E covering E then, setting $E_n := \langle A_n \rangle$ for each $n \in \mathbb{N}$, $\{E_n, n \in \mathbb{N}\}$ is an increasing sequence of subspaces of E covering E and some of them, say E_p, must be barrelled and dense in E. Hence A_p is a neighbourhood of the origin in E and E is Baire-like. On the other hand it is clear that $\mathscr{C}_s \supseteq \mathscr{C}_{s+1}$ for each $s \in \mathbb{N}$. Later on we shall see that all these inclusions are strict.

Proposition 3.2.3 *A space* E *is suprabarrelled if and only if given any increasing sequence of subspaces of* E *covering* E, *there is one that is Baire-like.*

Proof. Assuming E is suprabarrelled and $\{E_n, n \in \mathbb{N}\}$ is an increasing sequence of subspaces of E covering E, there is some $p \in \mathbb{N}$ such that E_p is dense and barrelled and then, by Proposition 3.1.2, E is Baire-like. Conversely, taking $E_n := E$ for every $n \in \mathbb{N}$, we deduce that E_p is Baire-like. So, if $\{E_n, n \in \mathbb{N}\}$ is an increasing sequence of subspaces of E covering E there is some $p \in \mathbb{N}$ such that E_p is dense, and by hypothesis Baire-like. Consequently E is suprabarrelled. ■

It is not difficult to notice that each totally barrelled space is barrelled of class s for every $s \in \mathbb{N}$. In any case, we recommend to the reader Exercise 1. The following result is easy to prove.

Proposition 3.2.4 *Given a positive integer* s, *a space* E *is barrelled of class* s *if, and only if, given any increasing sequence of subspaces of* E *covering* E, *there is one that is barrelled of class* s - 1.

Because of the previous result, this proposition becomes true for s = 1 if we consider as $\mathscr{C}_0$ the class of Baire-like spaces. As $\mathscr{C}_s \subseteq \mathscr{C}_0$ $\forall s \in \mathbb{N}$, the subspaces above may be chosen to be dense. Consequently, we have

Corollary 3.2.5. *Given* $s \in \mathbb{N}$, *a space* E *is barrelled of class* s *if, and only if, given any* s-*net* W *in* E, *there is some* $F \in W_s$ *that is Baire-like (and dense in* E*).*

Next we shall show that all classes $\mathscr{C}_s$ enjoy the same stability properties as those of $\mathscr{C}_0$.

Proposition 3.2.6 *Given* $s \in \mathbb{N}$, *if* $E \in \mathscr{C}_s$ *and* F *is a closed subspace of* E, *then* $E/F \in \mathscr{C}_s$.

Proof. Let $W = \{G_{m_1 m_2 \ldots m_p}, m_r \in \mathbb{N}, 1 \le r \le p \le s\}$ be an s-net in E/F. Let k be the canonical mapping of E onto E/F, then $k^{-1}(W) := \{k^{-1}(G_{m_1 m_2 \ldots m_p}), m_r \in \mathbb{N}, 1 \le r \le p \le s\}$ is an s-net in E and there is some $G \in W_s$ such that $k^{-1}(G)$ is Baire-like. Hence $G = k(k^{-1}(G)) \in W_s$ and it is Baire-like. ■

Proposition 3.2.7. *Given* $s \in \mathbb{N}$, *if* $F \in \mathscr{C}_s$ *and* F *is a dense subspace of* E, *then* $E \in \mathscr{C}_s$.

Proof. Let $W = \{E_{m_1m_2...m_p},\ m_r \in \mathbb{N},\ 1 \le r \le p \le s\}$ be an s-net in E. Then $W \cap F := \{E_{m_1m_2...m_p} \cap F,\ m_r \in \mathbb{N},\ 1 \le r \le p \le s\}$ is an s-net in F and there must be some $G \in W_s$ such that $G \cap F$ is Baire-like and dense in F. Since F is dense in E, $G \cap F$ is a dense Baire-like subspace of G and, consequently, G is Baire-like. ∎

Proposition 3.2.8 *Given* $s \in \mathbb{N}$, *if* $E \in \mathscr{C}_s$ *and* F *is a countable-codimensional subspace of* E, *then* $F \in \mathscr{C}_s$.

Proof. Let $W = \{F_{m_1m_2...m_p},\ m_r \in \mathbb{N},\ 1 \le r \le p \le s\}$ be an s-net in F. Let H be an algebraic complement of F in E, then $\{F_{m_1m_2...m_p} + H,\ m_r \in \mathbb{N},\ 1 \le r \le p \le s\}$ is an s-net in E and there is some $G \in W_s$ such that G + H is Baire-like and, consequently, G is Baire-like. ∎

For the next lemma consider the case that $\{F_{m_1m_2...m_p},\ m_r \in \mathbb{N},\ 1 \le r \le p \le s\}$, $s \ge 1$, is an s-net in the product space $E = \Pi\{E_n,\ n \in \mathbb{N}\}$. Suppose that $T_{m_1...m_s}$ is a barrel of $F_{m_1...m_s}$, let $B_{m_1...m_s}$ be its closure in E and $L_{m_1...m_s} := \langle B_{m_1...m_s} \rangle$ for $m_1, \ldots, m_s \in \mathbb{N}$. If $s = 1$, set $G_{m_1} := \bigcap\{L_m,\ m \ge m_1\}$, and if $s > 1$ define by decreasing recurrence the linear subspaces $G_{m_1...m_{p+1}} := \bigcap\{L_{m_1...m_p m},\ m \ge m_p\}$, $L_{m_1...m_p} := \bigcup\{G_{m_1...m_p m},\ m \in \mathbb{N}\}$ for $p = s-1, \ldots, 1$ and $G_{m_1} := \bigcap\{L_m,\ m \ge m_1\}$.

Lemma 3.2.9 *There is a positive integer* q *with* $G_q \supseteq E(q+1, q+2, \ldots)$.

Proof. If this were not true, then for each $j \in \mathbb{N}$ there would be some $x_j \in E(j+1, j+2, \ldots) \setminus G_j$. Now, $A := \overline{\Gamma\{x_j,\ j \in \mathbb{N}\}}$ is a Banach disk in E, and given that $\{G_m,\ m \in \mathbb{N}\}$ covers E_A, there is some $m'_1 \in \mathbb{N}$ such that $G_{m_1} \cap E_A$ is a dense Baire subspace of E_A $\forall m_1 \ge m'_1$. If $s > 1$, suppose by finite induction we have found m'_1, and the functions $m'_i(m_1, \ldots, m_{i-1})$, $2 \le i \le p \le s$, such that for any positive integers $m_1 \ge m'_1$. $\ldots$, $m_p \ge m'_p(m_1, \ldots, m_{p-1})$, $G_{m_1...m_i} \cap E_A$ is a dense Baire subspace of E_A, $1 \le i \le$

p. Then, for any $m_1 \geq m'_1$, ..., $m_p \geq m'_p(m_1,\dots,m_{p-1})$, since $\{G_{m_1\dots m_p m}$, $m \in \mathbb{N}\}$ covers $G_{m_1\dots m_p}$, there exists some $m'_{p+1}(m_1,\dots,m_p) \in \mathbb{N}$ such that $G_{m_1\dots m_{p+1}} \cap E_A$ is a (dense) Baire subspace of E_A for each $m_{p+1} \geq m'_{p+1}(m_1,\dots,m_p)$. So, $A \subseteq L_{m_1\dots m_s}$ if $m_1 \geq m'_1$, ..., $m_s \geq m'_s(m_1,\dots,m_{s-1})$ since $B_{m_1\dots m_s} \cap L_{m_1\dots m_s} \cap E_A$ is a barrel and hence a neighbourhood of the origin in the Baire space $L_{m_1\dots m_s} \cap E_A$ for each $m_1 \geq m'_1$, ..., $m_s \geq m'_s(m_1,\dots,m_{s-1})$. So it follows that $A \subseteq G_{m_1\dots m_s}$ for $m_1 \geq m'_1$, ..., $m_s \geq m'_s(m_1,\dots,m_{s-1})$ and, therefore, $A \subseteq L_{m_1\dots m_{s-1}}$ if $m_1 \geq m'_1$, ..., $m_{s-1} \geq m'_{s-1}(m_1,\dots,m_{s-2})$, which implies that $A \subseteq G_{m_1\dots m_{s-1}}$ for $m_1 \geq m'_1$, ..., $m_{s-1} \geq m'_{s-1}(m_1,\dots,m_{s-2})$. Going on in this way, we obtain that $A \subseteq G_{m_1}$ for $m_1 \geq m'_1$, and hence $x_{m_1} \in G_{m_1}$, a contradiction. On the other hand if $s = 1$ we modify easily the argument to get a similar contradiction. ■

Proposition 3.2.10 *Given* $s \in \mathbb{N}$, *if* $\{E_i,\ i \in I\}$ *is a family of barrelled spaces of class* s, *then* $E := \Pi\{E_i,\ i \in I\} \in \mathscr{C}_s$.

Proof. We will carry out the proof in three steps,

(i) $I = \{1, 2\}$. Suppose $W = \{F_{m_1 m_2\dots m_p},\ m_r \in \mathbb{N},\ 1 \leq r \leq p \leq s\}$ is an s-net in E, then $\{F_{m_1 m_2\dots m_p} \cap E_i,\ m_r \in \mathbb{N},\ 1 \leq r \leq p \leq s\}$ is an s-net in E_i for $i = 1, 2$ and there must be some $G \in W_s$ such that $G \cap E_i$ is a dense Baire-like subspace of E_i for $i = 1, 2$. Thus $G \cap E_1 \times G \cap E_2$ is a dense Baire-like subspace of G and G is Baire-like.

(ii) $I = \mathbb{N}$. We already know the property is true for $s = 0$. So we assume this property holds for some $s \geq 0$ and $E_i \in \mathscr{C}_{s+1}$, $i \in I$, but $E \in \mathscr{C}_s \setminus \mathscr{C}_{s+1}$. This means that there is an (s+1)-net in E, say $\{F_{m_1 m_2\dots m_p}$, $m_r \in \mathbb{N},\ 1 \leq r \leq p \leq s+1\}$, formed by dense subspaces such that no $F_{m_1\dots m_{s+1}}$ is barrelled. For each m_1, ..., $m_{s+1} \in \mathbb{N}$, let $T_{m_1\dots m_{s+1}}$ be a barrel of $F_{m_1\dots m_{s+1}}$ which is not a neighbourhood of the origin in

$F_{m_1 \dots m_{s+1}}$, let $B_{m_1 \dots m_{s+1}}$ be its closure in E and set $L_{m_1 \dots m_{s+1}} :=$ $\langle B_{m_1 \dots m_{s+1}} \rangle$. We define by decreasing recurrence for $p = s+1, \dots, 1$, $G_{m_1 \dots m_p}$, $L_{m_1 \dots m_p}$ and G_{m_1} as in the previous lemma. Notice that $\{G_m, m \in \mathbb{N}\}$ and (if $s \geq 1$) $\{G_{m_1 \dots m_p m}, m \in \mathbb{N}\}$ are 1-nets in E and $L_{m_1 \dots m_p}$, respectively, for each $m_r \in \mathbb{N}$, $1 \leq r \leq p \leq s+1$. Besides $F_{m_1 m_2 \dots m_p} \subseteq G_{m_1 m_2 \dots m_p}$ for each $m_r \in \mathbb{N}$, $1 \leq r \leq p \leq s+1$.

By Lemma 3.2.9 there is some $q \in \mathbb{N}$ such that $G_q \supseteq E(q+1, q+2, \dots)$.

On the other hand, $W := \{G_m \cap E(1, \dots, q), m \in \mathbb{N}\}$ is a 1-net in $E(1, \dots, q)$ and either by (i) if $s > 1$, or by the stability properties of the Baire-like spaces if $s = 0$, there exists $m_1 \in \mathbb{N}$ such that $G_{m_1} \cap E(1, \dots, q)$ is dense in $E(1, \dots q)$ and belongs to $\mathscr{C}_s$. Now as $G_{m_1} \supseteq G_m \cap E(1, \dots, q) + E(q+1, q+2, \dots)$, which belongs to $\mathscr{C}_s$ and is dense in E, it follows from Proposition 3.2.7 if $s > 1$ or by the stability properties of Baire-like spaces if $s = 0$ that $G_{m_1} \in \mathscr{C}_s$. Besides as G_{m_1} is dense in E, it follows that $L_{m_1} \in \mathscr{C}_s$. But this means there is an $m_2 \in \mathbb{N}$ such that $G_{m_1 m_2} \in \mathscr{C}_{s-1}$ and is dense in E. In this way we obtain a $G_{m_1 \dots m_{s+1}} \in \mathscr{C}_0$ which is dense in E. So $B_{m_1 \dots m_{s+1}}$ is a neighbourhood of the origin, a contradiction.

(iii) I is uncountable. As in (ii) we know this property is true for $s = 0$. Assuming $E_i \in \mathscr{C}_{s+1}$ for each $i \in I$, and $E \in \mathscr{C}_s$ for some $s \geq 0$, we show that $E \in \mathscr{C}_{s+1}$. If $E \notin \mathscr{C}_{s+1}$ there exists an (s+1)-net in E, say $W = \{F_{m_1 m_2 \dots m_p}, m_r \in \mathbb{N}, 1 \leq r \leq p \leq s+1\}$, formed by dense subspaces such that no $F_{m_1 \dots m_{s+1}}$ is a barrelled subspace. For each $m_1, \dots, m_{s+1} \in \mathbb{N}$, let $T_{m_1 \dots m_{s+1}}$ be a barrel of $F_{m_1 \dots m_{s+1}}$ which is not a neighbourhood of the origin in $F_{m_1 \dots m_{s+1}}$ and let $B_{m_1 \dots m_{s+1}}$ be its closure in E.

Now some $\langle B_{m_1 \dots m_{s+1}} \rangle$ contains the dense subspace E_0 of E formed by all those vectors of E with countable support. Indeed, if for each

$m_1, \ldots, m_{s+1} \in \mathbb{N}$ there exists some $x_{m_1 \ldots m_{s+1}} \in E_0 \setminus \langle B_{m_1 \ldots m_{s+1}} \rangle$, the support of the $x_{m_1 \ldots m_{s+1}}$ being countable, there is a countable subset H of I such that $\{x_{m_1 \ldots m_{s+1}}, m_r \in \mathbb{N}, 1 \le r \le s+1\} \subseteq E(H)$. On the other hand, $W \cap E(H)$ is an (s+1)-net in E(H) and, by (ii), there exists some barrelled subspace $F_{m_1 \ldots m_{s+1}} \cap E(H)$ of E(H) dense in E(H). Therefore $T_{m_1 \ldots m_{s+1}} \cap E(H)$ is a neighbourhood of the origin in $F_{m_1 \ldots m_{s+1}} \cap E(H)$ and so $B_{m_1 \ldots m_{s+1}}$ contains a neighbourhood of the origin in E(H). Hence $x_{m_1 \ldots m_{s+1}} \in \langle B_{m_1 \ldots m_{s+1}} \rangle$, a contradiction. The barrelledness of E_0 (see Exercise 2 (ii)) implies that $B_{m_1 \ldots m_{s+1}}$ is a neighbourhood of the origin in E, a contradiction. ■

Lemma 3.2.11 *Let* F *be a closed subspace of* E *and let* k *be the canonical mapping of* E *onto* E/F. *Suppose* U *is a balanced subset of* E. *If there is an absolutely convex neighbourhood of the origin* V *in* E *such that* $V \cap F \subseteq U$ *and* $\overline{k(V \cap U)}$ *is a neighbourhood of the origin in* E/F, *then* $\overline{U + U}$ *is a neighbourhood of the origin in* E.

Proof. By hypothesis $\overline{\frac{1}{3}k(V \cap U)}$ is a neighbourhood of zero in E/F and is contained in the subset $k\left(\overline{\frac{1}{3}(V \cap U)+F}\right)$ of E/F. Therefore the set Z:= $\frac{1}{3}V \cap \left(\overline{\frac{1}{3}(V \cap U)+F}\right)$ is a neighbourhood of the origin in E. Now, given any neighbourhood of the origin W in E contained in $\frac{1}{3}V$, we have

$$Z \subseteq \frac{1}{3}V \cap \left(\frac{1}{3}(V \cap U) + F + W\right) \subseteq \frac{1}{3}(V \cap U) + V \cap F + W \subseteq U + U + W.$$

Hence $Z \subseteq \overline{U + U}$, and the proof is complete. ■

Proposition 3.2.12 *Given* $s \in \mathbb{N}$, *if* F *is a closed subspace of* E *such that* F *and* $E/F \in \mathscr{C}_s$ *then* $E \in \mathscr{C}_s$.

Proof. We already know that the result is true for s = 0. Now assume by recurrence this result is true for $s \ge 0$ but not for s+1, i.e. that F and $E/F \in \mathscr{C}_{s+1}$ but $E \in \mathscr{C}_s \setminus \mathscr{C}_{s+1}$. Then there is an (s+1)-net in E, $W = \{E_{m_1 m_2 \ldots m_p}, m_r \in \mathbb{N}, 1 \le r \le p \le s+1\}$ formed by dense subspaces such that no $E_{m_1\ldots m_{s+1}}$ is barrelled. For each $m_1, \ldots, m_{s+1} \in \mathbb{N}$, let $T_{m_1\ldots m_{s+1}}$ be a barrel of $E_{m_1\ldots m_{s+1}}$ which is not a neighbourhood of the origin in $E_{m_1\ldots m_{s+1}}$ and let $B_{m_1\ldots m_{s+1}}$ be its closure in E. Let k be the canonical mapping of E onto E/F. Now we extract a sub-(s+1)-net W^* from W such that if $E_{m_1\ldots m_{s+1}} \in W^*_{s+1}$, then $E_{m_1\ldots m_{s+1}} \cap F$ and $k(E_{m_1\ldots m_{s+1}})$ are barrelled and dense in F and E/F, respectively.

The barrelledness of $E_{m_1\ldots m_{s+1}} \cap F$ implies the existence of a neighbourhood of the origin U in E such that $U \cap F \subseteq B_{m_1\ldots m_{s+1}}$. On the other hand, since $B_{m_1\ldots m_{s+1}} \cap U$ is absorbing in $E_{m_1\ldots m_{s+1}}$ and $k(E_{m_1\ldots m_{s+1}})$ is barrelled and dense in E/F, then $\overline{k(B_{m_1\ldots m_{s+1}} \cap U)}$ is a neighbourhood of the origin in E/F, contradicting the previous lemma. ■

3.3 DISTINGUISHING BETWEEN THE DIFFERENT CLASSES

In order to distinguish the different classes of barrelled spaces of class $s \in \mathbb{N}$ we are going to see that each non-normable Fréchet space E contains a dense subspace H that belongs to $\mathscr{C}_{s-1}$ but not to $\mathscr{C}_s$. As usual, ω will denote the topological product of infinite countably many one-dimensional spaces.

Proposition 3.3.1 *Given* $s \in \mathbb{N}$ *the space* ω *contains a dense subspace* $F \in \mathscr{C}_{s-1} \setminus \mathscr{C}_s$.

Proof. For each $n \in \mathbb{N}$, set $F_n := \omega^n \times (\ell^1)^{\mathbb{N}}$, ρ_n the product topology of F_n and consider the dense proper subspace of $\omega^{\mathbb{N}}$, $F = \bigcup\{F_n,\ n \in \mathbb{N}\}$ with the relative topology τ. To show $F(\tau)$ is barrelled, we need the following

<u>Claim</u>. $F(\tau)$ *is the inductive limit of the sequence* $\{F_n(\rho_n),\ n \in \mathbb{N}\}$

<u>Proof of the claim</u>. Denoting by ρ the inductive limit topology of the sequence $\{F_n(\rho_n),\ n \in \mathbb{N}\}$ pick a continuous linear functional u on $F(\rho)$. Fixing some $n \in \mathbb{N}$, the restriction u_n of u on F_n belongs to

$$F_n(\rho)' = \varphi \oplus \overset{n}{\ldots\ldots} \oplus \varphi \oplus \ell^\infty \oplus \ell^\infty \oplus \ldots .$$

Let $m \in \mathbb{N}$ be such that $u_n \in \varphi \oplus \overset{n}{\ldots\ldots} \oplus \varphi \oplus \ell^\infty \oplus \overset{m}{\ldots\ldots} \oplus \ell^\infty$. Since u is ρ_{n+m}-continuous on F_{n+m}, u_n is ρ_{n+m}/F_n-continuous on F_n, and since F_n is ρ_{n+m}-dense in F_{n+m}, it follows that

$$u_n \in F_n(\rho_{n+m}/F_{n+m})' = F_{n+m}(\rho_{n+m})' = \varphi \oplus \overset{n+m}{\cdots\cdots} \oplus \varphi \oplus \ell^\infty \oplus \ell^\infty \oplus \ldots$$

and therefore $u_n \in \varphi \oplus \varphi \oplus \ldots$. Given that $F(\tau)' = \varphi \oplus \varphi \oplus \ldots$, by the Hahn-Banach theorem u_n defines a τ-continuous linear functional v on the whole space F. But v = u, because both coincide in the τ-dense subspace F_n of F. Hence $u \in F(\tau)'$. Since $F(\tau)' = F(\rho)'$ we conclude that $F(\tau)$ coincides with $F(\rho)$ since they are two Mackey spaces with the same topological dual.

We continue with the proof of Proposition 3.3.1. In fact, if $F(\tau)$ were suprabarrelled, we would be able to assume that the F_n with the topology induced by τ are barrelled. Then according to the closed graph theorem, ρ_n would coincide with τ in F_n and hence $F(\tau)$ would be complete, a contradiction. Hence ω contains a barrelled dense subspace F which is not suprabarrelled, and by Proposition 3.1.2, the property becomes true for s = 1.

Now, assuming that for some $s \geq 1$ ω contains a dense subspace $F \in \mathscr{C}_{s-1} \setminus \mathscr{C}_s$, we are going to find a dense subspace of ω that belongs to $\mathscr{C}_s \setminus \mathscr{C}_{s+1}$. Consider the dense subspaces of $\omega^{\mathbb{N}}$, $E_r := \omega^r \times F^{\mathbb{N}}$, $r \in \mathbb{N}$, and $E := \bigcup\{E_r,\ r \in \mathbb{N}\}$. Then $E \notin \mathscr{C}_{s+1}$ since no E_r is barrelled of class s. We are going to show that $E \in \mathscr{C}_s$. Otherwise, $E \in \mathscr{C}_{s-1} \setminus \mathscr{C}_s$ and we would be able to find an s-net in E, $W = \{F_{m_1 m_2 \ldots m_p},\ m_r \in \mathbb{N},\ 1 \leq r \leq p \leq s\}$ of dense subspaces such that no $F_{m_1 \ldots m_s}$ is barrelled. For each $m_1, \ldots, m_s \in \mathbb{N}$, let $T_{m_1 \ldots m_s}$ be a barrel of $F_{m_1 \ldots m_s}$ which is not a neighbourhood of the origin in $F_{m_1 \ldots m_s}$, let $B_{m_1 \ldots m_s}$ be its closure in E and set $L_{m_1 \ldots m_s} := \langle B_{m_1 \ldots m_s} \rangle$. By decreasing recurrence, if $s-1 \geq p \geq 1$, define $G_{m_1 \ldots m_{p+1}} := \bigcap\{L_{m_1 \ldots m_p m},\ m \geq m_{p+1}\}$, $L_{m_1 \ldots m_p} := \bigcup\{G_{m_1 \ldots m_p m},\ m \in \mathbb{N}\}$, and $G_{m_1} := \bigcap\{L_m,\ m \geq m_1\}$. Now, $\{G_m,\ m \in \mathbb{N}\}$ and $\{G_{m_1 m_2 \ldots m_p},\ m_p \in \mathbb{N}\}$ are 1-nets in E and $L_{m_1 \ldots m_{p-1}}$, respectively, and $F_{m_1 m_2 \ldots m_p} \subseteq G_{m_1 m_2 \ldots m_p}$, for each $m_r \in \mathbb{N}$, $1 \leq r \leq p \leq s$.

As in Lemma 3.2.9 there is some $p \in \mathbb{N}$ such that $G_p \supseteq \{0\}^p \times F^{\mathbb{N}}$. On the other hand, since $\{G_s, s \geq p\}$ covers the Baire space (and hence, barrelled of class s, for every $s \in \mathbb{N}$) ω^p, there is some $m_1 \geq p$ such that $G_{m_1} \cap \omega^p$ is a dense subspace of ω^p and belongs to $\mathscr{C}_{s-1}$. This implies that G_{m_1} contains the subspace $(G_{m_1} \cap \omega^p) \times F^{\mathbb{N}}$ which belongs to $\mathscr{C}_{s-1}$ and is dense in E. Therefore $G_{m_1} \in \mathscr{C}_{s-1}$ and a fortiori $L_{m_1} \in \mathscr{C}_{s-1}$. Thus, there is some $m_2 \in \mathbb{N}$ such that $G_{m_1 m_2} \in \mathscr{C}_{s-2}$ and is dense in E. Continuing in this way we obtain some $G_{m_1 \ldots m_s}$ which is barrelled and dense in E. This implies that $B_{m_1 \ldots m_s}$ is a neighbourhood of the origin in E and, consequently, $T_{m_1 \ldots m_s}$ is a neighbourhood of the origin in $F_{m_1 \ldots m_s}$, a contradiction. Now the conclusion follows since $\omega^{\mathbb{N}} \cong \omega$. ■

Lemma 3.3.2 *Let* E *be a Fréchet space. If* E *is not a Banach space, then* E *has a closed linear subspace* H *such that* $E/H \cong \omega$.

Proof. Let $\{U_n, n \in \mathbb{N}\}$ be a strictly decreasing sequence of non-bounded neighbourhoods of the origin in E and let $\{B_n, n \in \mathbb{N}\}$ be the increasing sequence of its corresponding polars. As U_n is not bounded, its polar B_n is not absorbing in E'. Hence, there exist a sequence $\{K_n, n \in \mathbb{N}\}$ consisting of the linear spans of the elements of a subsequence of $\{B_n, n \in \mathbb{N}\}$ such that K_n is strictly contained in K_{n+1}. Set $u_n \in K_n \setminus K_{n-1}$ and denote by G the linear span of the sequence $\{u_n, n \in \mathbb{N}\}$ of linearly independent vectors of E'.

As $\{B_n, n \in \mathbb{N}\}$ is a fundamental sequence of weak* bounded sets in E', then $G \cap B$ is weakly* closed in E' for every weakly* bounded subset B of E' since the bounded sets of G in the weak* relative topology are finite-dimensional. Then by a well-known fact concerning the convex sets of the duals of Fréchet spaces G is weakly* closed in E'. Setting $H = G^{\perp}$, it is obvious that the quotient topology of E/H is the Mackey topology $\mu(E/H,H^{\perp}) = \mu(E/H,G)$, since G is weakly* closed in E'. But as the weakly* compact subsets of G are finite-dimensional, it follows that $\mu(E/H,G) = \sigma(E/H,H')$. Consequently E/H is weakly complete and so it is isomorphic to ω. ■

Theorem 3.3.3 *Given* $s \in \mathbb{N}$ *and any non-normable Fréchet space* E, E *contains a dense subspace* $F \in \mathscr{C}_{s-1} \setminus \mathscr{C}_s$.

Proof. By the previous lemma there is a closed subspace G of E such that $E/G \cong \omega$ and, by the previous result, ω contains a dense subspace H $\in \mathscr{C}_{s-1} \setminus \mathscr{C}_s$. Let k be the canonical mapping of E onto E/G and $F := k^{-1}(H)$. As $F/G \cong H$ and $\mathscr{C}_s$ satisfies the three-space problem, $F \in \mathscr{C}_{s-1} \setminus \mathscr{C}_s$. ■

3.4 EXERCISES

1. *On totally barrelled spaces.*

(i) Assuming the union of two families $\mathscr{A}$ and $\mathscr{B}$ of linear subspaces of a linear space E, with $|\mathscr{A}|$, $|\mathscr{B}| < c$, covers E, then one of them covers E.

(ii) A space E is said to satisfy condition (G) if, given a sequence $\{E_n, n \in \mathbb{N}\}$ of subspaces of E covering E, there is some E_m which is barrelled. If E is a separable space that satisfies condition (G) and does not contain a complemented copy of φ, then E is totally barrelled.

(iii) If a space E saisfies condition (G) and does not contain φ, then E is totally barrelled.

2. *A barrelled space of class s.* Let $\{E_i, i \in I\}$ be a family of spaces and let E_0 denote the linear subspace of $E = \Pi\{E_i, i \in I\}$ formed by all those elements of E having at most countable support, i.e., a countable number of coordinates different from zero.

(i) If T is a barrel in E_0, then T absorbs the bounded sets in E_0.

(ii) If for each $i \in I$, E_i is barrelled, then E_0 is barrelled.

(iii) If for each $i \in I$, E_i is barrelled of class s, E_0 is a barrelled space of class s.

3. *b spaces.* A space E is said to be a *b* space if, given an increasing sequence of subspaces of E covering E, one of them is barrelled.

(i) Let E be the locally convex hull of a family of *b* spaces and let F be an (LF) space. If f is a linear mapping from E into F with closed graph, then f is continuous.

(ii) Let $\{E_n, n \in \mathbb{N}\}$ be a sequence of spaces such that $E(1,2,\dots,n)$ is a *b* space for each $n \in \mathbb{N}$. Then $\Pi\{E_n, n \in \mathbb{N}\}$ is a *b* space.

(iii) $\varphi^{\mathbb{N}}$ is a *b* space which is not *db*.

(iv) Every non-complete (LF)-space is a barrelled space which is not *b*.

4. *d spaces.* A space E is said to be a *d* space if it is barrelled and given any increasing sequence of subspaces of E covering E, at least one of the subspaces is dense in E.

(i) Every Baire-like space is a *d* space.

(ii) Every $(LF)_1$-space, i.e., every (LF)-space with a defining sequence none of whose members is dense, is a barrelled space which is not *d*.

(iii) Every $(LF)_2$-space, i.e., every non-metrizable (LF)-space E with a defining sequence whose members are dense in E, is a *d* space which is not Baire-like.

5. *A d space.* Let ω be the vector space of all real sequences and let $\{A_i, i \in \mathbb{N}\}$ be an infinite partition of $\mathbb{N}$ such that each set A_i has infinitely many elements. Let $E_0 := \ell^1$ and define

$$E_i := \{x \in \omega: x_j = 0 \textit{ for each } j \notin A_i\}$$

for i = 1, 2, ... Denote by E the locally convex hull of the family $\{E_n, n = 0, 1, \dots\}$.

(i) E is barrelled.

(ii) E is even a *d* space.

(iii) E is not Baire-like. *[Hint: Define* $A_n := \{x \in \omega: |x_k| \le n$ *for each* $k \in A_i$ *with* $i > n\}$ *then show that* $\{A_n, n \in \mathbb{N}\}$ *is an increasing sequence of closed nowhere-dense absolutely convex subsets of* E *covering* E.*]*

6. *Distinguishing between Baire-like spaces and suprabarrelled spaces.* Every $(LF)_3$-space, i.e. each metrizable proper (LF)-space, is a Baire-like space which is not suprabarrelled.

7. *A Baire-like space which is not db and is not dominated by an (LF)-space.* Consider in the product space $\omega^{\mathbb{N}}$ the sequence $\{E_n, n \in \mathbb{N}\}$ with $E_n = \omega^n \times \varphi_\omega \times \varphi_\omega \times \dots$ where here φ_ω denotes the space φ provided with the topology induced by ω. Define $E := \cup\{E_n, n \in \mathbb{N}\}$.

(i) E is a dense subspace of $\omega^{\mathbb{N}}$ which is not *db*.

(ii) E is Baire-like.

(iii) If τ is a locally convex topology strictly finer than the one of E, then $E(\tau)$ is not an (LF)-space.

(iv) Suppose that E is the inductive limit of an increasing sequence $\{F_n(\rho_n), n \in \mathbb{N}\}$ of Fréchet spaces. Then there is some $m \in \mathbb{N}$ such that $\{0\} \times \overset{m}{\dots} \times \{0\} \times \varphi \times \varphi \times \dots \subseteq F_m$. *[Hints: Use Exercises 1 and 2 and closed graph theorems to show that* $\varphi^{\mathbb{N}} \cap F_n$ *is a closed barrelled subspace of* $\varphi^{\mathbb{N}}$ *for infinitely many indexes n and that there exists some m such that* $\varphi^{\mathbb{N}} \cap F_m$ *contains* $\{0\} \times \overset{m}{\dots} \times \{0\} \times \varphi \times \varphi \times \dots$ *.]*

(v) There is some $s \geq m$ such that $E_s \supseteq F_m$.

(vi) The canonical injections from $\{0\} \times \overset{m}{\dots} \times \{0\} \times \varphi \times \varphi \times \dots$ into $F_m(\rho_m)$ and from $F_m(\rho_m)$ into $\{0\} \times \overset{s}{\dots} \times \{0\} \times \varphi \times \varphi \times \dots$ are continuous so $\{0\} \times \overset{m}{\dots} \times \{0\} \times \varphi \times \varphi \times \dots$ must be metrizable, a contradiction.

8. *Each separable infinite-dimensional Fréchet space* E *contains a dense totally barrelled subspace* F *which is not unordered Baire-like.* Let us consider a biorthogonal system (x_n, u_n) for E such that (x_n) is total in E, let $\mathcal{F}$ be the filter in $\mathbb{N}$ of all cofinite subsets and let $\mathcal{U}$ be an ultrafilter that refines $\mathcal{F}$. Let us denote by L(U) the closure in E of $\langle\{x_n, n \notin U\}\rangle$ and set $L = \cup\{L(U), U \in \mathcal{U}\}$.

(i) L is a dense linear subspace of E.

(ii) L is not unordered Baire-like.

(iii) Let $\{L_n, n \in \mathbb{N}\}$ be a sequence of subspaces of L covering L, and let us define F_n as the linear hull of the set $L_n \cup \{x_j, j \in \mathbb{N}\}$ for each $n \in \mathbb{N}$. Assuming L_n is not barrelled for each $n \in \mathbb{N}$, then no F_n is barrelled.

(iv) Let T_n be a barrel in F_n which is not a neighbourhood of zero in F_n, and let B_n denote the closure of T_n in L. Then, there is some $m \in \mathbb{N}$

such that B_m absorbs each bounded closed absolutely convex subset of H $= \langle\{x_n, n \in \mathbb{N}\}\rangle$. *[Hint: Choose a bounded closed absolutely convex set* B *in* L *and assume by contradiction that no* B_n *absorbs* $B \cap H$. *Construct a countable set* $\{y_{jk}: j, k \in \mathbb{N}, k \le j\}$ *in* $B \cap H$ *and two sets* M_1, M_2 *in* $\mathbb{N}$, *so that* $y_{jk} \notin j \cdot B_k$ *for* $k \le j$ *while* $M_1 \cup M_2 = \mathbb{N}$ *and* $M_1 \cap M_2 = \emptyset$, *in such a way that if* $M_1 \in \mathcal{U}$ *then* $y_{2j,k} \in L(M_1)$ *for* $2j \ge k$ *with* $j, k \in \mathbb{N}$, *while if* $M_2 \in \mathcal{U}$ *then* $y_{2j+1,k} \in L(M_2)$ *for* $2j+1 \ge k$ *with* $j, k \in \mathbb{N}$. *Assuming for instance that* $M_1 \in \mathcal{U}$, *as* $L(M_1)$ *is a Fréchet space, there must be some* m *in* $\mathbb{N}$ *such that* $y_{2j,m} \in 2jB_m$ *for some* $j \in \mathbb{N}$, *a contradiction.]*

(v) There is a positive integer r such that B_r absorbs each bounded set of the subspace H.

(vi) As H is a bornological space, B_r is a neighbourhood of the origin in H, hence B_r is a neighbourhood of the origin in L, a contradiction.

(vii) Consequently, there is a $p \in \mathbb{N}$ such that L_p is Baire-like, and so L is totally barrelled.

9. *Spaces having property (L).* As we have seen in section 1.8 a space E is said to have property (L) if given a sequence $\{A_n, n \in \mathbb{N}\}$ of absolutely convex subsets of E covering E, with $2A_n \subseteq A_{n+1}$, then each absolutely convex set U, meeting A_n in a neighbourhood of the origin in A_n for all $n \in \mathbb{N}$, is a neighbourhood of zero in E (see Chapter 1).

(i) Every barrelled space has property (L).

(ii) Let F be a dense subspace of a space E having property (L). If $\{A_n, n \in \mathbb{N}\}$ is an increasing sequence of absolutely convex subsets of F covering F with $2A_n \subseteq A_{n+1}$, then $E = \bigcup\{\overline{A_n}, n \in \mathbb{N}\}$.

(iii) The strong dual of a non distinguished Fréchet space has property (L) but it is not barrelled.

3.5 NOTES AND REMARKS

Suprabarrelled spaces were introduced by M. Valdivia in [V7]. Independently, W.J. Robertson, I. Tweddle and F.E. Yeomans also consider this last class of spaces in [RTY], calling them spaces with property (db), without a study of its properties.

Other strong barrelledness conditions have been investigated by M. Valdivia in [V6], where Baire-hyperplane spaces are defined, and in the book [V9], where convex-Baire spaces are studied with great detail. Actually, a space is a Baire-hyperplane if it cannot be covered by any sequence of closed hyperplanes and a space is called convex-Baire if it cannot be covered by any sequence of nowhere-dense closed convex sets. Both classes of spaces have good permanence properties, in particular both are stable under arbitrary products. Ordered suprabarrelled spaces have been studied by J.C. Ferrando and M. López-Pellicer [FL1], quasi-Baire spaces have been introduced by S.A. Saxon [S1] and studied later on by Narayanaswami and Saxon in [NS]. Spaces satisfying condition (G) have been defined by P. Pérez Carreras and J. Bonet in [PB1], quasi-suprabarrelled spaces have been studied by Ferrando and López-Pellicer [FL2], quasi-totally barrelled spaces have been introduced by J.C. Ferrando and J. Más [FM], and countably quasi-suprabarrelled spaces have been investigated by J.C. Ferrando and L.M. Sánchez Ruiz in [FS4]. A space E is called quasi-Baire if E is barrelled and it cannot be covered by any increasing sequence of nowhere-dense subspaces. As we have seen above in Exercise 1, a space E is said to satisfy condition (G) if, given any sequence of subspaces of E covering E, then one of them is barrelled. A space E is said to be (countably) quasi-suprabarrelled if, given an increasing sequence of subspaces of E covering E, there is one that is barrelled (and its closure has countable codimension). Finally, E is a quasi-totally barrelled space if, given a sequence of subspaces of E covering E, there exists one of them that is barrelled and its closure has countable codimension in E.

While quasi-Baire, quasi-totally barrelled and countably quasi-suprabarrelled spaces enjoy good permanence properties, spaces satisfying condition (G) and quasi-suprabarrelled spaces are not stable under finite products, although the question for arbitrary products is reduced to this one. Three-space problems for some classes of barrelled spaces have been studied by Roelcke and Dierolf [RD], Bonet and Pérez Carreras [BP1] as well as by Valdivia [V13].

Barrelled spaces of class $n \in \mathbb{N}$ were defined for the first time by Rodríguez Salinas [Ro1-2] in a more general context. These spaces have been studied by Ferrando and López-Pellicer in [FL4]. As we have noticed above, suprabarrelled spaces coincide with barrelled spaces of class 1, and ordered suprabarrelled spaces are precisely barrelled spaces of class 2.

Some of these classes of strong barrelled spaces have been renamed in a more convenient way by S. Saxon in [S4]. So quasi-barrelled spaces are denoted by *d*, spaces with condition (G) by *ub*, quasi-suprabarrelled spaces by *b*, Baire-hyperplane spaces by *ud* and unordered Baire-like spaces by *udb*. In order to have a general scheme containing all classes considered we shall also denote convex-Baire by CB, totally barrelled by TB, quasi-totally barrelled by QTB, countably quasi-suprabarrelled by *cb*, Baire-like by BL and barrelled space of class n by BC(n)

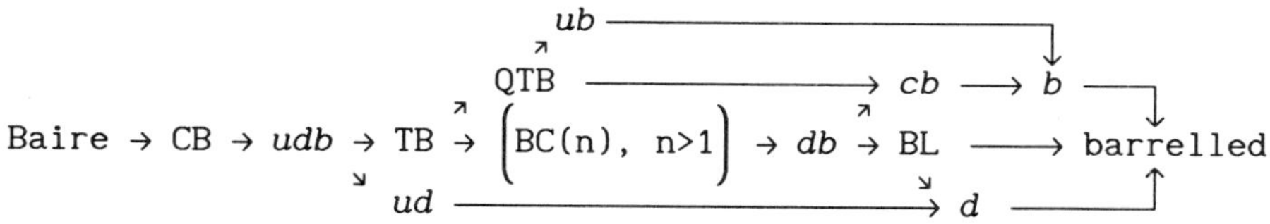

As far as the exposition of the chapter is concerned, Lemma 3.1.1 is based on I.§3.(3) of Valdivia [V9], with a different argument (see also Valdivia [V1] and De Wilde and Houet [DWH]). Definition 3.2.1 is taken from Ferrando and López-Pellicer [FL7], and for the exposition of the permanence properties of barrelled spaces of class n as well as for the proof of Propositions 3.2.12 and 3.3.1 we have followed with minor modifications Ferrando and López-Pellicer's paper [FL4]. Lemma 3.2.11 is taken from Valdivia [V13, Proposition 4] and Lemma 3.3.2 is a well-known result of M. Eidelheit, [E], which we have taken from §31.4.(1) of Köthe [K1].

Exercise 1 (i) is based upon Todd and Saxon's Lemma 1.3.1 [TS], while part (ii) is from Pérez Carreras and Bonet [PB1]. Part (iii) is inspired by Example 1 of Valdivia and Pérez Carreras [VP1]. Exercise 2 is from Valdivia [V9], pp. 101-102. More information is available in Todd and Saxon [TS] Theorem 4.11. Exercise 3 can be found in Ferrando and López Pellicer [FL2]. Our Exercise 4 has been taken from Saxon and Narayanaswami [SN2]. Exercise 5 comes from Narayanaswami and Saxon [NS]. Exercise 6 is straightforward but it can be found in Valdivia and Pérez Carreras [VP1, Thm. 2], Narayanaswami and Saxon [NS, Thm. 3], and in Saxon and Narayanaswami [SN1, Corollary of Thm. 1]. Exercise 7 has been taken from Ferrando and López-Pellicer [FL2]. Exercise 8 is chosen from [V7], and Exercise 9 has been adapted from Ruess [Ru].

We provide below a brief outline of some examples and counter-examples which have not been included in the exercises. So, in Valdivia [V12], locally convex Baire spaces whose product is not Baire (but is obviously convex-Baire) are constructed. Assuming c-additivity (c-A), i.e., assuming that the union of less than c subsets of ℝ, each of (Lebesgue) measure zero, has itself measure zero, a condition weaker than Martin's Axiom (MA), Arias de Reyna [Ar1] proved that there are dense hyperplanes of first category (hence convex-Baire subspaces) in separable Banach spaces, showing that the Klee and Wilansky conjecture [KW] was not true. Also assuming c-A, Saxon [S5] has shown that each locally convex space E such that E' is infinite-dimensional and $E' \neq E^*$ admits a dense non-Baire hyperplane, extending Arias de Reyna's result to the general locally convex case and answering in the positive a question raised in [PB2]. On the other hand, Valdivia [V10] has shown that if E and F are two infinite-dimensional Fréchet spaces, then the injective tensor product of E and F is an unordered Baire-like space which is not Baire. In fact, one has

Theorem 3.5.1 $E\otimes_\pi F$ *is a Baire space if and only if one of the following two conditions is satisfied*

1. dim E < ∞ *and* $F^{\dim E}$ *is a Baire space,*
2. dim F < ∞ *and* $E^{\dim F}$ *is a Baire space.*

In Saxon [S3] a norm $|\cdot|$ stronger than the usual operator norm is defined in the space $\mathcal{F}(E,F)$ of all finite-rank operators from E into F, E and F being infinite-dimensional Banach spaces, in such a way that $(\mathcal{F}(E,F),|\cdot|)$ is an unordered Baire-like space which is not Baire. In Valdivia [V6] it is proved that each separable infinite-dimensional Fréchet space contains a totally barrelled subspace which is not an inductive limit of Baire-hyperplane subspaces. In Valdivia and Pérez Carreras [VP1] the following theorem is proved. The proof will be given in the next chapter.

Theorem 3.5.2 *Let* E *be a totally barrelled space which is not unordered Baire-like and let* F *be a space. Then* $E\otimes_\pi F$ *is totally barrelled if and only if* F *is finite dimensional.*

In Pérez Carreras and Bonet's paper [PB1] a class of metrizable suprabarrelled spaces which are not the inductive limit of totally barrelled spaces is given. On the other hand, φ is a very simple example of a quasi-totally barrelled space which is not totally barrelled while, as can be found in Ferrando and Más [FM], $\varphi(I)$ with $|I| = c$ is an example of a space satisfying condition (G) which is not quasi-totally barrelled. Other distinguishing examples concerning countably quasi-suprabarrelled spaces are also included in [FS4].

We include now some remarkable results which are related to the (LF)-space incidence in infinite-dimensional Fréchet spaces. Indeed, in Saxon and Narayanaswami [SN1] it is shown that a given dense Baire-like subspace F of a Fréchet space E is not a suprabarrelled space if and only if there exists a subspace M of E such that $M \supseteq F$ and M with a topology stronger than the relative topology is an (LF)-space. As one can easily deduce from Exercise 6 above sometimes this subspace M needs to be different from F.

Also, in Saxon and Narayanaswami [SN1] it is shown that if E is a Fréchet space, E has a separable (infinite-dimensional) quotient if and only if E has a dense subspace which, with a topology stronger than the relative topology, is a metrizable (LF)-space. In Valdivia and Pérez Carreras [VP2] it is proved that every non-normable Fréchet space contains a (proper) dense subspace which is an (LF)-space, hence each non-normable Fréchet space has a separable quotient. The major step in this direction has been given by Saxon and Narayanaswami [SN2]:

Theorem 3.5.3 *A Fréchet space has a dense (LF)-subspace if it splits into infinitely many parts, each of which has a separable quotient.*

Finally, we end these notes with some comments about closed graph theorems involving strong barrelled spaces in the domain. The following closed graph theorem, which applies to suprabarrelled spaces in the domain, is taken from Valdivia [V7].

Theorem 3.5.4 *Let* E *be a suprabarrelled space and* F *be the inductive limit of an increasing sequence* (F_n) *of* Γ_r*-spaces contained in* F. *If* T *is a linear mapping from* E *into* F *with closed graph,* T *is continuous and there is some* $p \in \mathbb{N}$ *so that* $T(E) \subseteq F_p$.

For quasi-suprabarrelled spaces we refer the reader to Exercise 3 (i), which, using Γ_r-spaces (see Exercise 1, Chapter 5), is

Theorem 3.5.5 *Let* E *be a locally convex hull of a family of quasi-suprabarrelled spaces and let* F *be the inductive limit of an increasing sequence of* Γ_r*-spaces contained in* F. *If* T *is a linear mapping from* E *into* F *with closed graph, then* T *is continuous.*

In the case of totally barrelled spaces, we choose from Valdivia [V8] the following closed graph theorem.

Theorem 3.5.6 *Let* E *be a totally barrelled space and let* F *be a space with an absolutely convex* $\mathscr{C}$*-web. If* T *is a linear mapping from* E *into* F *with closed graph, there exists a subspace* H *of* F *and a locally convex topology* τ *on* H *stronger than the relative topology such that* $H(\tau)$ *is a Fréchet space,* $T(E) \subseteq H$ *and* $T : E \longrightarrow H(\tau)$ *is continuous.*

4. Barrelled spaces of class $\aleph_0$

4.1 DEFINITION OF BARRELLED SPACES OF CLASS $\aleph_0$

In this chapter we are going to study the class of strongly barrelled spaces defined as the intersection of all the classes $\mathcal{C}_s$ considered in Chapter 3. This new class of barrelled spaces will be of particular interest from now on.

Definition 4.1.1 *A space* E *is called barrelled of class* $\aleph_0$ *if* $E \in \mathcal{C}_s$ *for each* $s \in \mathbb{N}$.

Naturally, barrelled spaces of class $\aleph_0$ inherit the properties shared by all the $\mathcal{C}_s$ that we have seen in Chapter 3. Besides it follows from the previous definition that every totally barrelled space is barrelled of class $\aleph_0$ and that, for each $s \in \mathbb{N}$, there exist barrelled spaces of class s which are not of class $\aleph_0$. Our interest in this chapter will be centred on showing examples of barrelled spaces of class $\aleph_0$ that are not totally barrelled.

4.2 NORMED BARRELLED SPACES OF CLASS $\aleph_0$ WHICH ARE NOT TOTALLY BARRELLED

In what follows we shall prove that if E is a Banach space containing a sequence $\{X_n, n \in \mathbb{N}\}$ of infinite-dimensional linear subspaces such that for each $n \in \mathbb{N}$ the closed linear hull of the sequence $\{X_m, m > n\}$ is a topological complement of $X_1 + X_2 + \ldots + X_n$, then E contains a dense barrelled subspace of class $\aleph_0$ which is not totally barrelled. For each $n \in \mathbb{N}$, let E_n denote the closed linear hull of $\{X_m, m \in \mathbb{N} \setminus \{n\}\}$.

Let $\mathcal{U}$ be an ultrafilter which refines the Fréchet filter on $\mathbb{N}$ formed by all cofinite subsets of $\mathbb{N}$. Given any $U \in \mathcal{U}$ let L(U) represent the closed linear hull of $\bigcup\{X_n, n \in \mathbb{N} \setminus U\}$ and denote by L the linear space $\bigcup\{L(U), U \in \mathcal{U}\}$. Since $\oplus\{X_n, n \in \mathbb{N}\} \subseteq L$, we have that L is a dense subspace of E. On the other hand, L is not totally barrelled since $E_n \cap L$ is an infinite-codimensional subspace of L for every $n \in \mathbb{N}$, and if $z \in L(U)$ and $p \in U$, then $z \in E_p$, which implies that the sequence $\{E_n, n \in \mathbb{N}\}$ is a covering of L. Next we show that L is barrelled of class $\aleph_0$.

For the next lemma we consider $\{F_n, n \in \mathbb{N}\}$ to be an increasing sequence of linear subspaces of L covering L, and for each $n \in \mathbb{N}$ let T_n be a barrel of F_n, V_n the closure of T_n in E and $S_n := \bigcap\{\langle V_m\rangle, n \le m\}$.

Lemma 4.2.1 *If* F *is a Banach subspace of* L*, then there is some* S_n *which contains* F.

Proof. The existence of a positive integer n such that $F \cap S_n$ is dense in F and barrelled leads to $F \subseteq \langle V_m\rangle$ for each $n \le m$, hence $F \subseteq S_n$. ■

In particular, for every $m \in \mathbb{N}$ there is some positive integer n such that $\oplus\{X_i, 1 \le i \le m\} \subseteq S_n$. We shall say that a partition $\{M_1, M_2\}$ of $\mathbb{N}$ is subordinated to a strictly increasing sequence $\{m_i, i \in \mathbb{N}\}$ of positive integers if $M_2 = \{m \in \mathbb{N}: m_{2i-1} < m \le m_{2i}, i \in \mathbb{N}\}$.

Lemma 4.2.2 *There is some* $n \in \mathbb{N}$ *such that* $\oplus\{X_m : m \in \mathbb{N}\} \subseteq S_n$.

Proof. If, for every pair (m,n) of positive integers $\oplus\{X_i, i \ge m\}$ is not contained in S_n, we can select a strictly increasing sequence (m_i) of positive integers such that, setting $m_0 = 0$, $\oplus\{X_m, m_{i-1} < m \le m_i\}$ is not contained in S_i. One of the sets M_1 or M_2 of the partition subordinated to the sequence $\{m_i, i \in \mathbb{N}\}$ belongs to the ultrafilter $\mathcal{U}$. If $M_1 \in \mathcal{U}$, by Lemma 4.2.1 there is an $i \in \mathbb{N}$ such that $L(M_1) \subseteq S_{2i}$. This implies that $\oplus\{X_m, m_{2i-1} < m \le m_{2i}\} \subseteq S_{2i}$, a contradiction. If $M_2 \in \mathcal{U}$ we obtain a similar contradiction. Therefore there are two positive integers p and q such that $\oplus\{X_m, p \le m\} \subseteq S_q$. This, together with Lemma 4.2.1, allows us to reach the conclusion. ■

So given an increasing sequence $\{F_n, n \in \mathbb{N}\}$ of linear subspaces of L covering L we can suppose, disregarding a finite number of subspaces F_n if necessary, that $\oplus\{X_m, m \in \mathbb{N}\} \subseteq S_n$ for every $n \in \mathbb{N}$. By the same argument we can assume that all the F_n are dense in L since, taking $T_n := F_n$ for every $n \in \mathbb{N}$, we have that $S_n = \overline{F_n}$ and $\oplus\{X_m, m \in \mathbb{N}\} \subseteq \overline{F_n}$.

Proposition 4.2.3 L *is suprabarrelled.*

Proof. If L is not suprabarrelled, we can choose an increasing covering $\{F_n, n \in \mathbb{N}\}$ of dense linear subspaces of L, none of them barrelled. Let T_n be a barrel of F_n which is not a neighbourhood of the origin in F_n, $V_n := \overline{T_n}$ and $S_n := \cap\{\langle V_m\rangle, n \le m\}$. Then we may suppose that $\oplus\{X_m, m \in \mathbb{N}\} \subseteq S_n$ for every $n \in \mathbb{N}$. Denoting by B the closed unit ball of E we can distinguish two cases.

Case 1. There are two positive integers p and n such that V_n absorbs $B \cap \oplus\{X_m, m \ge p\}$. Then V_n absorbs $B \cap \oplus\{X_m, m \in \mathbb{N}\}$ and hence, as $\oplus\{X_n, n \in \mathbb{N}\}$ is dense in E, V_n absorbs B. This implies that T_n is a neighbourhood of the origin in F_n, which is a contradiction.

Case 2. For every pair (p,n) of positive integers $B \cap \oplus\{X_m, m \ge p\}$ is not absorbed by V_n. Then, we can select m_1 and m_2 and a vector

$$y_{1,1} \in (B \cap \oplus\{X_m, m_1 < m \le m_2\}) \setminus 2V_1.$$

Suppose we have determined the positive integers $m_1 < m_2 < \ldots < m_k$, and, for $i+j \le k$, the vectors

$$y_{i,j} \in (B \cap \oplus\{X_m, m_{i+j-1} < m \le m_{i+j}\}) \setminus (i+j)V_i .$$

Then we can select the vectors $y_{1,k}, y_{2,k-1}, \ldots, y_{k,1}$ such that

$$y_{r,k+1-r} \in (B \cap \oplus\{X_m, m_k < m\}) \setminus (k+1)V_r$$

with $r = 1, 2, \ldots, k$. Hence, there is some positive integer $m_{k+1} > m_k$ such that, for $1 \le r \le k$,

$$y_{r,k+1-r} \in (B \cap \oplus\{X_m, m_k < m \le m_{k+1}\}) \setminus (k+1)V_r.$$

By recurrence we can determine a strictly increasing sequence $\{m_k, k \in \mathbb{N}\}$ of positive integers and a family of vectors $\{y_{i,j}, (i,j) \in \mathbb{N}^2\}$ such that

$$y_{r,k-r} \in (B \cap \oplus\{X_m, m_{k-1} < m \le m_k\}) \setminus kV_r$$

with $k = 2, 3, \ldots$ and $r = 1, 2, \ldots, k-1$.

Let $\{M_1, M_2\}$ be the partition of $\mathbb{N}$ subordinated to the sequence $\{m_i, i \in \mathbb{N}\}$. As $\mathcal{U}$ is an ultrafilter, one of the two sets M_1 or M_2 is an element of $\mathcal{U}$. If $M_1 \in \mathcal{U}$, according to Lemma 4.2.1, there is some $r \in \mathbb{N}$ such that $L(M_1) \subseteq S_r$. Consequently, V_r absorbs $B \cap L(M_1)$. But this is a contradiction since $y_{r,2k-r} \in B \cap L(M_1) \setminus 2kV_r$. A similar contradiction is obtained if $M_2 \in \mathcal{U}$. ■

Given a positive integer p, consider a p-net $\{F_{n_1 n_2 \ldots n_i}, n_r \in \mathbb{N}, 1 \le r \le i \le p\}$ of L. Let $T_{n_1 \ldots n_p}$ be a barrel in $F_{n_1 \ldots n_p}$, $V_{n_1 \ldots n_p}$ the closure of $T_{n_1 n_2 \ldots n_p}$ in E, $Z_{n_1 n_2 \ldots n_p} := \langle V_{n_1 n_2 \ldots n_p} \rangle$, and $S_{n_1 n_2 \ldots n_p} := \bigcap\{Z_{n_1 n_2 \ldots n_{p-1} m}, m \ge n_p\}$. By decreasing induction, for $1 \le i \le p - 1$ we define $Z_{n_1 \ldots n_i} := \bigcup\{S_{n_1 \ldots n_i m}, m \in \mathbb{N}\}$ and $S_{n_1 \ldots n_i} := \bigcap\{Z_{n_1 \ldots n_{i-1} m}, m \ge n_i\}$. Then, $F_{n_1 \ldots n_i} \subseteq S_{n_1 \ldots n_i}$, $\{S_{n_1}, n_1 \in \mathbb{N}\}$ is an increasing sequence whose union contains L and $\{S_{n_1 \ldots n_i}, n_i \in \mathbb{N}\}$ is an increasing sequence whose union contains $S_{n_1 \ldots n_{i-1}}$ for $2 \le i \le p$.

Lemma 4.2.4 *If* F *is a Banach subspace of* L *then there exists a positive integer* n_1 *such that* $F \subseteq S_{n_1}$. *Moreover, if* $F \subseteq S_{n_1 \ldots n_i}$ *with* $1 \le i < p$, *there is some* $n_{i+1} \in \mathbb{N}$ *such that* $F \subseteq S_{n_1 \ldots n_{i+1}}$.

Proof. We are going to demonstrate by decreasing induction that if the subspace $F \cap S_{n_1 \ldots n_i}$ is of second category in F, then $F \subseteq S_{n_1 \ldots n_i}$. This, together with the inclusion relationships given above, allows us to reach the conclusion. Note that the first statement is a consequence of this property taking i = 1, as $\{S_n, n \in \mathbb{N}\}$ is an increasing covering of the Baire space F, and hence there is some $n_1 \in \mathbb{N}$ such that $F \cap S_{n_1}$ is of second category in F.

Assuming $F \cap S_{n_1 \ldots n_{p-1} n_p}$ is of second category in F, we have that $F \cap Z_{n_1 \ldots n_{p-1} n}$ is of second category in F for every $n \ge n_p$, and thus $V_{n_1 \ldots n_{p-1} n}$ contains a neighbourhood of the origin in F for $n \ge n_p$. So, $F \subseteq Z_{n_1 \ldots n_{p-1} n}$ for $n \ge n_p$, which implies that $F \subseteq S_{n_1 \ldots n_p}$.

Suppose now that, if $F \cap S_{n_1 \ldots n_{i+1}}$, with $1 \le i < p$, is of second category in F, then $F \subseteq S_{n_1 \ldots n_{i+1}}$. Assuming $F \cap S_{n_1 \ldots n_i}$ is of second category in F, then $F \cap Z_{n_1 \ldots n_{i-1} n}$ is of second category in F for each $n \ge n_i$. Hence, for every positive integer $n \ge n_i$, there is some $m \in \mathbb{N}$ such that $F \cap S_{n_1 \ldots n_{i-1} nm}$ is of second category in F. By the induction hypothesis this implies that $F \subseteq S_{n_1 \ldots n_{i-1} nm}$. So, $F \subseteq Z_{n_1 \ldots n_{i-1} n}$ for every $n \ge n_i$ and therefore

$$F \subseteq \bigcap \{Z_{n_1 \ldots n_{i-1} n},\ n \ge n_i\} = S_{n_1 \ldots n_{i-1} n_i}. \ \blacksquare$$

In particular, for each $m \in \mathbb{N}$, there exists some $n_1 \in \mathbb{N}$ such that $\oplus\{X_i,\ 1 \le i \le m\} \subseteq S_{n_1}$, and if $\oplus\{X_i,\ 1 \le i \le m\} \subseteq S_{n_1 \ldots n_i}$, then there is some $n_{i+1} \in \mathbb{N}$ such that $\oplus\{X_i,\ 1 \le i \le m\} \subseteq S_{n_1 \ldots n_{i+1}}$.

Definition 4.2.5 *Given a positive integer* s, *a subset* A *of* $\mathbb{N}^s$ *is said to be of class* s *if there are infinitely many* $b_1 \in \mathbb{N}$ *such that there is some* $(b_1, b_2, \ldots, b_s) \in A$; *and given any element* $(a_1, a_2, \ldots, a_s) \in A$ *and any natural number* i, $1 \le i \le s-1$, *there are infinitely many* $b_{i+1} \in \mathbb{N}$ *such that there is some* $(a_1, a_2, \ldots, a_i, b_{i+1}, b_{i+2}, \ldots, b_s) \in A$.

Note that if P_i denotes the projection of $\mathbb{N}^s$ onto the *i*-th factor, a set $A \subseteq \mathbb{N}^s$ is of class s if it satisfies the following conditions.

(1) $|P_1(A)| = \aleph_0$

(2) Given $1 \le i < s$, let

$$A(n_1, \ldots, n_i) = \{\mathbf{x} \in A\colon P_j(\mathbf{x}) = n_j \text{ for } 1 \le j \le i\}.$$

If $A(n_1, \ldots, n_i) \ne \emptyset$, then $|P_{i+1}(A(n_1, \ldots, n_i))| = \aleph_0$.

Lemma 4.2.6 *The set*

$$\{(n_1, n_2, \ldots, n_p) \in \mathbb{N}^p : \oplus\{X_m,\ m \in \mathbb{N}\} \subseteq S_{n_1 \ldots n_p}\}$$

contains a set of class p.

Proof. We prove first that the set

$$\{n_1 \in \mathbb{N} : \oplus\{X_m, m \in \mathbb{N}\} \subseteq S_{n_1}\}$$

contains a set of class 1. According to Lemma 4.2.4 it suffices to find a pair (m, n) of positive integers such that $\oplus\{X_r, m \le r\} \subseteq S_n$.

If, for each pair (m, n) of positive integers, $\oplus\{X_r, m \le r\}$ is not contained in S_n, there is a strictly increasing sequence $\{m_i, i = 0, 1, \ldots\}$ of integers with $m_0 = 0$ such that $\oplus\{X_m, m_{i-1} < m \le m_i\}$ is not contained in S_i for each $i \in \mathbb{N}$. Now one of the two sets M_1 or M_2 of the partition subordinated to the sequence $\{m_i, i = 1, 2, \ldots\}$ belongs to $\mathcal{U}$. If $M_1 \in \mathcal{U}$, according to Lemma 4.2.4 there is some $i \in \mathbb{N}$ such that $L(M_1) \subseteq S_{2i}$, hence $\oplus\{X_m, m_{2i-1} < m \le m_{2i}\} \subseteq S_{2i}$, a contradiction. There is a similar contradiction assuming $M_2 \in \mathcal{U}$. Hence, using again Lemma 4.2.4 and since $\{S_n, n \in \mathbb{N}\}$ is increasing there exists some $p_1 \in \mathbb{N}$ such that $\oplus\{X_m, m \in \mathbb{N}\} \subseteq S_{n_1}$ for each $n_1 \ge p_1$, and $D_1 = \{n_1 \in \mathbb{N} : n_1 \ge p_1\}$ is a set of class 1 contained in $\{n_1 \in \mathbb{N} : \oplus\{X_m, m \in \mathbb{N}\} \subseteq S_{n_1}\}$.

For each $n_1 \in D_1$ apply Lemma 4.2.4 and reason as above to obtain a $p_2(n_1) \in \mathbb{N}$ such that $\oplus\{X_m, m \in \mathbb{N}\} \subseteq S_{n_1 n_2}$ for each $n_2 \ge p_2(n_1)$. Clearly $D_2 = \{(n_1,n_2) \in \mathbb{N}^2 : n_1 \in D_1, n_2 \ge p_2(n_1)\}$ is a set of class 2 which is contained in $\{(n_1,n_2) \in \mathbb{N}^2 : \oplus\{X_m, m \in \mathbb{N}\} \subseteq S_{n_1 n_2}\}$. Now, for every pair $(n_1,n_2) \in D_2$, we obtain $p_3(n_1,n_2) \in \mathbb{N}$ such that $n_3 \ge p_3(n_1,n_2)$ implies $\oplus\{X_n, n \in \mathbb{N}\} \subseteq S_{n_1 n_2 n_3}$. Again,

$$D_3 = \{(n_1,n_2,n_3) \in \mathbb{N}^3, (n_1,n_2) \in D_2, n_3 \ge p_3(n_1,n_2)\}$$

is a subset of class 3 of $\{(n_1,n_2,n_3) \in \mathbb{N}^3, \oplus\{X_m, m \in \mathbb{N}\} \subseteq S_{n_1 n_2 n_3}\}$.

We continue inductively until we obtain a set D_p of class p which is contained in $\{(n_1,n_2,\ldots,n_p) \in \mathbb{N}^p : \oplus\{X_m, m \in \mathbb{N}\} \subseteq S_{n_1 \ldots n_p}\}$, which concludes the proof. ∎

In what follows we shall always tacitly consider in $\mathbb{N}^s$ the total ordering defined by $(a_1, a_2, \ldots, a_s) < (b_1, b_2, \ldots, b_s)$ if either $a_1 + a_2 + \ldots + a_s < b_1 + b_2 + \ldots + b_s$ or $a_1 + a_2 + \ldots + a_s = b_1 + b_2 +$

... + b_s and there is some index r such that $a_r < b_r$ and $a_i = b_i$ for $1 \le i < r$. This ordering allows us to write the elements of the set $\mathbb{N}^s$ as a sequence which, from now on, we will denote by $\{(m_1(k), m_2(k), \ldots, m_s(k)), k \in \mathbb{N}\}$. This generalized diagonal ordering of $\mathbb{N}^s$ induces in $\mathbb{N}$ the usual order.

Theorem 4.2.7 *L is a barrelled space of class* $\aleph_0$.

Proof. Suppose that L is barrelled of class p - 1 but not barrelled of class p with $p \ge 2$. Then we can determine a p-net $\{F_{n_1 \ldots n_i}, n_r \in \mathbb{N}, 1 \le r \le i \le p\}$ of dense linear subspaces of L such that no $F_{n_1 \ldots n_p}$ is barrelled. Assume that $T_{n_1 \ldots n_p}$ is a barrel in $F_{n_1 \ldots n_p}$ which is not a neighbourhood of the origin in $F_{n_1 \ldots n_p}$, and define $V_{n_1 \ldots n_p}$, $Z_{n_1 \ldots n_p}$ and $S_{n_1 \ldots n_p}$ as we did before Lemma 4.2.4. Now Lemma 4.2.6 allows us to select in $\mathbb{N}^p$ a set $\mathcal{P}$ of class p such that $\oplus\{X_m, m \in \mathbb{N}\}$ is contained in every $S_{n_1 \ldots n_p}$ whenever $(n_1, \ldots, n_p) \in \mathcal{P}$. Considering in $\mathcal{P}$ the induced ordering we write $\mathcal{P}$ as $\{(n_1(r), \ldots, n_p(r)), r \in \mathbb{N}\}$. It is obvious that $L \subseteq \bigcup\{S_{n_1(r)n_2(r)\ldots n_p(r)}, r \in \mathbb{N}\}$ and that $V_{n_1(r)n_2(r)\ldots n_p(r)}$ absorbs $B \cap \oplus\{X_i, 1 \le i \le q\}$ for every q, $r \in \mathbb{N}$. Now two cases are in order.

Case 1. There exists a pair of positive integers q and r such that $V_{n_1(r)n_2(r)\ldots n_p(r)}$ absorbs $B \cap \oplus\{X_m, m \ge q\}$. Then, $V_{n_1(r)n_2(r)\ldots n_p(r)}$ absorbs $B \cap \oplus\{X_m, m \in \mathbb{N}\}$, and therefore it absorbs B. This implies that $T_{n_1(r)\ldots n_p(r)}$ is a neighbourhood of the origin in $F_{n_1(r)\ldots n_p(r)}$, a contradiction.

Case 2. For every pair (q,r) of positive integers, $V_{n_1(r)\ldots n_p(r)}$ does not absorb $B \cap \oplus\{X_m, m \ge q\}$. Then, there are m_1 and m_2, and some

$$y_{1,1} \in (B \cap \oplus\{X_m, m_1 < m \le m_2\}) \setminus 2V_{n_1(1)n_2(1)\ldots n_p(1)}.$$

Suppose we have determined the positive integers $m_1 < m_2 < \ldots < m_k$ and, for $i+j \le k$, the vectors

$$y_{i,j} \in (B \cap \{X_m, m_{i+j-1} < m \le m_{i+j}\}) \setminus (i+j)V_{n_1(i)n_2(i)\ldots n_p(i)}.$$

By the hypothesis, we can determine the vectors $y_{1,k}$, $y_{2,k-1}$, ..., $y_{k,1}$ such that, for r = 1, 2, ..., k,

$$y_{r,k+1-r} \in (B \cap \oplus\{X_m, m > m_k\}) \setminus (k+1)V_{n_1(r)n_2(r)\ldots n_p(r)}.$$

Hence there is some positive integer $m_{k+1} > m_k$ such that, for r = 1, 2, ..., k, we have

$$y_{r,k+1-r} \in (B \cap \oplus\{X_m, m_k < m \le m_{k+1}\} \setminus (k+1)V_{n_1(r)n_2(r)\ldots n_p(r)}.$$

Proceeding by recurrence we can determine a strictly increasing sequence $\{m_k, k \in \mathbb{N}\}$ in $\mathbb{N}$ and a family of vectors $\{y_{i,j}, (i,j) \in \mathbb{N}^2\}$, such that

$$y_{r,k-r} \in (B \cap \oplus\{X_m, m_{k-1} < m \le m_k\}) \setminus kV_{n_1(r)n_2(r)\ldots n_p(r)}$$

with r = 1, 2, ..., k-1 and k = 2, 3, ...

Let $\{M_1, M_2\}$ be the partition of $\mathbb{N}$ subordinated to the sequence $\{m_k, k \in \mathbb{N}\}$. As $\mathcal{U}$ is an ultrafilter, either M_1 or M_2 belongs to $\mathcal{U}$. If $M_1 \in \mathcal{U}$, proceeding as in Lemma 4.2.4, there is some positive integer r such that $L(M_1) \subseteq S_{n_1(r)\ldots n_p(r)}$. Thus it follows that $V_{n_1(r)\ldots n_p(r)}$ absorbs $B \cap L(M_1)$. This contradicts

$$y_{r,2k-r} \in B \cap L(M_1) \setminus 2kV_{n_1(r)n_2(r)\ldots n_p(r)}.$$

An analogous contradiction is obtained if $M_2 \in \mathcal{U}$.

So, if L is barrelled of class p - 1 we have that L is barrelled of class p. This, together with Proposition 4.2.3, allows us to reach the conclusion.■

Example 4.2.8 Assume that p is a real number greater or equal than 1, Σ is a σ-algebra defined in some non-empty set Ω and μ a positive measure defined on Σ. We suppose that Ω admits an infinite countable partition $\{\Omega_n, n \in \mathbb{N}\}$, where each Ω_n belongs to Σ and for every $n \in \mathbb{N}$ the space $X_n := \{f \in L^p(\mu), f_{|\Omega\setminus\Omega_n} = 0 \text{ a.e.}\}$ is infinite-dimensional.

Clearly, X_n is a closed subspace of $L^p(\mu)$ since, if $\{f_i, i \in \mathbb{N}\}$ is a sequence of X_n which converges in norm to some $f \in L^p(\mu)$, there is a subsequence $\{f_{i_j}, j \in \mathbb{N}\}$ which converges almost everywhere to f and therefore $f_{|\Omega\setminus\Omega_n} = 0$ almost everywhere.

For each $n \in \mathbb{N}$ the closed linear hull of the family $\{X_m, m > n\}$ is the set $M = \{f \in L^p(\mu), f_{|\cup\{\Omega_r, r \le n\}} = 0 \text{ a.e.}\}$. Indeed, if χ_s stands for the indicator function of $\cup\{\Omega_r, r \le s\}$, $f \in M$ and $f_q := f \circ \chi_{q+n}$ for every $q \in \mathbb{N}$, then the sequence $\{f_q, q \in \mathbb{N}\}$ belongs to the linear hull of $\{X_m, m > n\}$ and then, as a consequence of the Lebesgue dominated convergence theorem, $\lim_q \int_\Omega |f_q - f| \, d\mu = 0$. Now, the closed linear hull of the family $\{X_m, m > n\}$ is a topological complement of $X_1 + X_2 + \ldots + X_n$ since the projection which maps f into $f \circ \chi_n$ has norm one, as it follows from the inequality $\int_\Omega |f \circ \chi_n|^p \, d\mu \le \int_\Omega |f|^p \, d\mu$. Therefore $L^p(\mu)$ contains a dense linear subspace which is barrelled of class $\aleph_0$, not totally barrelled.

In particular, if $\Omega = \mathbb{N}$ and $\{\Omega_n, n \in \mathbb{N}\}$ is a partition of $\mathbb{N}$ into infinitely many sets of infinite cardinality, it follows that each ℓ^p contains a dense linear subspace which is barrelled of class $\aleph_0$ but not totally barrelled. This is also true even if $p = \infty$ since, as we will see in the following chapter, the linear subspace ℓ_0^∞ of all simple functions, provided with the supremum norm, is barrelled of class $\aleph_0$.

Example 4.2.9 We already know that if E is a space and F is a closed linear subspace of E such that F and E/F are both barrelled of class $\aleph_0$, then E is also barrelled of class $\aleph_0$. Furthermore, if E/F is not totally barrelled, then E cannot be totally barrelled because of the stability of totally barrelled spaces under the formation of quotients.

On the other hand, let Σ denote an infinite σ-algebra on a set Ω. Then, if $\ell_0^\infty(\Sigma)$ denotes the linear space of all Σ-simple functions with the supremum norm and $\ell^\infty(\Sigma)$ denotes its completion, then $\ell^\infty(\Sigma)$ contains a copy of ℓ^∞. As ℓ^1 can be topologically embedded into ℓ^∞, it follows that $\ell^\infty(\Sigma)$ contains a copy of ℓ^1, whereupon its strong dual $ba(\Sigma)$, the Banach space of all bounded finitely additive scalar measures provided with the variation norm, contains a closed subspace H such that $ba(\Sigma)/H$ is topologically isomorphic to ℓ^∞. Applying the results mentioned above, ℓ_0^∞ is barrelled of class $\aleph_0$ (Chapter 5) but not totally barrelled (see Chapter 6), then if k denotes the quotient mapping from

$ba(\Sigma)$ onto ℓ^∞ and we set $G := k^{-1}(\ell_0^\infty)$, then G is a dense linear subspace of $ba(\Sigma)$ that is barrelled of class $\aleph_0$ but not totally barrelled.

4.3 PROJECTIVE TENSOR PRODUCT OF BARRELLED SPACES OF CLASS $\aleph_0$

We will show in section 4.5 that every non-normable Fréchet space contains a dense linear subspace of class $\aleph_0$ that is not totally barrelled. But to obtain this result we will need more machinery, which we are going to develop in this and the next section, and which are also interesting in themselves.

As we have seen in Chapter 1 the following property is known to be true for Baire-like and unordered Baire-like spaces.

Proposition 4.3.1 *Let* E *and* F *be two barrelled spaces of class* $\aleph_0$. *If* E *is metrizable, then* $E \otimes_\pi F$ *is barrelled of class* $\aleph_0$.

Proof. Given that $E \otimes_\pi F$ is barrelled of class 0, we are going to prove by induction that $E \otimes_\pi F$ is barrelled of class $\aleph_0$.

If $E \otimes_\pi F$ is barrelled of class p - 1 but not of class p ≥ 1, we can determine a p-net $\{L_{n_1 \ldots n_i},\ n_r \in \mathbb{N},\ 1 \le r \le i \le p\}$ in $E \otimes_\pi F$ formed by dense linear subspaces such that no $L_{n_1 \ldots n_p}$ is barrelled.

Let $T_{n_1 \ldots n_p}$ be a barrel in $L_{n_1 \ldots n_p}$ which is not a neighbourhood of the origin in $L_{n_1 \ldots n_p}$ and let $V_{n_1 \ldots n_p}$ be its closure in $E \otimes_\pi F$. Setting $Z_{n_1 \ldots n_p} := \langle V_{n_1 \ldots n_p} \rangle$ and $S_{n_1 \ldots n_p} := \bigcap \{Z_{n_1 \ldots n_p n},\ n \ge n_p\}$, we define by decreasing induction $Z_{n_1 \ldots n_k} := \bigcup \{S_{n_1 \ldots n_k n},\ n \in \mathbb{N}\}$ and $S_{n_1 \ldots n_k} := \bigcap \{Z_{n_1 \ldots n_{k-1} n},\ n \ge n_k\}$ for $1 \le k \le n-1$. Then, $L_{n_1 \ldots n_k}$ is contained in $S_{n_1 \ldots n_k}$, $\{S_n,\ n \in \mathbb{N}\}$ is increasing and covers $E \otimes F$, and for $1 < k \le n$ the sequence $\{S_{n_1 \ldots n_{k-1} n},\ n \in \mathbb{N}\}$ is increasing and its union contains the subspace $S_{n_1 \ldots n_{k-1}}$.

Now, if $\{U_r,\ r \in \mathbb{N}\}$ stands for a decreasing base of absolutely convex neighbourhoods of the origin in E we put

$$S^F_{n_1\dots n_k} := \left\{y \in F,\ U_{n_k} \otimes y \subseteq S_{n_1\dots n_k}\right\}$$

for $1 \le k \le n$. Clearly, $\{S^F_{n_1\dots n_k},\ n_k \in \mathbb{N}\}$ is an increasing sequence of subspaces of F whose union contains $S^F_{n_1\dots n_{k-1}}$ for $1 < k \le n$. Now if we show that $\{S^F_{n_1},\ n_1 \in \mathbb{N}\}$ covers F then, F being barrelled of class $\aleph_0$, we can determine $n_1, n_2, \dots, n_p \in \mathbb{N}$ such that $S^F_{n_1\dots n_p}$ is dense in F and barrelled, which is a contradiction. In fact, since $U_{n_p} \otimes S^F_{n_1\dots n_p} \subseteq S_{n_1\dots n_p}$, $E \otimes S^F_{n_1\dots n_p} \subseteq S_{n_1\dots n_p}$ and given that E and $S^F_{n_1\dots n_p}$ are barrelled, E is metrizable and $S^F_{n_1\dots n_p}$ is dense in F, we obtain that $E \otimes_\pi S^F_{n_1\dots n_p}$ is barrelled and, consequently, $S_{n_1\dots n_p}$ and $Z_{n_1\dots n_p}$ are barrelled. Then, $V_{n_1\dots n_p}$ is a neighbourhood of the origin in $Z_{n_1\dots n_p}$ and hence $T_{n_1\dots n_p}$ is a neighbourhood of the origin in $L_{n_1\dots n_p}$, which is not true. Consequently, we have only to show $\{S^F_{n_1},\ n_1 \in \mathbb{N}\}$ covers F. Given $z \in F$, for $1 \le k \le n$, let

$$S^E_{n_1\dots n_k} := \left\{x \in E : x \otimes z \in S_{n_1\dots n_k}\right\},$$

$$Z^E_{n_1\dots n_k} := \left\{x \in E : x \otimes z \in Z_{n_1\dots n_k}\right\}$$

Now, it is easy to check that $\{S^E_{n_1\dots n_{k-1}n},\ n \in \mathbb{N}\}$ is increasing and covers $S^E_{n_1\dots n_{k-1}}$ for $1 < k \le p$ and $S^E_{n_1\dots n_p}$ is the intersection of the family $\{Z^E_{n_1\dots n_{p-1}n},\ n \ge n_p\}$.

As E is barrelled of class $\aleph_0$ it follows that there is some $k_1 \in \mathbb{N}$ such that $S^E_{r_1}$ is barrelled of class $p - 1$ and dense in E for every $r_1 \ge k_1$. For each $s_1 \ge k_1$ there is some $k_2(s_1)$ such that $S^E_{s_1 s_2}$ is barrelled of class $p - 2$ and dense in E for every $s_2 \ge k_2(s_1)$. By induction we can determine k_1, $k_2(r_1)$, $k_3(r_1,r_2)$, ..., $k_p(r_1,r_2,\dots,r_{p-1})$ with $r_1 \ge k_1$, $r_2 \ge k_2(r_1)$, ..., $r_{p-1} \ge k_{p-1}(r_1,r_2,\dots,r_{p-2})$ such that every subspace $S^E_{s_1 s_2\dots s_p}$ with $s_1 \ge k_1$, $s_2 \ge k_2(s_1)$, ..., $s_p \ge k_p(s_1,s_2,\dots,s_{p-1})$ is barrelled and dense in E. Each of these subspaces coincides with E, since if $u \in E \setminus S^E_{s_1 s_2 \ldots s_p}$ then $u \otimes z \notin S_{s_1 s_2 \ldots s_p}$ and therefore there is some $s \geq s_p$ such that $u \otimes z \notin Z_{s_1 \ldots s_{p-1} s}$ and so, defining the closed absolutely convex subsets of E

$$W_{s_1 \ldots s_{p-1} s} := \left\{ x \in E : x \otimes z \in V_{s_1 \ldots s_{p-1} s} \right\}$$

and since $S^E_{s_1 \ldots s_{p-1} s} \subseteq \langle W_{s_1 \ldots s_{p-1} s} \rangle$, it follows that $W_{s_1 \ldots s_{p-1} s}$ is a neighbourhood of the origin in E, and we can find some $\lambda > 0$ such that $\lambda u \in W_{s_1 \ldots s_{p-1} s}$, which shows that $u \otimes z \in \langle V_{s_1 \ldots s_{p-1} s} \rangle = Z_{s_1 \ldots s_{p-1} s}$, a contradiction.

Finally, from $S^E_{s_1 \ldots s_p} = E$ for $s_1 \geq k_1, \ldots, s_p \geq k_p(s_1, \ldots, s_{p-1})$ it follows that $Z^E_{s_1 \ldots s_{p-1}} = \bigcup \{ S_{s_1 \ldots s_{p-1} s}, \ s \geq s_p \} = E$ for each $s_{p-1} \geq k_{p-1}(s_1, \ldots, s_{p-2})$, which implies that $S^E_{s_1 \ldots s_{p-2} k_{p-1}(s_1 \ldots s_{p-2})}$ and, consequently, $Z^E_{s_1 \ldots s_{p-2}}$ coincides with E for $s_1 \geq k_1, \ \ldots, \ s_{p-2} \geq k_{p-2}(s_1, s_2, \ldots, s_{p-3})$. By recurrence we obtain that $Z^E_{s_1} = E$ if $s_1 \geq k_1$, from where it follows that $S^E_{k_1} = \bigcap \{ Z^E_{s_1}, \ s_1 \geq k_1 \} = E$. Hence $E \otimes z \subseteq S_{k_1}$ and hence $U_{k_1} \otimes z \subseteq S_{k_1}$. So, $z \in S_{k_1}$, i.e., $\{ S^F_{n_1}, \ n_1 \in \mathbb{N} \}$ covers F.■

Remark 4.3.2 If E and F are barrelled spaces of class s with $s \in \mathbb{N}$ and E is metrizable, the argument used in the proof of the previous proposition also shows that $E \otimes_\pi F$ is a barrelled space of class s.

4.4 DENSE TOTALLY BARRELLED SUBSPACES OF NON-NORMABLE FRECHET SPACES

We are going to show that each infinite-dimensional Fréchet space has a dense totally barrelled subspace which is not unordered Baire-like. This is a deep result, which we prove in the next proposition.

Theorem 4.4.1 *Let* E *be a Fréchet space. If* E *is infinite-dimensional, there is a dense linear subspace* F *of* E *which is totally barrelled but not unordered Baire-like.*

Proof. Let $\mathbb{K}$ be the field of the real or complex numbers. If $\mathbb{K}$ is real,

$\mathbb{H}$ will denote the field of rational numbers, while if $\mathbb{K}$ is complex, $\mathbb{H}$ will denote the subfield of all numbers a + bi with a and b rational numbers. Assume $\{(x_i,u_i) \in E \times E', i \in \mathbb{N}\}$ is a biorthogonal system, and denote by A the subset $\{x \in E, u_i x \in \mathbb{H}, \forall i \in \mathbb{N}\}$ of E. We are going to show that $F := \langle A \rangle$ is a dense subspace of E which is totally barrelled but not unordered Baire-like. We shall always consider a base $\{U_n, n \in \mathbb{N}\}$ of absolutely convex neighbourhoods of the origin in E such that $2U_{n+1} \subseteq U_n$ for each $n \in \mathbb{N}$.

We proceed in three steps.

Step 1. F *is a dense linear subspace of* E.
Pick some $x \in E$ and choose any closed absolutely convex neighbourhood of the origin U in E. We are going to find a $y \in U$ such that $x + y \in A$. This will show that A, and hence F, is dense in E. Indeed, if $p \in \mathbb{N}$ is such that $U_p \subseteq U$, for each $j \in \mathbb{N}$ we select some scalar a_j verifying that $a_j x_j \in U_{p+j}$ and $u_j(x + a_j x_j) \in \mathbb{H}$. Clearly the Cauchy series $\sum_{j=1}^{\infty} a_j x_j$ converges to some y in $\overline{U_p} \subseteq U$. Since $u_j(x + y) = u_j(x + a_j x_j) \in \mathbb{H}$, it follows that $x + y \in A$.

Step 2. F *is totally barrelled.*
Let $\{F_n, n \in \mathbb{N}\}$ be a sequence of subspaces of F covering F and let L be the linear hull of the sequence $\{x_i, i \in \mathbb{N}\}$. It suffices to show that there is some positive integer p such that $F_p + L$ is barrelled, since L has countable dimension.

Suppose no subspace $F_n + L$ is barrelled and let T_n be a barrel in $F_n + L$ which is not a neighbourhood of the origin in this subspace. For each $n \in \mathbb{N}$, denote by V_n the closure of T_n in E, and by G_n the linear hull of V_n. Since E is a Baire space, the sequence $\{G_n, n \in \mathbb{N}\}$ cannot cover E. So there exists some $x \in E$ which does not belong to any G_n.

Let a_1 be a scalar such that $a_1 x_1 \in U_1$ and $u_1(x + a_1 x_1) \in \mathbb{H}$. Since $a_1 x_1 \in G_n$ for each $n \in \mathbb{N}$, it follows that $x + a_1 x_1 \notin G_n$ for each $n \in \mathbb{N}$. In particular, $x + a_1 x_1 \notin V_1$, and consequently there is some $v_{11} \in V_1^{\circ}$ such that $|v_{11}(x + a_1 x_1)| > 1$. Now we choose a scalar a_2 satisfying the following three conditions.

(1) $a_2x_2 \in U_2$

(2) $u_2(x + a_2x_2) \in \mathbb{H}$

(3) $|v_{11}(x + a_1x_1 + a_2x_2)| > 1$

As before, $x + a_1x_1 + a_2x_2 \notin G_n$ for each $n \in \mathbb{N}$. Hence, $x + a_1x_1 + a_2x_2$ neither belongs to $2V_1$ nor to $2V_2$, and there exists some $v_{12} \in V_1^o$ and $v_{21}, v_{22} \in V_2^o$ such that $|v_{ij}(x + a_1x_1 + a_2x_2)| > j$ for $1 \le i, j \le 2$. Again we are able to select a scalar a_3 satisfying

(1') $a_3x_3 \in U_3$

(2') $u_3(x + a_3x_3) \in \mathbb{H}$

(3') $|v_{ij}(x + a_1x_1 + a_2x_2 + a_3x_3)| > j$ for $1 \le i,j \le 2$

Proceeding by recurrence we obtain a sequence $\{a_i, \; i \in \mathbb{N}\}$ of scalars such that $u_i(x + a_ix_i) \in \mathbb{H}$ for each $i \in \mathbb{N}$, $\sum_{i=1}^{\infty} a_ix_i$ is a Cauchy series converging to some y in E and $|v_{ij}(x + a_1x_1 + \ldots + a_nx_n)| > j$ for each $1 \le i, j \le n$.

Making $n \longrightarrow \infty$ in the relation above we obtain

$$|v_{ij}(x + y)| \ge j \tag{1}$$

for each $i, j \in \mathbb{N}$. But, on the other hand, as $x + y \in A$ there is some $p \in \mathbb{N}$ such that $x + y \in G_p$. So, there exists some $q \in \mathbb{N}$ with $x + y \in qV_p$, and hence

$$|v_{pj}(x + y)| \le q \tag{2}$$

for each $j \in \mathbb{N}$.

Taking $i = p$ and $j = q+1$ in (1) we obtain

$$|v_{p,q+1}(x + y)| \ge q+1 > q$$

while, taking $j = q+1$ in (2),

$$|v_{p,q+1}(x + y)| \le q.$$

But this is a contradiction.

<u>Step 3</u>. F *is not unordered Baire-like.*

Let B_n represent the subset of F formed by all those vectors of F which are a linear combination of at most $n - 1$ elements of A. Clearly, we have $F = \bigcup\{B_n, \; n \ge 2\}$.

For each $n > 1$ define the continuous projection $P_n : F \longrightarrow F$ such that $P_nx = \sum_{i=1}^{n} u_i(x)x_i$ for every $x \in F$. Since $|P_n(A)| \le \aleph_0$, the family of all subsets of $P_n(A)$ of less than n elements is at most countable. We denote this family by $\{Q_{nm}, m \in \mathbb{N}\}$.

As each subspace $\langle Q_{nm}\rangle$ is finite-dimensional and P_n is continuous, then the subspaces $F_{nm} := P_n^{-1}\{\langle Q_{nm}\rangle\}$ are closed in F. Furthermore F_{nm} is a proper subspace of F for each n, m $\in \mathbb{N}$, since we have $\dim\{P_n(F)\} = n$ while $\dim\{\langle Q_{nm}\rangle\} \le n-1$.

Finally, $B_n \subseteq \bigcup\{F_{nm}, m \in \mathbb{N}\}$ since given $y = b_1y_1 + \ldots + b_{n-1}y_{n-1}$ with $y_i \in A$ for $1 \le i < n$, there is some $k \in \mathbb{N}$ such that $P_n(y_i) \in Q_{nk}$ for $1 \le i < n$. Hence $y \in F_{nk}$.

This completes the proof. ■

4.5 DENSE BARRELLED SPACES OF CLASS $\aleph_0$ IN NON-NORMABLE FRECHET SPACES

As promised in section 4.3 now we are prepared to prove that each non-normable Fréchet space contains a dense barrelled subspace of class $\aleph_0$ which is not totally barrelled. First we need two additional lemmas.

Lemma 4.5.1 *Let* E *be a totally barrelled space which is not unordered Baire-like and let* F *be a space. Then* $E\otimes_\pi F$ *is totally barrelled if and only if* F *is finite-dimensional.*

Proof. If F is finite-dimensional, $E \otimes_\pi F$ is isomorphic to $E^{\dim F}$ which is totally barrelled because of the well-known product permanence of totally barrelled spaces. Assume now that F is infinite-dimensional. Taking advantage of the fact that E is not unordered Baire-like but it is totally barrelled we can select a sequence $\{E_n, n \in \mathbb{N}\}$ of subspaces of E covering E such that E_n is barrelled and $\overline{E_n}$ has finite but non-null codimension in E for each $n \in \mathbb{N}$. Now denote by L_n some topological complement of $\overline{E_n}$ in E. Since every L_n is finite-dimensional and $L_n \otimes_\pi F$ is isomorphic to $F^{\dim L_n}$, we have

$$E \otimes_\pi F \cong (\overline{E_n} \times L_n) \otimes_\pi F \cong (\overline{E_n} \otimes_\pi F) \times (L_n \otimes_\pi F) \cong (\overline{E_n} \otimes F) \times F$$

where $\cong$ indicates topological isomorphism.

Setting $G_n := \overline{E_n}\otimes F$, the sequence $\{G_n, n \in \mathbb{N}\}$ covers $E\otimes F$. Indeed, if $z = \sum_{i=1}^{m} x_i\otimes y_i \in E \otimes F$ and L denotes the linear span of $\{x_1, x_2, \ldots, x_m\}$, there exists some $p \in \mathbb{N}$ with $\overline{E_n} \supseteq L$ since L is a Baire subspace and the sequence $\{E_n \cap L,\ n \in \mathbb{N}\}$ covers L. Consequently $z \in G_p$.

As each G_n is a closed subspace of $E \otimes_\pi F$ of infinite codimension in $E \otimes F$ and, as we have seen above, the sequence $\{G_n,\ n \in \mathbb{N}\}$ covers $E \otimes F$, it follows that $E \otimes_\pi F$ is not totally barrelled. ■

Lemma 4.5.2 *If* E *is an infinite-dimensional Fréchet space, then* $E \otimes_\pi E$ *contains a dense barrelled linear subspace of class* $\aleph_0$ *which is not totally barrelled.*

Proof. According to Theorem 4.4.1, E contains a dense linear subspace F which is totally barrelled but not unordered Baire-like. Then, by Lemma 4.5.1, $F \otimes_\pi E$ is not totally barrelled and finally, by Proposition 4.3.1, $F \otimes_\pi E$ is barrelled of class $\aleph_0$. ■

Theorem 4.5.3 *Each non-normable Fréchet space* E *contains a dense linear subspace* F *of class* $\aleph_0$ *which is not totally barrelled.*

Proof. According to Lemma 4.5.2, $\omega \otimes_\pi \omega$ contains a dense linear subspace H, barrelled of class $\aleph_0$, which is not totally barrelled. Since

$$\omega \hat{\otimes}_\pi \omega = K^{\mathbb{N}} \hat{\otimes}_\pi K^{\mathbb{N}} \cong [K \hat{\otimes}_\pi K]^{\mathbb{N}} \cong K^{\mathbb{N}} = \omega,$$

H is a dense barrelled subspace of class $\aleph_0$ of ω, not totally barrelled. Now, by Lemma 3.3.2 of Chapter 3, E has a closed linear subspace L such that E/L is topologically isomorphic to ω. If k is the quotient mapping from E onto ω then, by virtue of what we have said in Example 4.2.9, we have that $F := k^{-1}(H)$ is a dense linear subspace of E which is barrelled of class $\aleph_0$ but not totally barrelled. ■

4.6 EXERCISES

1. *Topological embedding of* ℓ^1 *in* ℓ^∞. Every separable normed space of infinite dimension is isometric to a closed linear subspace of ℓ^∞. Furthermore, ℓ^∞ contains $\ell^1(\mathbb{R})$ isometrically.

2. *Dense barrelled subspaces of class* $\aleph_0$ *that are not totally barrelled in some vector-valued function spaces.*

(i) Let Ω be a compact Hausdorff topological space and let $C(\Omega)$ be the Banach space of all scalar continuous functions on Ω provided with the supremum norm. Suppose that E is a nuclear space. If both $C(\Omega)$ and E are of infinite dimension, the space $C(\Omega,E)$ of all continuous E-valued functions on Ω endowed with the uniform convergence topology contains a dense subspace of class $\aleph_0$ which is not totally barrelled. *[Hint: Use the fact that* $C(\Omega)\otimes_{\varepsilon}E$ *is isomorphic to a dense subspace of* $C(\Omega,E)$*.]*
(ii) Assume that (Ω,Σ,μ) is a positive measure space such that $L^1(\mu)$ is infinite-dimensional and let X be an infinite-dimensional Banach space. Then the Banach space $L^1(\mu,X)$ of all (classes of) X-valued μ-measurable Bochner integrable functions defined on Ω endowed with its usual norm (see Chapter 8) contains a dense subspace of class $\aleph_0$ which is not totally barrelled. *[Hint: Use the fact that* $L^1(\mu)\otimes_{\pi}X$ *is isomorphic to a dense subspace of* $L^1(\mu,X)$*.]*

3. *The space* $\ell^\infty(X)$. Let X be a normed space and let us denote by $\ell^\infty(X)$ the linear space of all bounded sequences (x_n) in X provided with the norm $\|x\| = \sup\{\|x_n\|, n \in \mathbb{N}\}$ for every $x = (x_n) \in \ell^\infty(X)$.
(i) The completion of the space $\ell^\infty(X)$ coincides with $\ell^\infty(\hat{X})$.
(ii) Assuming T is a barrel in $\ell^\infty(X)$, there exists some $p \in \mathbb{N}$ such that $\ell^\infty(\mathbb{N}\backslash\{1,2,\ldots,p\},X) \subseteq pT$.
(iii) $\ell^\infty(X)$ is isomorphic to $\ell^\infty(X) \times X$.
(iv) $\ell^\infty(X)$ is barrelled if and only if X is barrelled.
(v) $\ell^\infty(X)$ is a barrelled space of class s if and only if X is barrelled of class s, and hence $\ell^\infty(X)$ is barrelled of class $\aleph_0$ if and only if X is barrelled of class $\aleph_0$. *[Hint: A modification of (ii) would be worth looking at here.]*

4. *The space* $c_0(X)$. If X is a normed space, denote by $c_0(X)$ the subspace of $\ell^\infty(X)$ formed by all null sequences.
(i) $c_0(X)$ is a closed subspace of $\ell^\infty(X)$.
(ii) The completion of the space $c_0(X)$ coincides with $c_0(\hat{X})$.
(iii) $c_0(X)$ is barrelled of class $\aleph_0$ if and only if X is.

5. *Baireled spaces.* A space E is said to be baireled if, whenever it contains a family $\{E_{n_1 \dots n_p},\ p, n_1, \dots, n_p \in \mathbb{N}\}$ of linear subspaces such that $E = \bigcup\{E_m,\ m \in \mathbb{N}\}$, $E_{n_1 \dots n_p} = \bigcup\{E_{n_1 \dots n_p m},\ m \in \mathbb{N}\}$ for each p, $n_1, \dots, n_p \in \mathbb{N}$ and with $E_n \subseteq E_{n+1}$ and $E_{n_1 \dots n_p n} \subseteq E_{n_1 \dots n_p n+1}$ for each p, n, $n_1, \dots, n_p \in \mathbb{N}$, there is a sequence $\{k_i,\ i \in \mathbb{N}\}$ in $\mathbb{N}$ such that $E_{k_1 \dots k_i}$ is Baire-like for each $i \in \mathbb{N}$.

(i) Every totally barrelled space is baireled and every baireled space is barrelled of class $\aleph_0$.

(ii) Every non-normable Fréchet space contains a dense barrelled subspace of class $\aleph_0$ which is not baireled. *[Hint: Prove first that this is true in ω and then proceed as usual.]*

(iii) Let E be a baireled space and F be unordered Baire-like. If E is metrizable, then $E \otimes_\pi F$ is baireled.

(iv) If E is a normed totally barrelled space which is not unordered Baire-like, then $E \otimes_\pi c_0$ is a separable normed baireled space which is not totally barrelled.

(v) ℓ^1 has a dense baireled subspace which is not totally barrelled.

6. *A barrelled space of class $\aleph_0$.* Assume that Ω is a set of uncountable cardinality. Denote by E the linear span of the indicator functions e(A) of the countable subsets A of Ω, provided with the supremum-norm $\|f\| = \sup\{|f(\omega)|,\ \omega \in \Omega\}$ for each $f \in E$.

(i) If B stands by the closed unit ball of E, then

$$(1/4)B \subseteq \Gamma\{e(A):\ A \subseteq \Omega,\ |A| \le \aleph_0\}$$

(ii) Every barrel in E absorbs the family $\{e(A):\ A \subseteq \Omega,\ |A| \le \aleph_0\}$

(iii) E is barrelled.

(iv) E is barrelled of class s for each $s \in \mathbb{N}$. *[Hint: An appropriate modification of (ii) is needed here.]*

7. *The Stone space of an algebra of sets.* Let Ω denote a non-empty set. Assume that Σ is an algebra of subsets of Ω. The Stone space S_Σ of the algebra Σ is defined as the collection of all non-trivial $\{0,1\}$-valued

additive measures defined in Σ provided with the relative topology of the product space $\{0,1\}^{\Sigma}$, where the set $\{0,1\}$ is assumed to be provided with the discrete topology.

(i) For each $A \in \Sigma$, denote by Π_A the canonical projection from S_Σ onto $\{0,1\}$, that is, $\Pi_A(\mu) = \mu(A)$ for $\mu \in S_\Sigma$. The family $\{\Pi_A^{-1}(\{1\}),\ A \in \Sigma\}$ defines a base of the topology of S_Σ formed by clopen sets.

(ii) S_Σ is a compact zero-dimensional Hausdorff topological space.

(iii) If $\mathcal{D}$ denotes the algebra of all clopen subsets of S_Σ, the mapping $\Phi : \Sigma \longrightarrow \mathcal{D}$ defined by $\Phi(A) = \Pi_A^{-1}(\{1\}) = \{\mu \in S_\Sigma : \mu(A) = 1\}$ for $A \in S_\Sigma$ is a homomorphism of algebras.

(iv) Denoting by $\ell_0^\infty(\mathcal{D})$ the linear space of all those scalar continuous functions defined on S_Σ taking only a finite number of different values provided with the supremum-norm topology, one has $\ell_0^\infty(\mathcal{D})$ is isometric to the space $\ell_0^\infty(\Sigma)$ of all Σ-simple functions on Ω with the supremum norm.

8. *The Stone-Čech compactification of a completely regular topological space.* Assume Ω stands for a Hausdorff topological space, and let $BC(\Omega)$ represent the Banach space of all real bounded and continuous functions defined in Ω, equipped with the supremum-norm. Denote by Q the closed unit ball of the strong dual $BC(\Omega)'$ of $BC(\Omega)$.

(i) The mapping $\Psi : \Omega \longrightarrow Q(\text{weak*})$ defined by $\langle\Psi(\omega),f\rangle = f(\omega)$ for each $f \in BC(\Omega)$ and each $\omega \in \Omega$ is continuous.

(ii) If Ω is completely regular, Ψ is a homeomorphism onto its image.

(iii) The closure S in Q of $\Psi(\Omega)$ under the $\sigma(BC(\Omega)',BC(\Omega))$-topology is a compact Hausdorff topological space called the Stone-Čech compactification of Ω, and it is denoted as $\beta\Omega$.

(iv) If Ω is completely regular, is a compact Hausdorff space and f is a continuous mapping from Ω into K, there exists a continuous extension $f^\beta : \beta\Omega \longrightarrow K$.

(v) If Ω is completely regular, then $BC(\Omega)' \cong C(\beta\Omega)' \cong \ell_0^\infty(\mathcal{D})'$, where $\mathcal{D}$ denotes the algebra of clopen sets of $\beta\Omega$.

(vi) If Ω is a discrete topological space and $\mathcal{B}$ is the σ-algebra of all subsets of Ω, then $\beta\Omega$ coincides with $S_{\mathcal{B}}$.

4.7 NOTES AND REMARKS

Our exposition of the chapter has been strongly motivated by Ferrando and López-Pellicer's papers [FL7] and [FL9], although our Lemma 4.2.6 is an appropriate modification of Lemma 4 there.

The definition of barrelled spaces of class $\aleph_0$ can be found in Ferrando and López Pellicer [FL6]. Sections 4.1 and 4.2 are taken from [FL9]. Proposition 4.3.1 can be found in [FL7], while Lemma 4.5.1 was enunciated in Theorem 3.5.2 and comes from Valdivia and Pérez Carreras [VP1]. Theorem 4.4.1 is a remarkable result of Saxon and Narayanaswami [SN1], although our proof differs slightly in some minor details from that given there. Lemma 4.5.2 and Theorem 4.5.3 are taken from [FL9].

The conditions imposed on the Banach space X in the first part of the chapter in order to guarantee that it contains a dense linear subspace which is barrelled of class $\aleph_0$ but not totally barrelled are analogous to those imposed by Valdivia in [V13] on a Banach space X to ensure that it contains a dense $\mathcal{L}$-suprabarrelled subspace which is not an inductive limit of a family of totally barrelled spaces. However our techniques based on ultrafilters, are inspired by Dierolf et al. [DDD] and Pérez Carreras and Bonet [PB1].

The first part of Exercise 1 is well known, while the second part is based upon Exercise 1 of Chapter XI of Diestel [D]. Exercise 2 comes from Ferrando and López-Pellicer [FL7] and all requirements needed can be found for example in Jarchow [J] or Köthe [K2]. See also the classic work of A. Grothendieck [G3]. The definitions of the spaces $\ell^\infty(E)$ and $c_0(E)$ are standard, and can be seen for instance in Chapter 4 of Pérez Carreras and Bonet [PB2] (for a study of the barrelledness of the space $c_0(E)$, E being locally convex, see Mendoza [Me2]). Exercise 5 has been taken from Ferrando and Sánchez Ruiz [FS2], where baireled spaces have been introduced. Another approach to the Stone-Čech compactification of a completely regular topological space to that which we have given in Exercise 8 may be found in Gillman and Jerison [GJ].

Barrelled spaces of class $\aleph_0$ and baireled spaces are located in the diagram of Chapter 3 as follows

$$\text{Baire} \longrightarrow \text{CB} \longrightarrow \textit{udb} \longrightarrow \text{TB} \longrightarrow \text{baireled} \longrightarrow \text{barrelled of class } \aleph_0 \longrightarrow \text{BC(n)}$$

Using techniques involving ultrafilters the following result is proved in Ferrando [Fe1], which should be compared with Theorem 4.4.1. Here, the demand for the existence of a basis has, as counterpart, the assumption that the totally barrelled subspace involved will be ultra-bornological.

Theorem 4.7.1 *Each (infinite-dimensional) Banach space with a basis has an ultrabornological totally barrelled subspace which is not unordered Baire-like.*

Let us now discuss the barrelledness of the space $\ell_0^\infty(\mathcal{A})$, $\mathcal{A}$ being a Boolean ring. Let us recall that a ring $(\mathcal{A},+,\cdot)$ is said to be a Boolean ring if $a + a = 0$ and $a \cdot a = a$ for each $a \in \mathcal{A}$. A Boolean algebra is a Boolean ring with unit element. If Σ is an algebra of subsets of a set Ω, then $(\Sigma,\Delta,\cap)$ is a Boolean algebra with Ω as unit element. By the Stone representation theorem, given a Boolean algebra $(\mathcal{A},+,\cdot)$ there exists a Hausdorff compact zero-dimensional topological space $S_{\mathcal{A}}$ — the Stone space of A — with the property that the algebra $\mathcal{D}$ of its clopen sets is homomorphic to $\mathcal{A}$. Hence the usual identification of the algebras $(\mathcal{A},+.\cdot)$ and $(\mathcal{D},\Delta,\cap)$. A Boolean algebra $\mathcal{A}$ is said to have property (N), [Sc], if each pointwise bounded family of finite measures defined on $\mathcal{A}$ is uniformly bounded. A Boolean algebra is said to have property (VHS), op. cit., if each sequence $\{\mu_n : n \in \mathbb{N}\}$ of bounded scalar measures defined on $\mathcal{A}$, such that there exists $\lim \mu_n(a)$ for each $a \in \mathcal{A}$, is uniformly exhaustive. A. Moltó [Mo1] introduces a class of algebras with property (VHS), namely algebras with property (f), for which for each pair of sequences of pairwise disjoint elements of $\mathcal{A}$, $\{A_n: n \in \mathbb{N}\}$ and $\{B_n: n \in \mathbb{N}\}$, with $A_i \cap B_j = \emptyset$ for every i, j $\in \mathbb{N}$, there is a strictly increasing sequence $\{n_k, k \in \mathbb{N}\} \subseteq \mathbb{N}$, a set $A \in \mathcal{A}$ and a $V_J \in \mathcal{A}$ for each $J \subseteq \mathbb{N}$, with

(i) $B_{n_k} \subseteq A$, $A \cap A_k = \emptyset$ for every $k \in \mathbb{N}$;

(ii) $B_{n_k} \subseteq V_J$ for every $k \in J$ and $B_{n_k} \cap V_J = \emptyset$ for every $k \notin J$.

Another class of Boolean algebra with property (VHS) defined by means of interpolation properties is given in [F]. On the other hand, given an algebra Σ of subsets of a set Ω, following the notation of Example 4.2.9, we denote by $\ell_0^\infty(\Sigma)$ the linear space of all Σ-simple functions equipped with the supremum norm. If $\mathcal{A}$ is a Boolean algebra and $\mathcal{D}$ is its Stone space, we will write $\ell_0^\infty(\mathcal{A})$ instead of $\ell_0^\infty(\mathcal{D})$. It is easy to check that $\ell_0^\infty(\mathcal{A})$ is barrelled if and only if $\mathcal{A}$ has property (N). So, since (VHS) $\Rightarrow$ (N), each $\ell_0^\infty(\mathcal{A})$ space, $\mathcal{A}$ being a Boolean algebra with property (f), is barrelled. Indeed in this case Moltó [Mo2] proved that $\ell_0^\infty(\mathcal{A})$ is even suprabarrelled. Since each σ-algebra of sets Σ has property (f), the aforementioned result implies that the space $\ell_0^\infty(\Sigma)$, Σ being a σ-algebra, is suprabarrelled. This had previously been proved by Valdivia [V5]. In the next chapter we will extensively study the properties of $\ell_0^\infty(\Sigma)$, Σ being a σ-algebra of subsets, and also include a proof of the suprabarrelledness of this space using Moltó's techniques. A class of algebras having property (f), named algebras with property (I), was studied in [FLS1] where it is shown that the corresponding ℓ_0^∞ space is barrelled of class $\aleph_0$.

Boolean algebras enjoying property (N) but not (VHS) have now been widely researched. The first known example of an algebra of this kind is due to Schachermayer [Sc] and consists of the algebra of all the Jordan measurable subsets of the closed interval [0,1]. Other examples may be seen in Graves and Weeler [GW]. M.T. Gassó [Ga] introduced a wider class of Boolean algebras, namely algebras with property (LI), which include algebras with property (f) and all the known examples of algebras with property (N) that do not have the (VHS) property. A class of algebras of sets with property (LI) that do not have property (VHS), namely algebras with the strong Nikodym property, has been introduced by Ferrando [Fe6] where it is shown that if $\mathcal{B}$ is an algebra with the strong Nikodym property, then $\ell_0^\infty(\mathcal{B})$ is barrelled of class $\aleph_0$.

If $\mathcal{A}$ is a ring of subsets of a set Ω we define $\ell_0^\infty(\mathcal{A})$ as the linear hull of the characteristic functions of the elements of $\mathcal{A}$. If we denote by $\mathcal{A}^*$ the algebra generated by a ring $\mathcal{A}$, it is easy to check that the space $\ell_0^\infty(\mathcal{A})$ is one-codimensional in $\ell_0^\infty(\mathcal{A}^*)$. Let us mention that a class of rings of sets with property (N), not necessarily having property (VHS), has been introduced by Freniche in [Fr2]. Finally, let us remark that Drewnowski et al. [DrFP4] have proved the ring $\mathcal{Z}$ of all subsets of $\mathbb{N}$ of density zero has property (N). This result may be obtained, with some effort, as a consequence of Exercise 4 of Chapter 1, or using some techniques from [NoS].

5. The space $\ell_0^\infty(\Sigma)$

5.1 THE NORMED SPACE $\ell_0^\infty(\Sigma)$

Let Σ be an algebra on a set Ω. For each $A \in \Sigma$, let $e(A)$ denote the characteristic function on A and let $\ell_0^\infty(\Omega,\Sigma)$ denote the linear space over the field $\mathbb{K}$ of real or complex numbers generated by the family $\{e(A),\ A \in \Sigma\}$. Whenever there is no doubt what the set Ω is, we will denote the space $\ell_0^\infty(\Omega,\Sigma)$ just as $\ell_0^\infty(\Sigma)$.

Every element μ of the algebraic dual $\ell_0^\infty(\Omega)^*$ of $\ell_0^\infty(\Omega)$ is settled by its restriction to the set $\{e(A),\ A \in \Sigma\}$. Therefore we can identify μ with the mapping, which we will also symbolize by μ, defined by

$$\mu(A) = \mu(e(A)),\ A \in \Sigma.$$

Obviously, this mapping μ is a finitely additive measure.

Conversely, if $\mu : \Sigma \longrightarrow \mathbb{K}$ is a finitely additive measure, the mapping from $\ell_0^\infty(\Omega)$ into $\mathbb{K}$, also denoted by μ, given by

$$\mu\Big(\sum_{i=1}^{n} a_i e(A_i)\Big) = \sum_{i=1}^{n} a_i \mu(A_i)$$

is linear. The linear map μ so defined does not depend on the representation of $f = \sum_{1\le i\le n} a_i e(A_i)$ considered since f has only one representation in the form $f = \sum_{1\le i\le m} b_i e(B_i)$, where the scalars b_i are the m different values of the range of f and $\{B_i,\ 1\le i\le m\}$ is a partition of Ω. In what follows we shall identify $\ell_0^\infty(\Sigma)^*$ with the space of all the finitely additive measures defined in Σ with values in $\mathbb{K}$.

In the space $\ell_0^\infty(\Sigma)$ we are going to consider the norm

$$|z| = \sup\{|z(\omega)|,\ \omega \in \Omega\}.$$

The Minkowski functional corresponding to the absolutely convex hull M of $\{e(A),\ A \in \Sigma\}$ will be denoted by $\|.\|$.

Proposition 5.1.1 *The norms* $|.|$ *and* $\|.\|$ *are equivalent.*

Proof. Clearly $|.| \le \|.\|$. Denoting by B be the closed unit ball of $|.|$, then the vectors of (1/2)B taking exactly two non-vanishing real values belong to the absolutely convex hull M of the family $\{e(A),\ A \in \Sigma\}$. Now, suppose by induction that for a positive integer $p \geq 2$ every vector of (1/2)B taking exactly p non-vanishing real values belongs to M. If the sets $A_1, A_2, \ldots, A_{p+1}$ are pairwise disjoint members of Σ and $z = \sum_{i=1}^{p+1} a_i e(A_i) \in (1/2)B$, with $a_i \neq a_j$ for $1 \leq i < j \leq p+1$, we are going to show that $z \in M$. In fact, as $p+1 \geq 3$, at least two different values of z will have the same sign. Without loss of generality we suppose that $0 < a_1 < a_2$. Then we define

$$z_1 = a_2 e(A_1) + a_2 e(A_2) + \ldots + a_{p+1} e(A_{p+1}),\ z_2 = z - a_1 e(A_1).$$

Given that $|z| \leq 1/2$, then $|z_1| \leq 1/2$ and $|z_2| \leq 1/2$. Since z_1 and z_2 have exactly p non-vanishing real values, we have by hypothesis that $z_1, z_2 \in M$. But

$$z = (a_1/a_2)z_1 + (1-a_1/a_2)z_2$$

with $0 < a_1/a_2 < 1$. Therefore, $z \in M$. In case $a_2 < a_1 < 0$, we would work with $-z$ instead of z and we are done.

In the complex case we have that $(1/4)B \subseteq M$. Indeed, taking $z \in (1/4)B$, we may write $z = z_1 + iz_2$ with z_1 and z_2 real vectors of (1/4)B. Now, as $2z_1$ and $2z_2$ are real vectors belonging to (1/2)B, they belong to M, because of the argument above. Hence the absolutely convex combination $z = (1/2)(2z_1) + (i/2)(2z_2)$ belongs to M. Thus we conclude that $\|.\| \leq 4|.|$. ■

The topological dual of $\ell_0^\infty(\Sigma)$ endowed with one of the two equivalent norms $|.|$ and $\|.\|$ will be denoted by $\ell_0^\infty(\Sigma)'$.

The dual norm of $|.|$ is given by

$$|\mu| = \sup\left\{ |\mu(\sum_{i=1}^{n} a_i e(A_i))|,\ \{A_1, A_2, \ldots, A_n\} \in \mathcal{F}(\Omega) \text{ with } \sup_{1 \leq i \leq n} |a_i| \leq 1 \right\},$$

where $\mathcal{F}(\Omega)$ denotes the family of all finite partitions of Ω with A_i in Σ. If, when possible, we put $a_i = \overline{(\mu(e(A_i)))}/\mu(e(A_i))$, then we obtain

$$|\mu| = \sup\left\{ \sum_{i=1}^{n} |\mu(A_i)|,\ \{A_1, A_2, \ldots, A_n\} \in \mathcal{F}(\Omega) \right\},$$

which is the expression of the variation of μ. So we shall write either $|\mu|$ or $|\mu|(\Omega)$ interchangeably. The dual norm of $\|.\|$ is given by

$$\|\mu\| = \sup\{|\mu(\sum_{i=1}^{n} a_i e(A_i))| : n \in \mathbb{N},\ A_i \in \Sigma,\ \sum_{i=1}^{n} |a_i| \le 1\}$$

which is obviously equal to

$$\|\mu\| = \sup\{|\mu(A)|,\ A \in \Sigma\}.$$

Therefore the elements of the space $\ell_0^\infty(\Sigma)'$ are the bounded finitely additive measures. So, we can denote $\ell_0^\infty(\Sigma)'$ by $ba(\Sigma)$ as usual. Notice that, on $\ell_0^\infty(\Sigma)'$ one has $\|.\| \le |.| \le 4\|.\|$.

The variation of μ in $A \in \Sigma$ is defined by

$$|\mu|(A) = \sup\left\{\sum_{i=1}^{n} |\mu(A_i)|,\ \{A_1, A_2, \ldots, A_n\} \in \mathcal{F}(A)\right\}$$

where $\mathcal{F}(A)$ denotes the family of all the finite partitions of A with A_i in Σ. It is easy to show that if A and B are disjoint elements of Σ, $|\mu|(A \cup B) = |\mu|(A) + |\mu|(B)$. Therefore, because of the equivalence of norms, we have that $\mu \in \ell_0^\infty(\Sigma)'$ implies $|\mu| \in \ell_0^\infty(\Sigma)'$.

Given $A \in \Sigma$, $\ell_0^\infty(A,\Sigma)$ will denote the subspace of $\ell_0^\infty(\Sigma)$ generated by $\{e(B),\ B \in \Sigma,\ B \subseteq A\}$, and given $\mu \in \ell_0^\infty(\Sigma)'$, $\mu[A]$ will stand for the restriction of μ to $\ell_0^\infty(A,\Sigma)$ and $\|\mu[A]\|$ for the norm of $\mu[A]$. Again note that $\|\mu[A]\| \le |\mu|(A) \le 4\|\mu[A]\|$ for each $\mu \in ba(\Sigma)$ and each $A \in \Sigma$.

5.2 BARRELLEDNESS OF $\ell_0^\infty(\Sigma)$

In section 5.3 we will prove that the space $\ell_0^\infty(\Sigma)$ is barrelled of class $\aleph_0$. This is going to involve considerable work but the reward will repay our effort. As a first step we are going to prove in this section that, whenever Σ is a σ-algebra on a set Ω, the space $\ell_0^\infty(\Sigma)$ is barrelled. With this target we will prove first the following two lemmas. In the following Σ is supposed to be a σ-algebra.

Lemma 5.2.1 *Let* H *be a* $\sigma(\ell_0^\infty(\Sigma)^*, \ell_0^\infty(\Sigma))$*-bounded subset of* $\ell_0^\infty(\Sigma)^*$ *such that* $\sup\{|\mu|,\ \mu \in H\} = \infty$. *Then we can determine a sequence* $\{\mu_n,\ n \in \mathbb{N}\}$ *in* H *and a sequence* $\{A_n,\ n \in \mathbb{N}\}$ *of pairwise disjoint elements of* Σ *such that* $|\mu_1(A_1)| > 1$ *and, for* $n > 2$,

$$|\mu_n(A_n)| > |\mu_n(A_1)| + |\mu_n(A_2)| + \ldots + |\mu_n(A_{n-1})| + n.$$

Proof. By hypothesis we also have that $\sup\{\|\mu\|, \mu \in H\} = \infty$. Now if $\sup\{|\mu(\Omega)|, \mu \in H\} = h_1$ we can determine some $\mu_1 \in H$ such that $\|\mu_1\| > h_1 + 1$ and, therefore, there is some $P_1 \in \Sigma$ such that $|\mu_1(P_1)| > h_1 + 1$.

We have that $|\mu_1(\Omega\setminus P_1)| > |\mu_1(P_1)| - |\mu_1(\Omega)| > 1$ and also that $|\mu_1(P_1)| > 1$. Now as $|\mu| = |\mu|(P_1) + |\mu|(\Omega\setminus P_1)$, setting A_1 either equal to P_1 or equal to $\Omega\setminus P_1$ at our convenience and putting $B_1 := \Omega\setminus A_1$, we can suppose that $\sup\{|\mu|(B_1), \mu \in H\} = \infty$. Therefore,

$$\sup\{\|\mu[B_1]\|, \mu \in H\} = \infty.$$

The lemma follows easily by induction since if $\sup\{|\mu(B_1)|, \mu \in H\} = h_2$ we determine $\mu_2 \in H$ whose restriction $\mu_2[B_1]$ to $\ell_0^\infty(B_1,\Sigma)$ satisfies $\|\mu_2(B_1)\| > h_2 + |\mu_2(A_1)| + 2$ and hence there is a subset $P_2 \in \Sigma$ of B_1 such that $|\mu_2(P_2)| > h_2 + |\mu_2(A_1)| + 2$. As before, by taking A_2 equal either to P_2 or to $B_1\setminus P_2$ depending on whether $\sup\{|\mu|(\Omega\setminus P_2), \mu \in H\} = \infty$ or whether $\sup\{|\mu|(P_2), \mu \in H\} = \infty$, we have $|\mu_2(A_2)| > |\mu_2(A_1)| + 2$, and we continue as in the first step. ■

Remark 5.2.2 If $H = \{\mu\} \subseteq \ell_0^\infty(\Sigma)^* \setminus \ell_0^\infty(\Sigma)'$ i.e. $|\mu|(\Omega) = \infty$, according to the preceding lemma, we can find a sequence $\{A_n, n \in \mathbb{N}\}$ of pairwise disjoint elements of Σ such that $\sum_n|\mu(A_n)|$ is divergent. On the other hand, if $\mu \in \ell_0^\infty(\Sigma)'$, then we have $\sum_n|\mu(A_n)| < |\mu|$, and we conclude that *a finitely additive measure* μ *is bounded, i.e.,* $\mu \in ba(\Sigma)$ *if and only if for every sequence* $\{A_n, n \in \mathbb{N}\}$ *of pairwise disjoint elements of* Σ, $\sum_n|\mu(A_n)| < \infty$.

Lemma 5.2.3 *Let* $\{\mu_n, n \in \mathbb{N}\}$ *be a sequence in* $ba(\Sigma)$ *and let* $\{A_n, n \in \mathbb{N}\}$ *be a sequence of pairwise disjoint elements of* Σ. *Given* $\varepsilon > 0$ *there is a subsequence* $\{A_{n_p}, p \in \mathbb{N}\}$ *of* $\{A_n, n \in \mathbb{N}\}$ *such that*

$$|\mu_{n_p}|(\cup\{A_{n_q}, q = p+1, p+2, \dots\}) < \varepsilon.$$

Proof. Take $n_1 = 1$ and let $\{N_1, N_2, \dots\}$ be a partition of $\mathbb{N}\setminus\{1\}$ such that each N_i is infinite. Setting $B_{1i} = \cup\{A_n, n \in N_i\}$ for each $i \in \mathbb{N}$ then there is some $j_1 \in \mathbb{N}$ such that $|\mu_{n_1}|(B_{1j_1}) < \varepsilon$ since $\sum\{|\mu_{n_1}|(B_{1i}), i \in \mathbb{N}\} < \infty$. Now take n_2 equal to the first element of N_{j_1}.

Repeating the above argument we find a partition $\{M_1, M_2, \ldots\}$ of $N_{j_1} \setminus \{n_2\}$ such that every M_k is infinite, and we find some $j_2 \in \mathbb{N}$ such that if $B_{2j_2} = \bigcup\{A_n, n \in M_{j_2}\}$, then $|\mu_{n_2}|(B_{2j_2}) < \varepsilon$. Now we continue taking n_3 equal to the first element of M_{j_2}.

The sequence determined in this way verifies that

$$|\mu_{n_1}|(\bigcup\{A_{n_q}, q = 2, 3, \ldots\}) < |\mu_{n_1}|(B_{1j_1}) < \varepsilon,$$
$$|\mu_{n_2}|(\bigcup\{A_{n_q}, q = 3, 4, \ldots\}) < |\mu_{n_2}|(B_{2j_2}) < \varepsilon, \ldots,$$

and the proof is complete. ■

Theorem 5.2.4 $\ell_0^\infty(\Sigma)$ *is barrelled.*

Proof. Let H be a $\sigma(ba(\Sigma),\ell_0^\infty(\Sigma))$-bounded subset of $ba(\Sigma)$. If $\sup\{|\mu|, \mu \in H\} = \infty$ and $\varepsilon = 1$, then by the preceding two lemmas we can obtain a sequence $\{\mu_n, n \in \mathbb{N}\}$ in H and a sequence $\{A_n, n \in \mathbb{N}\}$ of pairwise disjoint elements of Σ such that

$$|\mu_n(A_n)| > |\mu_n(A_1)| + |\mu_n(A_2)| + \ldots + |\mu_n(A_{n-1})| + n$$

and

$$|\mu_n|(\bigcup\{A_q, q = n+1, n+2, \ldots\}) < \varepsilon.$$

Hence if $A := \bigcup\{A_p, p \in \mathbb{N}\}$ we have

$$|\mu_n(A)| \geq |\mu_n(A_n)| - |\mu_n(A_1)| - |\mu_n(A_2)| - \ldots - |\mu_n(A_{n-1})| -$$
$$|\mu_n|(\bigcup\{A_q, q = n+1, n+2, \ldots\}) > n - 1,$$

in contradiction with $\sup\{|\mu_n(A)|, n \in \mathbb{N}\} < \infty$.

Therefore, H is an equicontinuous subset of $ba(\Sigma)$ and $\ell_0^\infty(\Sigma)$ is barrelled. ■

Remark 5.2.5 Using Phillips's lemma, another proof of the last result may be given. Indeed let H be a $\sigma(ba(\Sigma),\ell_0^\infty(\Sigma))$-bounded subset of $ba(\Sigma)$. If $\sup\{|\mu|, \mu \in H\} = \infty$, then according to Lemma 5.2.1, we can determine a sequence $\{\mu_n, n \in \mathbb{N}\}$ in H and a sequence $\{A_n, n \in \mathbb{N}\}$ of pairwise disjoint elements of Σ such that $|\mu_n(A_n)| > n$. Since $\{\mu_n/n, n \in \mathbb{N}\}$ converges to 0 in $ba(\Sigma)(\sigma(ba(\Sigma),\ell_0^\infty(\Sigma)))$, Phillips' lemma shows that $\lim_n \sum_p|\mu_n(A_p)|/n = 0$, a contradiction with $\sum_p|\mu_n(A_p)|/n \geq |\mu_n(A_n)|/n > 1$ for every $n \in \mathbb{N}$.

Remark 5.2.6 The barrelledness of $\ell_0^\infty(\Sigma)$ is equivalent to the sequential completeness of the space $ba(\Sigma)(\sigma(ba(\Sigma),\ell_0^\infty(\Sigma)))$. In fact, if W is a closed bounded absolutely convex subset of $ba(\Sigma)(\sigma(ba(\Sigma),\ell_0^\infty(\Sigma)))$, then W is a Banach disk and hence W is strongly bounded. Now because of the metrizability of $\ell_0^\infty(\Sigma)$, W is equicontinuous. Conversely, Phillips lemma can help us to show that $ba(\Sigma)(\sigma(ba(\Sigma),\ell_0^\infty(\Sigma)))$ is sequentially complete. Indeed if $\{\mu_n,\ n \in \mathbb{N}\}$ is a $\sigma(ba(\Sigma),\ell_0^\infty(\Sigma))$ Cauchy sequence in $ba(\Sigma)$ and μ is its limit in $\ell_0^\infty(\Sigma)^*$, suppose $\mu \in \ell_0^\infty(\Sigma)^* \setminus ba(\Sigma)$. Then we can determine a sequence $\{A_n,\ n \in \mathbb{N}\}$ of pairwise disjoint members of Σ such that $|\mu(A_n)| > n\ \forall n \in \mathbb{N}$. Given that $\lim_m \mu_m(A_n) = \mu(A_n)$, then for each $n \in \mathbb{N}$ we can find some m_n such that $|\mu_{m_n}(A_n)| > n$. Now, writing μ_n instead of μ_{m_n} we get a similar contradiction to the previous remark.

5.3 BARRELLEDNESS OF CLASS $\aleph_0$ OF $\ell_0^\infty(\Omega,\Sigma)$

Now, we will give some results with the aim of subsequently smoothing the proof of the barrelledness of class $\aleph_0$ of the space $\ell_0^\infty(\Sigma)$.

Lemma 5.3.1 *Let* $\mu \in \ell_0^\infty(\Sigma)'$. *If* $A \in \Sigma$ *and* $\{A_1, A_2, \dots, A_n\} \in \mathcal{F}(A)$ *with* $A_k \in \Sigma$ *for* $1 \le k \le n$, *then* $\sum_{1\le i\le n}|\mu|(A_i) = |\mu|(A)$.

Proof. This is obvious, since if $\mu \in \ell_0^\infty(\Sigma)'$, then $|\mu| \in \ell_0^\infty(\Sigma)'$. ∎

Lemma 5.3.2 *Let* V *be an absolutely convex subset of a space* F *which is not a neighbourhood of the origin in its linear hull* L *and let* B *be a finite subset of* F. *Then the absolutely convex hull of* $V \cup B$ *is not a neighbourhood of the origin in* F.

Proof. If L is infinite-codimensional in F the conclusion is clear since $\Gamma(V \cup B)$ is not absorbing.

If $\{z_1, z_2, \dots, z_p\}$ is a cobasis of L in F and $M := \Gamma\{z_1, z_2, \dots, z_p\}$, then the set $W = V + M$ is absorbing and verifies that $W \cap L = V$. Hence W is not a neighbourhood of the origin in F. It follows from $B \subseteq nW$ that $\Gamma(V \cup B) \subseteq V + nW \subseteq (n+1)W$, so we have that $\Gamma(V \cup B)$ is not a neighbourhood of the origin in F. ∎

Corollary 5.3.3 *Let* U *be a barrel in the subspace* E *of* $\ell_0^\infty(\Sigma)$ *and let* B *be a finite subset of* $\ell_0^\infty(\Sigma)$. *If* U *is not a neighbourhood of the origin in* E, *then* $\sup\{|\mu|, \mu \in (U \cup B)^\circ\} = \infty$, *the polar being taken in* $\ell_0^\infty(\Sigma)'$.

Proof. Let V be the closure of U in $\ell_0^\infty(\Sigma)$. As V is not a neighbourhood of the origin in its linear hull, by the previous lemma we have that $\{\Gamma(V \cup B)\}^\circ = (V \cup B)^\circ = (U \cup B)^\circ$ is not an equicontinuous subset of $ba(\Sigma)$. Hence $\sup\{|\mu|, \mu \in (U \cup B)^\circ\} = \infty$. ■

If U is a closed absolutely convex subset of $\ell_0^\infty(\Sigma)$ and $A \in \Sigma$, then one says that A *has property* U if there is a finite set B in $\ell_0^\infty(\Sigma)$ such that $\Gamma(U \cup B)$ contains a neighbourhood of the origin in $\ell_0^\infty(A,\Sigma)$. If U is a barrel in the subspace E of $\ell_0^\infty(\Sigma)$ and $A \in \Sigma$ we shall say that A does not have property U if $\sup\{|\mu|(A), \mu \in (U \cup B)^\circ\} = +\infty$ for every finite subset B of $\ell_0^\infty(\Sigma)$. So, the preceding corollary states that if U is a barrel in the subspace E of $\ell_0^\infty(\Sigma)$ and U is not a neighbourhood of the origin in $\ell_0^\infty(\Sigma)$, then Ω does not have property U.

In the following lemmas we shall suppose that $A \in \Sigma$ and U is a barrel in the subspace E of $\ell_0^\infty(\Sigma)$.

Lemma 5.3.4 *If* A *does not have property* U *and* $A_1, A_2, \ldots, A_n$ *is a partition of* A *whose elements belong to* Σ, *then there is some* A_q *that does not have property* U.

Proof. The conclusion follows from $\sup\{|\mu|(A_i), \mu \in (U \cup B_i)^\circ\} = h_i$ for $i = 1, 2, \ldots, n$, and, consequently, $\sup\{|\mu|(A), \mu \in (U \cup B)^\circ\} \le h_1 + \ldots + h_n$, where $B := B_1 \cup \ldots \cup B_n$. ■

Lemma 5.3.5 *Suppose that* A *does not have property* U. *Given a positive integer* $p \ge 2$, *the elements* $x_1, x_2, \ldots, x_n$ *of* $\ell_0^\infty(\Sigma)$ *and a positive real number* a, *there are* p *elements* $A_1, A_2, \ldots, A_p$ *of* Σ *which are a partition of* A *and* p *bounded measures* $\mu_1, \mu_2, \ldots, \mu_p$ *belonging to* U° *such that* $|\mu_i(A_i)| > a$ *and* $\sum_{1\le j\le n}|\mu_i(x_j)| \le 1$.

Proof. Let $B = \{e(A), nx_1, nx_2, \ldots, nx_n\}$. Since

$$\sup\{|\mu|(A), \mu \in (U \cup B)^\circ\} = \infty,$$

there is a subset of A, $C_1 \in \Sigma$, and some $\mu_1 \in (U \cup B)^o$ such that $|\mu_1(C_1)| > a + 1$ and $|\mu_1(A)| \le 1$. So,

$$|\mu_1(A \setminus C_1)| \ge |\mu_1(C_1)| - |\mu_1(A)| > a + 1 - 1 = a.$$

On the other hand, as $\mu_1 \in (U \cup B)^o$, $|\mu_1(nx_j)| \le 1$ for $j = 1, 2, \dots, n$ and, consequently, $\sum_{1 \le j \le n} |\mu_i(x_j)| \le 1$.

We have obtained that $|\mu_1(C_1)| > a$ and $|\mu_1(A \setminus C_1)| > a$. By the preceding lemma, either C_1 or $A \setminus C_1$ does not have property U. Should the first of these happen, set $A_1 := C_1$ and replace $A \setminus C_1$ with A; otherwise set $A_1 := A \setminus C_1$ and replace C_1 with A. Then we conclude the proof by finite induction. ■

Lemma 5.3.6 *Suppose that* $A \in \Sigma$ *does not have property* U_n *for* $n = n_1, n_2, \dots, n_p$ *and for infinitely many values of* n. *Then, given a positive real number* a *and the vectors* $x_1, x_2, \dots, x_r$ *in* $\ell_0^\infty(\Sigma)$ *there are* p *pairwise disjoint subsets* $M_1, M_2, \dots, M_p$ of A *belonging to* Σ *and* p *measures* $\mu_1, \mu_2, \dots, \mu_p \in ba(\Sigma)$ *such that, for every* $i = 1, 2, \dots, p$,

$$|\mu_i(M_i)| > a \qquad \sum_{j=1}^{r} |\mu_i(x_j)| \le 1 \qquad \mu_i \in U_{n_i}^o$$

and $A \setminus \bigcup\{M_i,\ i = 1, 2, \dots, p\}$ *does not have property* U_n *for* $n = n_1, n_2, \dots, n_p$, *and for infinitely many values of* n.

Proof. According to the previous lemma we can find a partition $Q_1, Q_2, \dots, Q_{p+2}$ of A formed by elements of Σ and $\nu_1, \nu_2, \dots, \nu_{p+2}$ in $ba(\Sigma)$ such that, for $i = 1, 2, \dots, p+2$

$$|\nu_i(Q_i)| > a \qquad \sum_{j=1}^{r} |\nu_i(x_j)| \le 1 \qquad \nu_i \in U_{n_1}^o$$

According to Lemma 5.3.4 we can suppose that Q_1 does not have property U_n for infinitely many values of n and that given a positive integer q, $1 \le q \le p$, some of the sets $Q_1, \dots, Q_{p+2}$ does not have property U_{n_q}. Then, as there are p + 2 elements, we can find some element Q_h, $1 < h \le p + 2$, such that $A \setminus Q_h$ does not have property U_n for $n = n_1, n_2, \dots, n_p$. Since $A \setminus Q_h$ contains Q_1 we have that $A \setminus Q_h$ does not have property U_n for infinitely many values of n. We set $M_1 = Q_h$, $\mu_1 = \nu_h$ and then

$$|\mu_1(M_1)| > a \qquad \sum_{j=1}^{r} |\mu_1(x_j)| \le 1 \qquad \mu_1 \in U^o_{n_1}$$

and $A \setminus Q_h$ does not have property U_n for $n = n_1, n_2, \ldots, n_p$ and for an infinity of values of n. Now we apply the above reasoning p times. ■

Recalling the definition of a set of class s which we introduced in the previous chapter, we prove next a property enjoyed by these kind of sets that we use repeatedly in what follows.

Lemma 5.3.7 *If the set* $A = A_1 \cup A_2 \cup \ldots \cup A_q$ *is of class* s, *then there exists some* A_i, $1 \le i \le q$, *which contains a set of class* s.

Proof. The result is obvious for $s = 1$. So, we will assume $s \ge 2$ and that the lemma is true for s-1. Let $B := \{(a_1, a_2, \ldots, a_{s-1}) \in \mathbb{N}^{s-1}$: *there are infinitely many* $(a_1, a_2, \ldots, a_{s-1}, a_s) \in A\}$ and let $B_i :=$ $\{(a_1, a_2, \ldots, a_{s-1}) \in \mathbb{N}^{s-1}$: *there are infinitely many* $(a_1, a_2, \ldots, a_{s-1}, a_s) \in A_i\}$, for each $1 \le i \le q$.

We are going to show that B is of class s-1. In fact, there are infinitely many $b_1 \in \mathbb{N}$ such that there is some $(b_1, b_2, \ldots, b_{s-1}, b_s) \in A$. But for each one of these, there are infinitely many c_s such that $(b_1, b_2, \ldots, b_{s-1}, c_s) \in A$, so $(b_1, b_2, \ldots, b_{s-1}) \in B$ and therefore there are infinitely many b_1 such that $(b_1, b_2, \ldots, b_{s-1}) \in B$. Now, given $(a_1, a_2, \ldots, a_{s-1}) \in B$ and $1 \le i \le s-2$, as there exists $(a_1, a_2, \ldots, a_{s-1}, a_s) \in A$, then there exist infinitely many b_{i+1} and, for each b_{i+1}, there exist $b_{i+2}, \ldots, b_s$ such that $(a_1, a_2, \ldots, a_i, b_{i+1}, b_{i+2}, \ldots, b_{s-1}, b_s) \in A$; but as there are infinitely many c_s such that $(a_1, a_2, \ldots, a_i, b_{i+1}, b_{i+2}, \ldots, b_{s-1}, c_s) \in A$, we have that $(a_1, a_2, \ldots, a_i, b_{i+1}, b_{i+2}, \ldots, b_{s-1}) \in B$.

It is clear that $B = B_1 \cup B_2 \cup \ldots \cup B_q$. Hence, by the induction hypothesis, there exists a set B_i which contains a set of class s-1 and, by construction, A_i contains a set of class s. ■

In the following lemma let us recall that $\{(m_1(i), m_2(i), \ldots, m_s(i)), i \in \mathbb{N}\}$ denotes $\mathbb{N}^s$ endowed with the diagonal ordering and consider $U_{m_1 m_2 \ldots m_s}$ to be closed absolutely convex subsets of $\ell_0^\infty(\Sigma)$.

Lemma 5.3.8 *Suppose that* $A \in \Sigma$ *does not have property* $U_{m_1 m_2 \ldots m_s}$ *for all* $(m_1, m_2, \ldots, m_s)$ *equal to* $(m_1(i), m_2(i), \ldots, m_s(i))$, $1 \le i \le k$; *for all* $(m_1, m_2, \ldots, m_s)$ *belonging to the cartesian products of* $\{(m_1(i), m_2(i), \ldots, m_r(i))\}$ *and a set* $S_i^{(s-r)}$ *of class* $s-r$, $1 \le r \le s-1$ *and* $1 \le i \le k$; *and for all* $(m_1, m_2, \ldots, m_s)$ *belonging to a set* S *of class* s. *Suppose besides that* $a > 0$ *and* $x_1, x_2, \ldots, x_n$ *are vectors of* $\ell_0^\infty(\Sigma)$. *Then we can determine* k *pairwise disjoint subsets* $A_1, A_2, \ldots, A_k$ *of* A, *all of them members of* Σ, *and* k *measures* $\mu_i \in ba(\Sigma)$ *such that*

$$|\mu_i(A_i)| > a, \ \sum_{k=1}^{n} |\mu_i(x_k)| \le 1, \ \mu_i \in \left(U_{m_1(i), m_2(i), \ldots, m_s(i)}\right)^\circ$$

for $1 \le i \le k$, *and* $A \setminus \bigcup\{A_i, 1 \le i \le k\}$ *does not have property* $U_{m_1 \ldots m_s}$ *for all* $(m_1, m_2, \ldots, m_s)$ *equal to* $(m_1(i), m_2(i), \ldots, m_s(i))$, $1 \le i \le k$; *for all* $(m_1, m_2, \ldots, m_s)$ *belonging to the cartesian products of* $\{(m_1(i), m_2(i), \ldots, m_r(i))\}$ and *a set* $S_i^{*(s-r)}$ *of class* $s-r$, $1 \le r \le s-1$ *and* $1 \le i \le k$; *and for all* $(m_1, m_2, \ldots, m_s)$ *belonging to a set* S^* *of class* s.

Proof. Applying Lemma 5.3.6 to the set A with $U_{m_1(1)m_2(1)\ldots m_s(1)}$ instead of U we can divide A in $sk+2$ pairwise disjoint subsets $B_1, B_2, \ldots, B_{sk+2}$, all of them members of Σ, and determine $sk+2$ bounded measures $\lambda_i \in \left(U_{m_1(1)m_2(1)\ldots m_s(1)}\right)^\circ$, $1 \le i \le sk+2$, with $|\lambda_i(B_i)| > a$, and $\sum_{1 \le j \le n} |\lambda_i(x_j)| \le 1$, for $i = 1, 2, \ldots, sk + 2$.

By Lemma 5.3.4 there exists some $B_{h(i)}$, with $1 \le h(i) \le sk + 2$, which does not have property $U_{m_1(i)m_2(i)\ldots m_s(i)}$ for $1 \le i \le k$.

Given an element $(m_1(i), m_2(i), \ldots, m_r(i), m_{r+1}, \ldots, m_s)$ of the cartesian product of $\{(m_1(i), m_2(i), \ldots, m_r(i))\}$ and $S_i^{(s-r)}$ we have, by hypothesis, that $A = B_1 \cup B_2 \cup \ldots \cup B_{sk+2}$ does not have property $U_{m_1(i)m_2(i)\ldots m_r(i)m_{r+1}\ldots m_s}$. Thus by Lemma 5.3.4 there exists at least some B_h which does not have property $U_{m_1(i)m_2(i)\ldots m_r(i)m_{r+1}\ldots m_s}$. So we can write

$$S_i^{(s-r)} = (S_i^{(s-r)})_1 \cup (S_i^{(s-r)})_2 \cup \ldots \cup (S_i^{(s-r)})_{sk+2}$$

assuming that, given any $(m_{r+1},\ldots,m_s) \in S_i^{(s-r)}$, if B_j does not have property $U_{m_1(i)m_2(i)\ldots m_r(i)m_{r+1}\ldots m_s}$, then

$$(m_{r+1}, \ \ldots, \ m_s) \in \left(S_i^{(s-r)}\right)_j, \ 1 \le j \le sk + 2.$$

By Lemma 5.3.7 we can suppose that there is some $\left(S_i^{(s-r)}\right)_{v(r,i)}$ with $1 \le v(r,i) \le sk + 2$ which contains some set $S'^{(s-r)}$ of class s-r. Hence there is a $B_{v(r,i)}$ which does not have property $U_{m_1m_2\ldots m_s}$ for all $(m_1, m_2, \ldots, m_s)$ belonging to the cartesian product of $\{(m_1(i), m_2(i), \ldots, m_r(i))\}$ and $S_i'^{(s-r)}$, for $1 \le r \le s-1$ and $1 \le i \le k$.

The same argument shows that there is some B_h, $1 \le h \le sk+2$, which does not have property $U_{m_1m_2\ldots m_s}$ for all $(m_1, m_2, \ldots, m_s)$ belonging to some set S' of class s.

In this process there must be at least one B_h, $1 \le h \le sk+2$, which has not been considered in any of the previous three stages since the number of sets B_i considered is smaller than or equal $k + (s-1)k + 1 = sk + 1$. Such a B_h and the corresponding λ_h will be denoted by A_1 and μ_1, respectively, satisfying

$$|\mu_1(A_1)| > a \qquad \text{and} \qquad \sum_{l=1}^{n} |\mu_1(x_l)| \le 1$$

and it is clear that $A \setminus A_1$ does not have property $U_{m_1m_2\ldots m_s}$ for all $(m_1, m_2, \ldots, m_s)$ equal to $(m_1(i), m_2(i), \ldots, m_s(i))$, with $1 \le i \le k$; for all $(m_1, m_2, \ldots, m_s)$ belonging to the cartesian products of $\{(m_1(i), m_2(i), \ldots, m_r(i))\}$ and a set $S_i'^{(s-r)}$ of class s-r for $1 \le r \le s-1$ and $1 \le i \le k$; and for all $(m_1, m_2, \ldots, m_s)$ belonging to some set S' of class s.

Applying again the previous reasoning to $A \setminus A_1$ we obtain A_2, and continuing in this way k times we determine all the desired sets A_i and the measures μ_i, $1 \le i \le k$. ∎

Remark 5.3.9 Lemma 5.3.8 applies without considering the vectors x_1, x_2, ..., x_n of the space $\ell_0^\infty(\Sigma)$. Then the measures μ_i verify that $|\mu_i(A_i)| > a$, for $i = 1, 2, \ldots, k$.

Lemma 5.3.10 *For each* $i \in \mathbb{N}$, *let us suppose that* $U_{m_1(i)m_2(i)\dots m_s(i)}$ *is a closed absolutely convex subset of the space* $\ell_0^\infty(\Sigma)$ *such that* Ω *does not have property* $U_{m_1(i)m_2(i)\dots m_s(i)}$. *Then we can determine a sequence* $\{(m'_1(i), m'_2(i), \dots, m'_s(i)), i \in \mathbb{N}\}$ *in* $\mathbb{N}^s$ *whose terms form a set of class* s, *a family* $\{A_{ij}, i, j \in \mathbb{N}\}$ *of pairwise disjoint elements of* Σ *and a family* $\{\mu_{ij}, i, j \in \mathbb{N}\}$ *of elements of* $\left(U_{m'_1(i)m'_2(i)\dots m'_s(i)}\right)^\circ$, *in such a way that*

$$|\mu_{ij}(A_{ij})| \geq i + j \quad \textit{and} \quad \sum\{|\mu_{ij}(A_{gh})|, g + h < i + j\} \leq 1$$

for all $i, j \in \mathbb{N}$.

Proof. We proceed by induction. So, choose $(m'_1(1), m'_2(1), \dots, m'_s(1)) = (1, \dots, 1)$. Then Ω does not have property $U_{m_1m_2\dots m_s}$ for $(m_1, \dots, m_s) = (m'_1(1), m'_2(1), \dots, m'_s(1))$; for all $(m_1, m_2, \dots, m_s)$ belonging to the cartesian products of $\{(m'_1(1), m'_2(1), \dots, m'_r(1))\}$ and a set of class s-r, for $1 \leq r \leq s-1$; and for all $(m_1, m_2, \dots, m_s)$ of some set of class s.

Applying Lemma 5.3.8 with k = 1, a = 2 and taking into account the last remark, we obtain a $\mu_{11} \in \left(U_{m'_1(1)m'_2(1)\dots m'_s(1)}\right)^\circ$ and some $A_{11} \in \Sigma$ such that $|\mu_{11}(A_{11})| \geq 2$ and $\Omega \setminus A_{11}$ does not have property $U_{m_1m_2\dots m_s}$ for $(m_1, m_2, \dots, m_s) = (1, 1, \dots, 1)$; for all $(m_1, m_2, \dots, m_s)$ belonging to the cartesian products of $\{(m'_1(1), m'_2(1), \dots, m'_r(1))\}$ and some set of class s-r, for $1 \leq r \leq s-1$; and for all $(m_1, m_2, \dots, m_s)$ of a set of class s.

Suppose that in the (k-1)-th stage of the inductive process we have obtained the element $(m'_1(k-1), m'_2(k-1), \dots, m'_s(k-1))$ of $\mathbb{N}^s$, the bounded measures $\mu_{ij} \in \left(U_{m'_1(i)m'_2(i)\dots m'_s(i)}\right)^\circ$ and the pairwise disjoint sets $A_{ij} \in \Sigma$ contained in $\Omega \setminus \bigcup\{A_{gh}, g + h < k\}$, for $1 \leq i \leq k-1$ and $i+j = k$, which satisfy the relations

$$|\mu_{ij}(A_{ij})| \geq i + j = k \quad \text{and} \quad \sum\{|\mu_{ij}(A_{gh})|, g + h < i + j\} \leq 1$$

in such a way that $\Omega \setminus \bigcup\{A_{ij}, i+j \leq k\}$ does not have property $U_{m_1\dots m_s}$ for all $(m_1, m_2, \dots, m_s)$ equal to $(m'_1(i), m'_2(i), \dots, m'_s(i))$, $1 \leq i \leq$

k-1; for all $(m_1, m_2, \ldots, m_s)$ belonging to the cartesian products of $\{(m'_1(i), m'_2(i), \ldots, m'_r(i))\}$ and some set $S_i^{(s-r)}$ of class s-r, for r = 1, 2, ..., s-1 and $1 \le i \le k-1$; and for all $(m_1, m_2, \ldots, m_s)$ of a set S of class s.

To determine $(m'_1(k), m'_2(k), \ldots, m'_s(k))$ we distinguish two cases:

Case 1. The first component of the k-th element of $\mathbb{N}^s$ is greater than the first one of all the first k-1 elements of $\mathbb{N}^s$.

Then we take $(m'_1(k), m'_2(k), \ldots, m'_s(k))$ equal to the first element of S such that $m'_1(k) > m'_1(k-1)$.

Case 2. The i-th element of $\mathbb{N}^s$, with $i<k$, is the last of those elements smaller than the k-th one with a maximum number r of first components respectively equal to those of the k-th element of $\mathbb{N}^s$.

Then we take $(m'_1(k), m'_2(k), \ldots, m'_r(k), m'_{r+1}(k), \ldots, m'_s(k))$ equal to $(m'_1(i), m'_2(i), \ldots, m'_r(i), m'_{r+1}(k), \ldots, m'_s(k))$, $(m'_{r+1}(k), \ldots, m'_s(k))$ being the first element of $S_i^{(s-r)}$ such that $m'_{r+1}(i) < m'_{r+1}(k)$.

Now, owing to the choice made, the definition of a set of class s and the supposition of the (k-1)-th stage, we have in both cases that $\Omega \setminus \bigcup\{A_{ij}, i+j \le k\}$ does not have property $U_{m_1 m_2 \ldots m_s}$ for $(m_1, m_2, \ldots, m_s)$ equal to $(m'_1(i), m'_2(i), \ldots, m'_s(i))$, i = 1, 2, ..., k; for all $(m_1, m_2, \ldots, m_s)$ belonging to the cartesian products of $\{(m'_1(i), m'_2(i), \ldots, m'_r(i)\}$ and some set of class s-r, for $1 \le r \le s-1$ and $1 \le i \le k$; and for all $(m_1, m_2, \ldots, m_s)$ of a set of class s.

Then Lemma 5.3.8 allows us to define the bounded measures $\mu_{ij} \in \left(U_{m'_1(i)m'_2(i)\ldots m'_s(i)}\right)^\circ$ and the pairwise disjoint sets A_{ij}, contained in $\Omega \setminus \bigcup\{A_{ij}, i+j \le k\}$, with $1 \le i \le k$ and $i+j = k+1$, such that

$$|\mu_{ij}(A_{ij})| \ge i+j = k+1 \quad \text{and} \quad \sum\{|\mu_{ij}(A_{gh})|, g+h < i+j\} \le 1$$

and $\Omega \setminus \bigcup\{A_{ij}, i+j \le k+1\}$ does not have property $U_{m_1 m_2 \ldots m_s}$ for all $(m_1, m_2, \ldots, m_s)$ equal to $(m'_1(i), m'_2(i), \ldots, m'_s(i))$, i = 1, 2, ..., k; for all $(m_1, m_2, \ldots, m_s)$ belonging to the cartesian products of the set $\{(m'_1(i), m'_2(i), \ldots, m'_r(i)\}$ and some set of class s-r, for $1 \le r \le s-1$ and $1 \le i \le k$; and for all $(m_1, m_2, \ldots, m_s)$ of a set of class s.

This completes the inductive process and establishes the lemma. ∎

In the following let $W = \{E_{m_1m_2\dots m_i}, m_r \in \mathbb{N}, 1 \le r \le i \le s\}$ be an s-net in $\ell_0^\infty(\Sigma)$, and reorder W as $\{E_{m_1(i)m_2(i)\dots m_s(i)}, i \in \mathbb{N}\}$, the subindexes following the diagonal ordering of $\mathbb{N}^s$.

Lemma 5.3.11 *Let* $\{(m'_1(i), m'_2(i), \dots, m'_s(i)), i \in \mathbb{N}\}$ *be a sequence in* $\mathbb{N}^s$ *whose terms form a set of class* s. *If* $x \in \ell_0^\infty(\Sigma)$, *then there exists some positive integer* p *such that* $x \in E_{m'_1(p)m'_2(p)\dots m'_s(p)}$.

Proof. Obviously there exists some natural number k_1 such that $x \in E_{m'_1(k_1)}$. Suppose there is some index k_{i-1}, $1<i\le s$, such that $x \in E_{m'_1(k_{i-1})\dots m'_{i-1}(k_{i-1})}$. Then, by the hypothesis on the sequence, the set M given by

$$M:= \{k \in \mathbb{N}: m'_1(k) = m'_1(k_{i-1}), \dots, m'_{i-1}(k) = m'_{i-1}(k_{i-1})\}$$

verifies that

$$E_{m'_1(k_{i-1})\dots m'_{i-1}(k_{i-1})} = \bigcup\{E_{m'_1(k_{i-1})\dots m'_{i-1}(k_{i-1})m'_i(k)}, k \in M\};$$

thus there is some index k_i such that $x \in E_{m'_1(k_i)\dots m'_{i-1}(k_i)m'_i(k_i)}$. By induction we conclude the proof. ∎

Lemma 5.3.12 *There exists some positive integer* i *such that* $E_{m_1(i)m_2(i)\dots m_s(i)}$ *is barrelled.*

Proof. Assume that on the contrary for each $i \in \mathbb{N}$ there exists some barrel $T_{m_1(i)m_2(i)\dots m_s(i)}$ in $E_{m_1(i)m_2(i)\dots m_s(i)}$ which is not a neighbourhood of the origin in $E_{m_1(i)m_2(i)\dots m_s(i)}$. If $U_{m_1(i)m_2(i)\dots m_s(i)}$ is the closure of $T_{m_1(i)m_2(i)\dots m_s(i)}$ in $\ell_0^\infty(\Sigma)$, given that $U_{m_1(i)m_2(i)\dots m_s(i)}$ is not a neighbourhood of the origin in its linear hull, it is easy to notice that Ω does not have property $U_{m_1(i)m_2(i)\dots m_s(i)}$ for any $i \in \mathbb{N}$. Now Lemma 5.3.10 allows us to determine a sequence $\{(m'_1(i), m'_2(i), \dots, m'_s(i)), i \in \mathbb{N}\}$, whose elements constitute a set of class s, a family $\{A_{ij}, i, j \in \mathbb{N}\}$ of pairwise disjoint elements of Σ and a family $\{\mu_{ij},\ i,\ j \in \mathbb{N}\}$ with $\mu_{ij} \in \left(U_{m_1'(i)m_2'(i)\ldots m_s'(i)}\right)^{\circ}$ such that

$$|\mu_{ij}(A_{ij})| \geq i+j, \text{ and } \sum\{|\mu_{ij}(A_{gh})|,\ g+h < i+j\} \leq 1 \qquad (1)$$

for all i, j $\in \mathbb{N}$.

By recurrence, following the ordering of $\mathbb{N}^2$, we are going to determine a family $\{P^{(i,j)},\ (i,j) \in \mathbb{N}^2\}$ of subsets of $\mathbb{N}^2$ which satisfies the following conditions:

α) If $(i,j) < (k,t)$ then $P^{(k,t)} \subseteq P^{(i,j)}$.

β) $P^{(i,j)} \subseteq \{(m,n),\ i+j < m+n\}$, for all i, j $\in \mathbb{N}$.

γ) For each natural number p there exist infinitely many pairs in $P^{(i,j)}$ whose first component is p.

δ)| $|\mu_{ij}|(\bigcup\{A_{gh},\ (g,h) \in P^{(i,j)}\}) < 1$, for all i, j $\in \mathbb{N}$. (2)

In fact, set $G := \bigcup\{A_{ij},\ i,\ j \in \mathbb{N}\}$ and let m be a positive integer such that $|\mu_{11}|(G) < m$. We make a partition of the set $\{(i,j) \in \mathbb{N}^2,\ i+j > 2\}$ in m parts P_k, $1 \leq k \leq m$, such that, given any natural number p, each P_k contains infinitely many pairs whose first component is p. A simple way of doing this is using the diagonals of $\mathbb{N}^2$. According to Lemma 5.3.1 we have

$$\sum_{k=1}^{m} |\mu_{11}|(\bigcup\{A_{ij},\ (i,j) \in P_k\}) \leq |\mu_{11}|(G) \leq m$$

and consequently there is some r, $1 \leq r \leq m$, such that

$$|\mu_{11}|(\bigcup\{A_{ij} : (i,j) \in P_r\}) < 1$$

Then we set $P^{(1,1)} := P_r$.

Suppose we have determined $P^{(ij)}$ and that (k,t) is the pair following (i,j) in the ordering of $\mathbb{N}^2$. Now, if $q \in \mathbb{N}$ is such that $|\mu_{kt}|(G) \leq q$, we make a partition of the set $\{(m,n) \in P^{(ij)},\ k+t < m+n\}$ in q parts Q_r, $1 \leq r \leq q$, so that, in each one of them, given any natural number p there are infinitely many pairs whose first component coincides with p.

Now, applying again Lemma 5.3.1 we have

$$\sum_{r=1}^{q} |\mu_{kt}|(\bigcup\{A_{ij},\ (i,j) \in Q_r\}) \leq |\mu_{kt}|(G) \leq q$$

and consequently there is some h, $1 \leq h \leq q$, such that

$$|\mu_{kt}|(\bigcup\{A_{ij},\ (i,j) \in Q_h\}) < 1.$$

Then we set $P^{(k,t)} := Q_h$.

Next we determine by recurrence a sequence $\{(i(k),j(k)),\ k \in \mathbb{N}\}$ in $\mathbb{N}^2$ whose terms form a set of class 2 and verify that for each $k \in \mathbb{N}$

$$(i(k+1),j(k+1)) \in P^{(i(k),j(k))} \tag{3}$$

We start by setting $(i(1),j(1)) = (1,1)$ and suppose that we have determined the first k-1 terms of this sequence. For determining the k-th term of the sequence we distinguish two cases:

<u>Case 1</u>. The first component of the k-th element of $\mathbb{N}^2$ is greater than the first components of the first k-1 elements of $\mathbb{N}^2$.

Then we take $(i(k),j(k))$ equal to some element of $P^{(i(k-1),j(k-1))}$ such that $i(k) = i(k-1) + 1$ and $j(k) > j(r)$ for $r = 1, 2, \ldots, k-1$.

<u>Case 2</u>. The first component of the k-th element of $\mathbb{N}^2$ is equal to the first component of the r-th element of $\mathbb{N}^2$, with $r < k$.

Then take $(i(k),j(k))$ equal to some element of $P^{(i(k-1),j(k-1))}$ such that $i(k) = i(r)$, $(i(r),j(r))$ being the element determined in the step r of the inductive process, and $j(k) > j(s)$ for $s = 1, 2, \ldots, k-1$.

Let $H := \bigcup\{A_{i(k)j(k)},\ k \in \mathbb{N}\}$. Given that the terms of the sequence $\{(m'_1(k), m'_2(k), \ldots, m'_s(k)),\ k \in \mathbb{N}\}$ form a set of class s, we apply Lemma 5.3.11 to find some positive integer p such that $e(H) \in E_{m'_1(p)m'_2(p)\ldots m'_s(p)}$. Hence, there is some $\lambda > 0$ such that $\lambda e(H) \in U_{m'_1(p)m'_2(p)\ldots m'_s(p)}$.

Now by the construction of the sequence $\{(i(k),j(k)),\ k \in \mathbb{N}\}$ there is a strictly increasing sequence $\{k_n,\ n \in \mathbb{N}\}$ of positive integers such that $i(k_n) = p$ for every $n \in \mathbb{N}$ and $\lim_n j(k_n) = \infty$.

Hence, for every $n \in \mathbb{N}$ we have

$$\mu_{(i(k_n),j(k_n))} = \mu_{(p,j(k_n))} \in \left(U_{m'_1(p)m'_2(p)\ldots m'_s(p)}\right)^{\circ},$$

and consequently, for every $n \in \mathbb{N}$,

$$|\mu_{(i(k_n),j(k_n))}(H)| \leq 1/\lambda \tag{4}$$

But, using (1) and (2), and taking into account (3), we obtain

$$|\mu_{(i(k_n),j(k_n))}(H)| = |\mu_{(i(k_n),j(k_n))}(\cup\{A_{i(k)j(k)},\ k \in \mathbb{N}\})| \geq$$
$$|\mu_{(i(k_n),j(k_n))}(\{A_{(i(k_n),j(k_n))}\})| - \sum_{k<k_n}|\mu_{(i(k_n),j(k_n))}(A_{i(k)j(k)})| -$$
$$|\mu_{(i(k_n),j(k_n))}(\cup\{A_{i(k)j(k)},\ k > k_n\})| \geq i(k_n) + j(k_n) - 1 - 1 =$$
$$j(k_n) + p - 2.$$

Therefore, $\lim_n|\mu_{(i(k_n),j(k_n))}(H)| = +\infty$, contradicting (4). ∎

Theorem 5.3.13 *The space* $\ell_0^\infty(\Sigma)$ *is barrelled of class* $\aleph_0$.

Proof. We have to show that $\ell_0^\infty(\Sigma)$ is barrelled of class s for any $s \in \mathbb{N}$. Clearly $\ell_0^\infty(\Sigma) \in \mathscr{C}_0$ because of Theorem 5.2.4 and Proposition 3.1.2, and given any $s \in \mathbb{N}$, if $\ell_0^\infty(\Sigma) \in \mathscr{C}_{s-1}\backslash\mathscr{C}_s$, then there would be some s-net W in $\ell_0^\infty(\Sigma)$ formed by dense subspaces such that no $F \in W_s$ is barrelled, a contradiction with the previous lemma. ∎

5.4 EXERCISES

1. Γ_r*-spaces.* A space E is said to be a Γ_r-space if given a quasicomplete linear subspace G of the algebraic dual H of E such that $G \cap H$ is weakly* dense in the topological dual E' of E, then $G \supseteq E'$.

(i) Let E be a linear space and τ and ρ be two Hausdorff locally convex topologies on E such that $E(\tau)$ is barrelled and $E(\rho)$ is a Γ_r-space. If the identity map i_E from $E(\tau)$ onto $E(\rho)$ has a closed graph, then i_E is continuous.

(ii) Let E be a barrelled space, F a Γ_r-space and T a linear mapping from E into F with closed graph. If K is an algebraic complement of T(E) in F provided with its finest locally convex topology, then the mapping $S(x,y) = T(x) + y$ from $F \times K$ into F is continuous and so T is continuous.

(iii) Let E be a barrelled space, F a Γ_r-space and T a linear continuous mapping from E into F. Then there exists a continuous linear extension U of T from the completion $\hat{E}$ of E into F. *[Hints: Take* U *as the continuous linear extension of* T *from* $\hat{E}$ *into* $\hat{F}$ *and show that* $U(\hat{E}) \subseteq$ F. *With this end pick an* $x \in \hat{E}\backslash E$*, take care that* $\langle E \cup \{x\}\rangle$ *is barrelled* *and use a well-known theorem of S. Iyahen concerning one-codimensional extensions of closed graph linear mappings.]*

(iv) Every B_r-complete space is a Γ_r-space.

2. *Λ_r-spaces.* A space E is said to be a Λ_r-space if, given a locally complete subspace G of the algebraic dual H of E such that $G \cap H$ is weakly* dense in the topological dual E' of E, then $G \supseteq E'$.

(i) Let E be a linear space and τ and ρ be two Hausdorff locally convex topologies on E such that $E(\tau)'$ is weakly* locally complete and $E(\rho)$ is a Λ_r-space. If the identity map i_E from $E(\tau)$ onto $E(\rho)$ has closed graph, then i_E is weakly continuous.

(ii) Let E be a space such that E' is weakly* locally complete, F be a Λ_r-space and T be a linear mapping from E into F with closed graph. Then T is weakly continuous.

(iii) Let E be a space with E' weakly* locally complete, F a Λ_r-space and T a weakly continuous linear mapping from E into F. Then there exists a weakly continuous linear extension U from the bounded closure $\check{E}$ of E into F.

(iv) Every reflexive Banach space and every Fréchet-Schwartz space is a Λ_r-space.

3. *A closed graph theorem for barrelled spaces of class p.* Let E be a barrelled space of class p and F be a space which contains a p-net W such that for each $L \in W_p$ there exists a Hausdorff locally convex topology τ_L on L, finer than the relative one, with the property that $L(\tau_L)$ is a Γ_r-space. Suppose that T is a linear mapping from E into F with closed graph.

(i) There is some $F \in W_p$ such that the restriction of T on $T^{-1}(F)$ is continuous.

(ii) $T(E) \subseteq F$ and T is a continuous mapping from E into $F(\tau_F)$.

4. *(BM)-convergent series in locally convex spaces.* A series $\sum_n x_n$ in a space E is said to be subseries convergent or (SF)-convergent if the series $\sum_{n\in J}x_n$ converges in E for every subset J of $\mathbb{N}$. A series $\sum_n x_n$ in E is called bounded multiplier convergent or (BM)-convergent if the series $\sum_n a_n x_n$ converges in E for every $(a_n) \in \ell^\infty$. Let E be a space and $\sum_n x_n$ a series in E which is subseries convergent. Suppose that E contains an s-net W such that each $F \in W_s$ is a Γ_r-space.

(i) The linear mapping $T : \ell_0^\infty(\sigma(\ell_0^\infty,\ell^1)) \to E$ defined by $T(e(A)) = \sum_{n\in A}x_n$ for each subset A of $\mathbb{N}$ has a continuous linear extension S from $\ell^\infty(\sigma(\ell^\infty,\ell^1))$ into the completion $\hat{E}$ of E.

(ii) There is some $F \in W_s$ such that the restriction of T on $T^{-1}(F)$ has a continuous linear extension on the whole E with values in F.

(iii) $S(\ell^\infty) \subseteq F$ and $\sum_n x_n$ is a (BM)-convergent series in F.

5. *If Ω is a discrete topological space, then $\beta\Omega$ does not contain a convergent sequence.* Assume Ω is a discrete topological space and denote by $\beta\Omega$ the Stone-Čech compactification of Ω.

(i) If $\{t_n,\ n \in \mathbb{N}\}$ is an injective sequence in $\beta\Omega$ and for each $n \in \mathbb{N}$ define $E_n := \{f \in \ell_0^\infty(\beta\Omega)\colon f(t_r) = f(t_s)\ \forall r,s \geq n\}$, then $\{E_n,\ n \in \mathbb{N}\}$ is an increasing sequence of closed linear subspaces of $\ell_0^\infty(\beta\Omega)$.

(ii) If $\{t_n,\ n \in \mathbb{N}\}$ converges to t in $\beta\Omega$, then $\{E_n,\ n \in \mathbb{N}\}$ covers $\ell_0^\infty(\beta\Omega)$, so in $\beta\Omega$ there are no non-trivial convergent sequences.

6. *Nikodym-Grothendieck boundedness theorem.* Let Ω be a set and Σ be a σ-algebra of subsets of Ω. Suppose that $\{\mu_i,\ i \in I\}$ is a family of elements of $ba(\Sigma)$ satisfying $\sup\{|\mu_i(E)|,\ i \in I\} < \infty$ for each $E \in \Sigma$. The barrelledness of the space $\ell_0^\infty(\Sigma)$ guarantees that $\sup\{|\mu_i(E)|,\ i \in I,\ E \in \Sigma\} < \infty$.

7. *The space $K(\Sigma)$.* If Ω is a set and Σ is an algebra of subsets of Ω, denote by $\ell^\infty(\Sigma)$ the completion of $\ell_0^\infty(\Sigma)$. We represent by $K(\Sigma)$ the subspace of $\ell^\infty(\Sigma)$ of all countably valued functions. These spaces are supposed to be provided with the supremum-norm $\|f\| = \sup\{|f(\omega)|,\ \omega \in \Omega\}$. Let Q stand for the absolutely convex hull of the indicator functions e(A) of all the elements A of the algebra Σ. Obviously, $\ell_0^\infty(\Sigma) \subseteq K(\Sigma) \subseteq \ell^\infty(\Sigma)$.

(i) If T is an absolutely convex absorbent set of $\ell_0^\infty(\Sigma)$ which does not absorb Q, there exists a sequence (A_i) of pairwise disjoint members of Σ and a strictly increasing sequence of positive integers (n_i) such that $e(A_i) \notin n_i T$ for each $i \in \mathbb{N}$.

(ii) $K(\Sigma)$ is a barrelled space. *[Hint: If B denotes the closed unit ball of $\ell_0^\infty(\Sigma)$, by Proposition 5.1.1 one has that $B \subseteq 4Q$.]*

(iii) Is $K(\Sigma)$ an unordered Baire-like space?

5.5 NOTES AND REMARKS

The barrelledness of the space $\ell_0^\infty(\Sigma)$, with Σ a σ-algebra on Ω, is a straightforward consequence of (in fact, equivalent to, as follows from Exercise 6) the Nikodym-Grothendieck boundedness theorem. A proof of the barrelledness of $\ell_0^\infty(\Sigma)$ given in these terms can be found in Valdivia's text [V9]. This also has been noted for ℓ_0^∞ in Saxon [S3]. We have provided two proofs of the barrelledness of $\ell_0^\infty(\Sigma)$ without using explicitly the Nikodym-Grothendieck theorem. The first is based upon Lemmas 5.2.1 and 5.2.3 (although Lemma 5.2.3 is taken from [V9]) and the second proof is based upon Phillips's lemma, which in turn is a consequence of the Nikodym-Grothendieck theorem. The barrelledness of the space ℓ_0^∞ has also been pointed out in Bennett and Kalton [BK3] to be a consequence of a result of Seever [Se].

The proof of the suprabarrelledness of the space $\ell_0^\infty(\Sigma)$ has been performed by M. Valdivia in [V5], and his methods of sliding humps have been exploited by us, along with some methods of our own, such as the introduction of the concept of set of class n in the previous chapter, in order to obtain Theorem 5.3.13.

As we mentioned in section 4.7, Valdivia's result was extended by A. Moltó in [Mo1] and [Mo2] to Boolean algebras satisfying property (f). As we stated there, in what follows we adapt the techniques used by A. Moltó to a σ-algebra Σ of subsets of a set Ω to provide another proof of the suprabarrelledness of the space $\ell_0^\infty(\Sigma)$.

Theorem 5.5.1 *Let* $\{\Sigma_n, n \in \mathbb{N}\}$ *be an increasing covering of* Σ. *Then there exists an* $n_0 \in \mathbb{N}$ *such that if* $\mu_i \in ba(\Sigma)$ *for each* $i \in \mathbb{N}$ *and* $\{\mu_i(E), i \in \mathbb{N}\}$ *is a Cauchy sequence for every* $E \in \Sigma_{n_0}$, *then* $\{\mu_i, i \in \mathbb{N}\}$ *is a* $\sigma(ba(\Sigma),\ell_0^\infty(\Omega,\Sigma))$*-Cauchy sequence.*

Proof. If the proposition were not true there would be some increasing covering $\{\Sigma_n: n \in \mathbb{N}\}$ of Σ, a sequence $\{B_n: n = 2, 3, \dots\}$ with $B_n \in \Sigma_n$ for each $n \in \mathbb{N}$, and a family $\{\mu_i^n: i, n \in \mathbb{N}\}$ such that for every $n \in \mathbb{N}$

(i) $\{\mu_i^n(E): i \in \mathbb{N}\}$ is a Cauchy sequence for every $E \in \Sigma_n$.

(ii) $\{\mu_i^n(B_{n+1}): i, n \in \mathbb{N}\}$ is not a Cauchy sequence.

If, for every $E \in \Sigma$ there exists some $n_E \in \mathbb{N}$ such that $n > n_E$ and $D \subseteq E$ imply that $\{\mu_i^n(D): i, n \in \mathbb{N}\}$ is a Cauchy sequence, then we can suppose besides that

(iii) $B_i \cap B_j = \emptyset$ if $i \neq j$ since the difference of two sequences, with only one of them of Cauchy, is not a Cauchy sequence.

If there exits some $E_0 \in \Sigma$ such that

$$\left\{n \in \mathbb{N} : \exists B \subseteq E_0,\ B \in \Sigma,\ \{\mu_i^n(B),\ i \in \mathbb{N}\}\ \textit{is not a Cauchy sequence}\right\}$$

is infinite, then we can also suppose (iii) since, given any $n_0 \in \mathbb{N}$, there is some $n > n_0$ and some $B \in \Sigma$, $B \subseteq E_0$, such that $E_0 \in \Sigma_n$ and $\{\mu_i^n(B): i \in \mathbb{N}\}$ is not a Cauchy sequence. Therefore $\{\mu_i^n(E_0\backslash B): i \in \mathbb{N}\}$ is not a Cauchy sequence, since $\{\mu_i^n(E_0),\ i \in \mathbb{N}\}$ is a Cauchy sequence. One of the sets B or $E_0 \setminus B$ has the same property as E_0 and hence we may suppose that

$$\left\{n \in \mathbb{N} : \exists\ C \subseteq E_0\backslash B,\ C \in \Sigma,\ \{\mu_i^n(C),\ i \in \mathbb{N}\}\ \textit{is not a Cauchy sequence}\right\}$$

is infinite. This enables us to prove the affirmation by induction.

We may also suppose the following condition:

(iv) All μ_i^n, with n, $i \in \mathbb{N}$, may be extended to countably additive measures, also denoted by μ_i^n, defined on the σ-algebra generated by the sequence $\{B_n: n = 2, 3, \dots\}$. In fact, let ρ be the bounded measure defined by $\rho(B) := \sum\left\{|\mu_i^n|(B)/\left(2^{n+i}|\mu_i^n|(\Omega)\right): i, n \in \mathbb{N}\right\}$. If $\{N_i,\ i \in \mathbb{N}\}$ is a partition of $\mathbb{N}$ and every N_i is infinite, there is some N_i such that $\rho(N_i) \le \varepsilon$, where $\varepsilon > 0$. Thus, by choosing an adequate subsequence of $\{B_n: n = 2, 3, \dots\}$ which we shall denote also by $\{B_n,\ n = 2, 3, \dots\}$, we can suppose that $\rho\left(\bigcup\{B_m: m = n, n + 1, \dots\}\right) \le 1/2^n$. Finally, the absolute convergence yields the countably additive extension.

Let $\mathcal{B}$ be the σ-algebra generated by $\{B_n,\ n = 2, 3, \dots\}$. The limit in $\mathcal{B}$ of a Cauchy sequence with respect to the pseudometric $\rho(A,B)$ defined by $\rho(A \Delta B)$ is the union of the B_n that are contained in infinitely many members of the Cauchy sequence. More precisely, if $\rho(B_n)$ is different from zero, then either B_n is contained only in a finite number of terms of a Cauchy sequence or it is contained in all but a finite number of terms of the Cauchy sequence. Let $\mathcal{B}_n := \{D \in \mathcal{B} : \{\mu_i^n(D),\ i \in \mathbb{N}\},\ \{\mu_i^{n+1}(D),\ i \in \mathbb{N}\},\ \dots$ *are Cauchy sequences*$\}$. According to (i), $\mathcal{B} = \bigcup\{\mathcal{B}_n: n \in \mathbb{N}\}$ and by the Baire category theorem there is some $p_0 \in \mathbb{N}$ such that, for $n \ge p_0$, $\mathcal{B}_n$ is of second category in $\mathcal{B}$. We can suppose $p_0 = 1$.

By (ii), for every $n \in \mathbb{N}$ there is an $\varepsilon_n > 0$ that contradicts the definition of Cauchy sequence for $\{\mu_i^n(B_{n+1}),\ i \in \mathbb{N}\}$.

Since $\mathcal{B}_n \subseteq \bigcup\left\{\{D \in \mathcal{B}: |\mu_i^n(D) - \mu_j^n(D)| \le \varepsilon_n/6,\ i,\ j \ge m\},\ m \in \mathbb{N}\right\}$, we are able to determine for each n some $m_n \in \mathbb{N}$ so that, if $m > m_n$ the set $\{D \in \mathcal{B} : |\mu_i^n(D)-\mu_j^n(D)| \le \varepsilon_n/6,\ i,\ j \ge m\}$ contains a ball in the pseudometric defined by ρ. Without loss of generality we can suppose that $m_n = 1$ for every $n \in \mathbb{N}$. Hence we can also assume that

(v) Given $n \in \mathbb{N}$ there exists some $D_n \in \mathcal{B}$ and a number $\delta_n > 0$ such that $\{D \in \mathcal{B}: \rho(D_n, D) < \delta_n\} \subseteq \{D \in \mathcal{B}: |\mu_i^n(D)-\mu_j^n(D)| \le \varepsilon_n/6,\ i,\ j \ge 1\}$.

A natural question is to ask what $D \in \mathcal{B}$ and $\rho(D) < \delta_n$ implies. It is obvious that $|\mu_i^n\{(D\backslash D_n)\ \Delta\ D_n\} - \mu_j^n\{(D\backslash D_n)\ \Delta\ D_n\}| \le \varepsilon_n/6$. Besides, $|\mu_i^n(D_n\backslash D) - \mu_j^n(D_n\backslash D)| \le \varepsilon_n/6$ for $i,\ j \ge 1$. From these two inequalities it follows immediately that

(vi) $D \in \mathcal{B}$ and $\rho(D) < \delta_n$ implies that $|\mu_i^n(D) - \mu_j^n(D)| \le \varepsilon_n/3$.

Finally, with the usual technique of making a partition of $\mathbb{N}$ in countably many parts, each of infinitely many elements, it is possible to obtain a subsequence of the sequence $\{B_n,\ n = 2, 3, \ldots\}$, which again we shall denote by $\{B_n: n = 2, 3, \ldots\}$, such that

(vii) $\rho\left(\bigcup\{B_m: m = n+1, n+2, \ldots\}\right) < \delta_n$.

Now we are going to reach a contradiction. Let us suppose that $B = \bigcup\{B_m: m = 2, 3, \ldots\} \in \mathcal{B}$ verifies that $B \in \mathcal{B}_n$. Then, by definition, $\{\mu_i^n(B),\ i \in \mathbb{N}\}$ is a Cauchy sequence. This contradicts the inequalities

$$|\mu_i^n(B)-\mu_j^n(B)| \ge -\left|\mu_i^n\left(\bigcup\{B_m: m= 2, 3, \ldots, n\}\right)-\mu_j^n\left(\bigcup\{B_m: m= 2, 3, \ldots, n\}\right)\right|$$
$$+ |\mu_i^n(B_{n+1})-\mu_j^n(B_{n+1})| - \left|\mu_i^n\left(\bigcup\{B_m: m = n+2, n+3, \ldots\}\right)\right.$$
$$\left.- \mu_j^n\left(\bigcup\{B_m: m = n+2, n+3, \ldots\}\right)\right|.$$

As there is some $i_0 \in \mathbb{N}$ such that for $i,\ j > i_0$ we have that

$$\left|\mu_i^n\left(\bigcup\{B_m: m = 2, 3, \ldots, n\}\right) - \mu_j^n\left(\bigcup\{B_m: m = 2, 3, \ldots, n\}\right)\right| < \varepsilon_n/3$$

and given $k > i_0$, by virtue of (ii), there are $i > j > k$ such that

$$|\mu_i^n(B_{n+1})-\mu_j^n(B_{n+1})| > \varepsilon_n$$

The last two inequalities along with (vi) and (vii) enables us to deduce that in the case of $i > j > k > i_0$, then

$$|\mu_i^n(B) - \mu_j^n(B)| \ge -\varepsilon_n/3 + \varepsilon_n - \varepsilon_n/3 = \varepsilon_n/3. \blacksquare$$

Corollary 5.5.2 *Let* $\{\Sigma_n, n = 1, 2, \ldots\}$ *be an increasing covering of* Σ. *Then there exists an* $n_0 \in \mathbb{N}$ *such that if* $\mu_i \in ba(\Sigma)$, $i = 1, 2, \ldots$, *and* $\{\mu_i(E), i \in \mathbb{N}\}$ *is a bounded sequence for every* $E \in \Sigma_{n_0}$ *then* $\{\mu_i, i \in \mathbb{N}\}$ *is a* $\sigma(ba(\Sigma),\ell_0^\infty(\Sigma))$*-bounded sequence.*

Proof. If the statement is not true, for each $n \in \mathbb{N}$ there exists a sequence $\{\mu_i^n, i \in \mathbb{N}\}$ in $ba(\Sigma)$ and some $E_n \in \Sigma$ such that for each $E \in \Sigma_n$ the sequence $\{\mu_i^n(E), i \in \mathbb{N}\}$ is bounded and $\mu_i^n(E_n) \geq i^2$. Then $\{(1/i)\mu_i^n(E), i \in \mathbb{N}\}$ is a Cauchy sequence for every $E \in \Sigma_n$, and $\{(1/i)\mu_i^n(E_n), i \in \mathbb{N}\}$ is unbounded for each $n \in \mathbb{N}$, contradicting 5.5.1.■

Corollary 5.5.3 $\ell_0^\infty(\Sigma)$ *is suprabarrelled.*

Proof. Let $\{E_n, n \in \mathbb{N}\}$ be an increasing covering of $\ell_0^\infty(\Sigma)$. As $\ell_0^\infty(\Sigma)$ is barrelled and metrizable and consequently Baire-like, we can suppose that every E_n is dense in $\ell_0^\infty(\Sigma)$. If we find some $n_0 \in \mathbb{N}$ such that every $\sigma(ba(\Sigma),E_{n_0})$-bounded subset is $\sigma(ba(\Sigma),\ell_0^\infty(\Sigma))$-bounded, we are done. The previous corollary provides an adequate positive integer n_0 corresponding to the sequence $\Sigma_n = \{A \in \Sigma: e(A) \in E_n\}$. ■

As we shall see in Chapter 6, the algebra $\mathcal{D}$ of clopen sets of the Stone-Čech compactification $\beta\mathbb{N}$ of the positive integers with its discrete topology is not a σ-algebra, but in Moltó [Mo1] it is proved that $\mathcal{D}$ satisfies (f), so $\ell_0^\infty(\beta\mathbb{N},\mathcal{D})$ is suprabarrelled.

Our presentation of the chapter has been strongly influenced by Valdivia [V5] and Ferrando and López Pellicer [FL3]. In fact, our Proposition 5.1.1 is a reformulation of Propositions 1 and 2 of [V5], Lemma 5.2.3 is the result I,§7.(17) of Valdivia [V9]. Lemma 5.3.1 is plain but it can be seen in §31.1.(1) of Köthe [K1]. Lemmas 5.3.2, 5.3.4, 5.3.5 and 5.3.6 correspond respectively to Propositions 7, 3, 4 and 5 of Valdivia [V5]. The rest of the chapter is based on Ferrando and López-Pellicer [FL3].

Exercise 1 is taken from Valdivia [V3] and from I.§6 of his book [V9]. Exercise 2 is based on Valdivia [V4] and also in I.§6 of [V9]. Exercise 4 is an extension of Theorem 5 of Valdivia's paper [V5]. A nice exposition of the Nikodym-Grothendieck boundedness theorem from the point of view of measure theory can be read in [D]. Indeed, an important extension of the Nikodym-Grothendieck theorem may be obtained from Theorem 5.3.13 but we shall postpone this to Chapter 7.

Several alternative proofs to those considered so far concerning strong barrelledness properties of the space $\ell_0^\infty(\Sigma)$, have been given in Ferrando and López Pellicer [FL5], Ferrando and Sánchez Ruiz [FS3] and also in Ferrer, López-Pellicer and Sánchez Ruiz [FLS1].

6. Troubles with some $\ell_0^\infty(\Sigma)$ spaces

6.1 INTRODUCTION

Many of the barrelled spaces occurring in analysis are known to enjoy strong barrelledness properties as we have seen to happen with $\ell_0^\infty(\Sigma)$ in Chapter 5, Σ being a σ-algebra of subsets of some set Ω. However, it is easy to see that the space $\ell_0^\infty(\Sigma)$ is not of second category when Σ is an infinite algebra. In fact, $\ell_0^\infty(\Sigma)$ may be covered by the nowhere dense closed subsets $X_n := \{f \in \ell_0^\infty(\Sigma),\ |f(\Omega)| \le n\}$. Indeed, assuming some X_n is somewhere dense, from $X_n - X_n \subseteq X_{n^2}$ and $aX_n \subseteq X_n$ for each $a \in \mathbb{K}$, it follows that $X_{n^2} = \ell_0^\infty(\Sigma)$, a contradiction.

In this chapter we shall look into some problems you may find for $\ell_0^\infty(\Sigma)$ to fulfil strong barrelledness properties. We will start with Σ being an infinite arbitrary algebra and, later on, we will deal with some kinds of particular algebras which are not a σ-algebra, showing in those cases that $\ell_0^\infty(\Sigma)$ may even fail to be barrelled, and deducing from this fact some consequences. Although in Chapter 5 we have already seen that the space $\ell_0^\infty(\Sigma)$, Σ being a σ-algebra, is barrelled of class $\aleph_0$, we shall end this chapter by showing that if Ω is infinite and Σ is the σ-algebra of all the subsets of Ω, $\ell_0^\infty(\Sigma)$ is not totally barrelled.

6.2 $\ell_0^\infty(\Sigma)$ AND ULTRABORNOLOGICAL SPACES

In this section we are going to show that if Σ is an infinite algebra of subsets of a set Ω, then $\ell_0^\infty(\Sigma)$ is not an ultrabornological space.

Lemma 6.2.1 *Let* E *be a subspace of* $\ell_0^\infty(\Sigma)$. *If* E *is contained in the set* $X_n := \{f \in \ell_0^\infty(\Sigma),\ |f(\Omega)| \le n\}$, *then* $\dim E \le n$.

Proof. Suppose that $\dim E > n$, set $m := \max\{|s(\Omega)|,\ s \in E\}$ and let g be a function belonging to E, say $s = \sum_{i=1}^{m} a_i e(A_i)$, such that $\{A_1, \ldots, A_m\}$

is a partition of Ω by members of Σ and $a_i \neq a_j$ for $i \neq j$. As dim $E > n \geq m$, there must be some $h \in E$ and some $j \in \{1, \ldots, m\}$ with $|h(A_j)| > 1$ since, otherwise, $E \subseteq \langle\{e(A_i), 1 \leq i \leq m\}\rangle$. Then choose some $\varepsilon > 0$ such that $\varepsilon h(\omega) < \min\{|a_i - a_j|, 1 \leq i < j \leq m\}$ for each $\omega \in \Omega$. Therefore, $|(g+\varepsilon h)(\Omega)| > m$, which is against the definition of m. ∎

Theorem 6.2.2 *If Σ is an infinite algebra, then the space $\ell_0^\infty(\Sigma)$ is not ultrabornological.*

Proof. Assume B is a Banach disk contained in $\ell_0^\infty(\Sigma)$. Defining $X_n := \{f \in \ell_0^\infty(\Sigma), |f(\Omega)| \leq n\}$, as $\ell_0^\infty(\Sigma) = \bigcup\{X_n, n \in \mathbb{N}\}$, we may find some $p \in \mathbb{N}$ such that $E_B \subseteq X_p$. Therefore, Lemma 6.2.1 ensures that E_B has finite dimension. If $\ell_0^\infty(\Sigma)$ were ultrabornological, then necessarily it would have the finest locally convex topology. Hence the topology induced by $\ell_0^\infty(\Sigma)$ in every countable dimensional subspace E would be the finest locally convex topology. So E would be a countable infinite-dimensional Fréchet space, which is a contradiction. ∎

Remark 6.2.3 From the previous theorem, it follows that the associated ultrabornological topology to $\ell_0^\infty(\Sigma)$ is the strongest locally convex topology. Next, we will see that this is true even when we endow $\ell_0^\infty(\Sigma)$ with the topology of pointwise convergence.

Proposition 6.2.4 *Let Σ be an algebra of subsets of a set Ω, and let τ and π denote, respectively, the topologies on $\ell_0^\infty(\Sigma)$ of uniform and pointwise convergence. Let $\{x_n, n \in \mathbb{N}\}$ be a sequence in $\ell_0^\infty(\Sigma)$.*

(i) If the (formal) series $\sum_{n=1}^{\infty} x_n$ is subseries convergent under the uniform convergence topology, then $\langle\{x_n, n \in \mathbb{N}\}\rangle$ is finite-dimensional.

(ii) If the series $\sum_{n=1}^{\infty} x_n$ is (BM)-convergent under the pointwise convergence topology, then $\langle\{x_n, n \in \mathbb{N}\}\rangle$ is finite-dimensional.

Proof. (i) Let us assume that dim $\langle\{x_n, n \in \mathbb{N}\}\rangle = \infty$. Then, given any $n_0 \in \mathbb{N}$, any $J \subseteq \{1, \ldots, n_0\}$ and any $\varepsilon > 0$, there is some $p > n_0$ such that

$$|(x_p + \sum\{x_n, n \in J\})(\Omega)| > |(\sum\{x_n, n \in J\})(\Omega)|$$

and since $\lim_{n\to\infty} \|x_n\| = 0$ we may assume $\|x_p\| < \varepsilon$. We shall denote this property by (P). Take $n_1 = 1$ and define either $\alpha_1 := 1$ if $|x_{n_1}(\Omega)| = 1$, or $\alpha_1 := (1/3)\cdot\min\{|a - b|: a, b \in x_{n_1}(\Omega), a \neq b\}$ if $|x_{n_1}(\Omega)| > 1$.

By (P) there is some $n_2 > n_1$ such that $\|x_{n_2}\| < (1/2^2)\min\{1,\alpha_1\}$ and

$$|(x_{n_1} + x_{n_2})(\Omega)| > |x_{n_1}(\Omega)|.$$

By recurrence, suppose we have obtained the positive integers $n_1 < n_2 < \dots < n_r$, such that for $i = 2, \dots, r$

(a) $|(\sum_{j=1}^{i-1} x_{n_j})(\Omega)| < |(\sum_{j=1}^{i} x_{n_j})(\Omega)|$ and

(b) $\|x_{n_i}\| < (1/2^i)\cdot\min\{1, \alpha_1, \alpha_2, \dots, \alpha_{i-1}\}$, with the previously determined α_1 and $\alpha_{i-1} = (1/3)\cdot\min\{|a - b|: a, b \in (\sum_{j=1}^{i-1} x_{n_j})(\Omega), a \neq b\}$ for $2 \leq i - 1 \leq r - 1$.

Then, by (P), there must be some $n_{r+1} > n_r$ such that

$$|(\sum_{j=1}^{r} x_{n_j})(\Omega)| < |(\sum_{j=1}^{r+1} x_{n_j})(\Omega)|$$

and $\|x_{n_{r+1}}\| < (1/2^{r+1})\cdot\min\{1, \alpha_1, \dots, \alpha_r\}$, where

$$\alpha_r = (1/3)\cdot\min\{|a - b| : a, b \in (\sum_{j=1}^{r} x_{n_j})(\Omega), a \neq b\}.$$

The sequence $\{x_{n_j}, j \in \mathbb{N}\}$ so obtained clearly satisfies

$$\lim_{r\to\infty} |(\sum_{j=1}^{r} x_{n_j})(\Omega)| = \infty.$$

Besides, for each $i \in \mathbb{N}$, if s and t are two different elements of $(\sum_{j=1}^{i} x_{n_j})(\Omega)$, then $B(s,\alpha_i) \cap B(t,\alpha_i) = \emptyset$.

If we prove that given any $i \in \mathbb{N}$ and $s \in (\sum_{j=1}^{i} x_{n_j})(\Omega)$ there is some $v \in \left(\sum\{x_{n_j}, j \in \mathbb{N}\}\right)(\Omega)$ such that $v \in B(s,\alpha_i)$, then we will have

$$|(\sum_{j=1}^{\infty} x_{n_j})(\Omega)| \geq |(\sum_{j=1}^{i} x_{n_j})(\Omega)|$$

since $\sum_{j=1}^{\infty} x_{n_j} \in \ell_0^\infty(\Sigma)$. Hence, by (a), $|(\sum_{j=1}^{\infty} x_{n_j})(\Omega)| = \infty$, a contradiction.

Let $h \in \Omega$ be such that $s = (\sum_{j=1}^{i} x_{n_j})(h)$ and let $v := (\sum_{j=1}^{\infty} x_{n_j})(h)$. Then

$$|v - s| = |(\sum_{j=i+1}^{\infty} x_{n_j})(h)| \le \sum_{j=i+1}^{\infty} \|x_{n_j}\| < \alpha_i \sum_{j=i+1}^{\infty} 2^{-j} < \alpha_i .$$

(ii) According to the Banach-Steinhaus theorem, the mapping T from ℓ^∞ into $(\ell_0^\infty(\Sigma),\pi)$ defined by $T(\{\xi_n, n \in \mathbb{N}\}) = \sum_{n=1}^{\infty} \xi_n x_n$ is continuous, since the mapping $T_i(\{\xi_n, n \in \mathbb{N}\}) = \sum_{n=1}^{i} \xi_n x_n$ is bounded for each $i \in \mathbb{N}$, and $\lim_{i\to\infty} T_i(\{\xi_n, n \in \mathbb{N}\}) = T(\{\xi_n, n \in \mathbb{N}\})$ under the topology π. Hence T, considered as a mapping from ℓ^∞ into $(\ell_0^\infty(\Sigma),\tau)$, is continuous, since the strong topology corresponding to $(\ell_0^\infty(\Sigma),\pi)$ is τ.

Since $T(e_n) = x_n$, the continuity of T implies that $\{x_n, n \in \mathbb{N}\}$ is τ-bounded and, thus, $\sum_{p=1}^{\infty} \xi_{n_p} x_{n_p}$ is a τ-Cauchy series in $\ell_0^\infty(\Sigma)$ for each $\{\xi_n, n \in \mathbb{N}\} \in \ell^1$ and every increasing sequence of positive integers $\{n_p, p \in \mathbb{N}\}$. Therefore, as τ has a base of π-closed neighbourhoods of the origin, $\{\xi_n x_n, n \in \mathbb{N}\}$ satisfies the hypothesis of (i) for each $\{\xi_n, n \in \mathbb{N}\} \in \ell^1$. Finally, choosing the sequence (ξ_n) such that $\xi_n \ne 0$ for each $n \in \mathbb{N}$, it follows from (i) that

$$\dim \langle\{\xi_n x_n, n \in \mathbb{N}\}\rangle = \dim \langle\{x_n, n \in \mathbb{N}\}\rangle < \infty.$$

This completes the proof. ∎

Corollary 6.2.5 *Let* B *be a Banach disk in* $(\ell_0^\infty(\Sigma),\pi)$*. Then, the linear hull* $\langle B\rangle$ *of* B *is finite-dimensional.*

Proof. Suppose the result is false and there is a sequence of linearly independent vectors in $\langle B\rangle$, say $\{z_n, n \in \mathbb{N}\}$. For each $n \in \mathbb{N}$, let $x_n := z_n/(2^n p_B(z_n))$, where p_B denotes the gauge of B. Then, the dimension of $\langle\{x_n, n \in \mathbb{N}\}\rangle$ is infinite. But it is obvious that, given an arbitrary vector $\{\xi_n, n \in \mathbb{N}\}$ of ℓ^∞, then the series $\sum_{n=1}^{\infty} \xi_n x_n$ is convergent in E_B and consequently in $(\ell_0^\infty(\Sigma),\pi)$, contradicting the previous theorem. ∎

Theorem 6.2.6 *The associated ultrabornological topology to* $(\ell_0^\infty(\Sigma),\pi)$ *is the finest locally convex topology.*

Proof. According to the preceding corollary, every absorbing absolutely convex subset of $(\ell_0^\infty(\Sigma),\pi)$ is a neighbourhood of the origin under the associated ultrabornological topology of π. ∎

6.3 SEPARABLE SUBSPACES OF $\ell_0^\infty(\Sigma)$

In this section we shall prove that the infinite-dimensional separable subspaces of $\ell_0^\infty(\Sigma)$, Σ being an algebra, are not barrelled.

Theorem 6.3.1 *If* E *is a separable subspace of* $\ell_0^\infty(\Sigma)$, E *has countable dimension.*

Proof. If D is a countable dense subset in E, then we can determine a countable subalgebra $\mathcal{B}$ of Σ such that D is contained in $\ell_0^\infty(\mathcal{B})$.

Let $x = \sum_{i=1}^{n} a_i e(A_i) \in \overline{\ell_0^\infty(\mathcal{B})}$, where $\{A_i,\ 1 \le i \le n\}$ is a partition of Ω formed by elements of Σ, with $a_i \ne a_j$ whenever $i \ne j$, and choose $\varepsilon = \min\{|a_i - a_j|,\ 1 \le i < j \le n\}$. Then, we can take $y = \sum_{j=1}^{m} b_j e(B_j)$, where $\{B_j,\ 1 \le j \le m\}$ is a partition of Ω formed by elements of $\mathcal{B}$ and $b_i \ne b_j$ when $i \ne j$, such that $\|x - y\| < \varepsilon$. By the triangular inequality each B_j must be contained in some A_i, and since $\{A_i,\ 1 \le i \le n\}$ and $\{B_j,\ 1 \le j \le m\}$ are partitions of Ω, we have that each A_i is the union of finitely many B_j's, i.e. $A_i \in \mathcal{B}$ and, therefore, $x \in \ell_0^\infty(\mathcal{B})$.

Thus, the closed subspace $\ell_0^\infty(\mathcal{B})$ contains E and the result follows from the countable dimension of $\ell_0^\infty(\mathcal{B})$. ∎

Corollary 6.3.2 *Every separable infinite-dimensional subspace of the space* $\ell_0^\infty(\Sigma)$ *is not barrelled.*

Proof. By the preceding theorem and the closed graph theorem we have that a separable barrelled subspace of $\ell_0^\infty(\Sigma)$ of infinite dimension is a countable infinite-dimensional linear space endowed with the finest locally convex topology, which contradicts the metrizability. ∎

Remark 6.3.3 An immediate consequence of this corollary is that if Σ is an infinite algebra then $\ell_0^\infty(\Sigma)$ does not contain any copy of c_0.

6.4 $\ell_0^\infty(\Sigma)$ AND $\mathscr{L}$-BARRELLED SPACES

Let us now recall some other results in the context of non-necessarily locally convex topological vector spaces. A sequence $\mathcal{U} = \{U_n, n \in \mathbb{N}\}$ in a topological vector space E is said to be a string if it is formed by absorbing balanced subsets of E and $U_{n+1} + U_{n+1} \subseteq U_n$ for each $n \in \mathbb{N}$. A string $\mathcal{U}$ is called closed if each U_n is closed, and topological if each U_n is a neighbourhood of zero. A topological vector space is called $\mathscr{L}$-barrelled or ultrabarrelled if each closed string is topological.

Every $\mathscr{L}$-barrelled locally convex space is barrelled. However there are barrelled locally convex spaces which are not $\mathscr{L}$-barrelled. Next, we shall see that this is the case with $\ell_0^\infty(\Sigma)$, which is not even countably $\mathscr{L}$-barrelled in spite of being barrelled of class $\aleph_0$.

Given an arbitrary infinite cardinal number $\mathfrak{m}$ a topological vector space $E(\mathcal{T})$ is called $\mathfrak{m}$-$\mathscr{L}$-barrelled if every closed string which is the intersection of $\mathfrak{m}$ topological strings is topological. Clearly, if $E(\mathcal{T})$ is $\mathscr{L}$-barrelled then it is $\mathfrak{m}$-$\mathscr{L}$-barrelled, and if $\mathfrak{m} = \aleph_0$ then $E(\mathcal{T})$ is called countably $\mathscr{L}$-barrelled.

A sequence $\mathscr{S} := \{B_n, n \in \mathbb{N}\}$ of balanced subsets of the linear space E such that $B_n + B_n \subseteq B_{n+1}$ for each $n \in \mathbb{N}$ is called absorbing if $\bigcup\{B_n, n \in \mathbb{N}\}$ is absorbing in E, i.e., if $E = \bigcup\{B_n, n \in \mathbb{N}\}$. Given an absorbing sequence $\mathscr{S} = \{B_n, n \in \mathbb{N}\}$ in the topological vector space $E(\mathcal{T})$, $\mathcal{T}_{\mathscr{S}}$ will denote the finest linear topology on E that induces on every B_n the same neighbourhoods of the origin as $\mathcal{T}$. In the following lemma we are going to give a set of strings generating $\mathcal{T}_{\mathscr{S}}$.

Lemma 6.4.1 *Assume* $\mathscr{S} = \{B_n, n \in \mathbb{N}\}$ *is an absorbing sequence in the topological vector space* $E(\mathcal{T})$ *and let*

$$\{\mathcal{U}^n, n \in \mathbb{N}\} = \{\{U_j^n, j \in \mathbb{N}\}, n \in \mathbb{N}\}$$

run through all the sequences of topological strings in $E(\mathcal{T})$.

Then the strings $\mathcal{W} = \{W_j, j \in \mathbb{N}\}$ *with*

$$W_j := \bigcap\{\overline{B_{n-j+1} + U_j^n}^{\mathcal{T}}, n \in \mathbb{N}\}$$

where $B_n = \{0\}$ *for* $n \leq 0$, *generate* $\mathcal{T}_{\mathscr{S}}$.

Proof. Each $\mathcal{W} = \{W_j,\ j \in \mathbb{N}\}$ is $\mathcal{T}_{\mathcal{S}}$-topological since for $n \geq m+j-1$,

$$B_m \subseteq \overline{B_{n-j+1} + U_j^n}^{\mathcal{T}}$$

and thus for any $m \in \mathbb{N}$,

$$W_j \cap B_m \supseteq B_m \cap \left(\bigcap\left\{ \overline{B_{n-j+1} + U_j^n}^{\mathcal{T}},\ 1 \leq n \leq m+j-1\right\}\right) \supseteq$$
$$B_m \cap \left(\bigcap\left\{ U_j^n,\ 1 \leq n \leq m+j-1\right\}\right),$$

and hence each W_j induces $\mathcal{T}$-neighbourhoods of the origin on B_m.

Let us show that the strings $\mathcal{W}$ generate $\mathcal{T}_{\mathcal{S}}$. Let $\{V_n,\ n \in \mathbb{N}\}$ be a topological string in $E(\mathcal{T}_{\mathcal{S}})$. Then for each $n \in \mathbb{N}$ there is a topological string $\mathcal{U}^n = \{U_j^n,\ j \in \mathbb{N}\}$ in $E(\mathcal{T})$ such that

$$B_n \cap U_j^n \subseteq V_{n+j} \cap B_n \subseteq V_{n+j} \quad \forall j \in \mathbb{N}.$$

We will assume $\{\mathcal{U}^n,\ n \in \mathbb{N}\}$ is decreasing, i.e. $U_j^{n+1} \subseteq U_j^n$ for each $n,\ j \in \mathbb{N}$ and build up the $\mathcal{W} = \{W_j,\ j \in \mathbb{N}\}$ corresponding to the sequence $\{\mathcal{U}^n,\ n \in \mathbb{N}\}$. Let us see that $W_{j+2} \subseteq V_j$ for each $j \in \mathbb{N}$. From

$$W_{j+2} = \bigcap\left\{ \overline{B_{n-j-1} + U_{j+2}^n}^{\mathcal{T}},\ n \in \mathbb{N}\right\} \subseteq \bigcap\left\{B_{n-j-1} + U_{j+2}^n + U_{j+2}^n,\ n \in \mathbb{N}\right\} \subseteq$$
$$\bigcap\left\{B_{n-j-1} + U_{j+1}^n,\ n \in \mathbb{N}\right\},$$

it follows that if $x \in W_{j+2}$ then

$$x = b_n + u_n,$$

with $b_n \in B_{n-j-1}$, $u_n \in U_{j+1}^n$ for each $n \in \mathbb{N}$.

Since $\mathcal{S}$ is an absorbing sequence, then there exists some positive integer $p \geq 3$ such that $x \in B_{p-j}$. Therefore

$$x = b_p + u_p = b_1 + \sum_{i=2}^{p}\{b_i - b_{i-1}\} + u_p$$

and since

$$u_p = x - b_p \in B_{p-j} + B_{p-j-1} \subseteq B_{p-j+1},$$
$$b_i - b_{i-1} \in B_{i-j-1} + B_{i-j-2} \subseteq B_{i-j},\quad i \geq 2,$$

and

$$b_i - b_{i-1} = u_{i-1} - u_i \in U_{j+1}^{i-1} + U_{j+1}^{i} \subseteq U_{j+1}^{i-1} + U_{j+1}^{i-1} \subseteq U_j^{i-1},\quad i \geq 2,$$

we conclude that

$$x \in B_{-j} + \sum_{i=2}^{p}\{B_{i-j} \cap U_j^{i-1}\} + B_{p-j+1} \cap U_{j+1}^{p} \subseteq \{0\} + \sum_{i=2}^{p}\{B_{i-1} \cap U_j^{i-1}\} +$$
$$B_p \cap U_{j+1}^{p} \subseteq \sum_{i=2}^{p} V_{i-1+j} + B_p \cap U_{j+1}^{p} \subseteq \sum_{i=2}^{p} V_{i-1+j} + V_{p+1+j} \subseteq V_j. \quad \blacksquare$$

Lemma 6.4.2 *Let* $\mathcal{S}$ *be an absorbing sequence in the topological vector space* $E(\mathcal{T})$. *If* $E(\mathcal{T})$ *is countably* $\mathcal{L}$*-barrelled, then* $\mathcal{T} = \mathcal{T}_{\mathcal{S}}$.

Proof. Clearly $\mathcal{T}_{\mathcal{S}}$ is always finer that $\mathcal{T}$. On the other hand, according to the previous lemma, $\mathcal{T}_{\mathcal{S}}$ is generated by a set of strings which are closed in $E(\mathcal{T})$ and are the intersection of sequences of $\mathcal{T}$-topological strings. Taking into account that $E(\mathcal{T})$ is countably $\mathcal{L}$-barrelled this implies that $\mathcal{T}_{\mathcal{S}} \subseteq \mathcal{T}$ and, consequently, $\mathcal{T}' = \mathcal{T}_{\mathcal{S}}$. ■

Lemma 6.4.3 *Let* H *be a dense* subspace *of the topological vector space* $E(\mathcal{T})$ *and let* $\mathcal{T}/H$ *be the topology on* H *induced by* $\mathcal{T}$. *Assume that* H *has an absorbing sequence* $\mathcal{S} = \{B_n,\ n \in \mathbb{N}\}$ *with* $\mathcal{T}/H = (\mathcal{T}/H)_{\mathcal{S}}$. *Then* $\overline{\mathcal{S}} = \{\overline{B_n}^{\mathcal{T}},\ n \in \mathbb{N}\}$ *is an absorbing sequence in* $E(\mathcal{T})$ *and* $\mathcal{T} = \mathcal{T}_{\overline{\mathcal{S}}}$.

Proof. If $\overline{\mathcal{S}}$ were not absorbing in $E(\mathcal{T})$, there would be some $x \notin \bigcup\{\overline{B_n}^{\mathcal{T}},\ n \in \mathbb{N}\}$, i.e. for each $n \in \mathbb{N}$ there would be some topological string $\mathcal{U}^n = \{U_j^n,\ j \in \mathbb{N}\}$ in $E(\mathcal{T})$ such that $x \notin B_n + U_1^n$.

Let $\mathcal{W} = \{W_j,\ j \in \mathbb{N}\}$ be defined by $W_j := \bigcap\{\overline{B_{n-j+1} + U_j^n}^{\mathcal{T}},\ n \in \mathbb{N}\}$, where $B_n = \{0\}$ for $n \le 0$. Then $\{W_j \cap H,\ j \in \mathbb{N}\}$ is a $(\mathcal{T}/H)_{\mathcal{S}}$-topological string and, by the hypothesis on $\mathcal{T}/H$ and $\mathcal{S}$, $(\mathcal{T}/H)$-topological.

Then $\{\overline{W_j \cap H}^{\mathcal{T}},\ j \in \mathbb{N}\}$ is a topological string in $E(\mathcal{T})$ and $\mathcal{W}$ is also a topological string in $E(\mathcal{T})$ since $\overline{W_j \cap H}^{\mathcal{T}} \subseteq W_j$ for each $j \in \mathbb{N}$.

Now from,

$$B_n + W_2 \subseteq B_n + \overline{B_n + U_2^{n+1}}^{\mathcal{T}} \subseteq B_n + B_n + U_2^{n+1} + U_2^{n+1} \subseteq B_{n+1} + U_1^{n+1}\ \forall n \in \mathbb{N}$$

it follows that $x \notin \overline{\bigcup\{B_n,\ n \in \mathbb{N}\}}^{\mathcal{T}}$, which is a contradiction since $\mathcal{S}$ is an absorbing sequence of H and H is a dense subspace of $E(\mathcal{T})$.

Therefore, $\overline{\mathcal{S}}$ is absorbing in $E(\mathcal{T})$ and $\mathcal{T} \subseteq \mathcal{T}_{\overline{\mathcal{S}}}$. Now let $\mathcal{W}$ be one of the $\mathcal{T}$-closed strings generating $\mathcal{T}_{\overline{\mathcal{S}}}$ provided by Lemma 6.4.1. Then, $\mathcal{W} \cap H$ is topological in $H((\mathcal{T}/H)_{\mathcal{S}})$ and consequently in $H(\mathcal{T}/H)$. Hence, $\{\overline{W_j \cap H}^{\mathcal{T}},\ j \in \mathbb{N}\}$ is a topological string in $E(\mathcal{T})$, as well as $\mathcal{W}$ since $\overline{W_j \cap H}^{\mathcal{T}} \subseteq W_j\ \forall j \in \mathbb{N}$. Thus $\mathcal{T}_{\overline{\mathcal{S}}} \subseteq \mathcal{T}$ and, therefore, $\mathcal{T} = \mathcal{T}_{\overline{\mathcal{S}}}$. ■

Theorem 6.4.4 *Let* E *be a subspace of* $\ell_0^\infty(\Sigma)$. *If* dim E *is infinite, then* E *is not countably* $\mathcal{L}$*-barrelled.*

Proof. If E were countably $\mathcal{L}$-barrelled, setting $X_n := \{f \in \ell_0^\infty(\Sigma), |f(\Omega)| \le n\}$, given that $E = \bigcup\{(X_n \cap E), n \in \mathbb{N}\}$, it would follow from Lemmas 10.4.2 and 10.4.3 that the completion of E would be contained in the closures of the sets $X_n \cap E$, a contradiction. ■

6.5 ALGEBRAS OF SETS GENERATING NON-BARRELLED SPACES

In this section we shall generate some classes of algebras of sets Σ whose corresponding $\ell_0^\infty(\Sigma)$ spaces are not barrelled. First of all note that, because of the metrizability, $\ell_0^\infty(\Sigma)$ is barrelled if and only if it is Baire-like. Hence, it is obvious that $\ell_0^\infty(\Sigma)$ cannot be barrelled whenever $\ell_0^\infty(\Sigma)$ has a strictly increasing sequence of closed subspaces covering it.

Proposition 6.5.1 *Let $\{\Sigma_n, n \in \mathbb{N}\}$ denote an increasing sequence of subalgebras of Σ such that $\Sigma = \bigcup\{\Sigma_n, n \in \mathbb{N}\}$. If each $\ell_0^\infty(\Sigma_n)$ is strictly contained in $\ell_0^\infty(\Sigma_{n+1})$, then $\ell_0^\infty(\Sigma)$ is not barrelled.*

Proof. As we have seen in the proof of Theorem 6.3.1, $\ell_0^\infty(\Sigma_n)$ is a closed subspace of $\ell_0^\infty(\Sigma) = \bigcup\{\ell_0^\infty(\Sigma_n), n \in \mathbb{N}\}$. So, the conclusion follows from the previous observation. ■

Example 6.5.2 Let $\{\alpha_n, n \in \mathbb{N}\}$ be an increasing sequence of infinite cardinal numbers. Assume that $\{\Sigma_n, n \in \mathbb{N}\}$ is an increasing sequence of subfamilies of Σ such that $|\Sigma_n| = \alpha_n$ and $\bigcup\{\Sigma_n, n \in \mathbb{N}\} = \Sigma$. Then, the cardinal of the algebra $\mathcal{B}_n$ generated by Σ_n is α_n, because if $\Sigma_n^1 = \Sigma_n$ and, by induction, Σ_n^{i+1} is the family of finite unions of elements of Σ_n^i and its complements, then $\mathcal{B}_n = \bigcup\{\Sigma_n^i, i \in \mathbb{N}\}$. Consequently, $\ell_0^\infty(\mathcal{B}_n)$ is strictly contained in $\ell_0^\infty(\mathcal{B}_{n+1})$ and, by Proposition 6.5.1, $\ell_0^\infty(\Sigma) = \bigcup\{\ell_0^\infty(\mathcal{B}_n), n \in \mathbb{N}\}$ is not barrelled. In particular, if Ω is a set whose cardinal α is the limit of an increasing sequence $\{\alpha_n, n \in \mathbb{N}\}$ of infinite cardinals, and if Σ is the algebra of the finite and cofinite subsets of Ω, then $\ell_0^\infty(\Sigma)$ is not barrelled.

Definition 6.5.3 *An algebra* Σ *of subsets of* Ω *is said to have property* g *whenever there exists a strictly increasing sequence* $\{A_n, n \in \mathbb{N}\}$ *of members of* Σ *such that, for each* $A \in \Sigma$ *either* A *or its complement* $\Omega \setminus A$ *is contained in* A_n *for some* $n \in \mathbb{N}$.

Example 6.5.4 Let Σ be the collection of all finite unions of disjoint intervals of the form (a, b], (-∞, b], (a, +∞). Taking $A_n = (-\infty, n]$ for every $n \in \mathbb{N}$ we obtain that Σ is an algebra in $\mathbb{R}$ with property g. Also Sorgenfrey's line, that is the real line $\mathbb{R}$ with the left half-open interval topology, has property g, with $A_n = (-\infty, n]$.

Proposition 6.5.5 *For every algebra* Σ *satisfying property* g *the normed space* $\ell_0^\infty(\Sigma)$ *is not barrelled.*

Proof. Let Σ_n denote the subalgebra of Σ such that an element A of Σ belongs to Σ_n when either A or $\Omega \setminus A$ is contained in A_n for some $n \in \mathbb{N}$, $\{A_n, n \in \mathbb{N}\}$ being the strictly increasing sequence provided by the definition above. By property g, then $\ell_0^\infty(\Sigma_n)$ is strictly contained in $\ell_0^\infty(\Sigma_{n+1})$. Now since each one of these subspaces is closed in $\ell_0^\infty(\Sigma)$ as the proof of Theorem 6.3.1 ensures, the conclusion follows. ■

In the next theorem and its corollary $\mathcal{D}$ will denote the algebra of clopen sets of a zero-dimensional Hausdorff topological space Ω. We are going to determine the compact spaces for which $\mathcal{D}$ satisfies property g.

Proposition 6.5.6 *A sufficient condition for* $\mathcal{D}$ *to have property* g *is that* Ω *has a point* ω *with a strictly decreasing countable neighbourhood base. Besides, if* Ω *is compact the condition is also necessary.*

Proof. Let $\{V_n, n \in \mathbb{N}\}$ be a strictly decreasing clopen neighbourhood base of $\omega \in \Omega$. If A is any clopen subset of Ω which is a neighbourhood of ω then there exists some V_n such that $V_n \subseteq A$, and if A is not a neighbourhood of ω, then there is some V_m such that $V_m \cap A = \varnothing$. Hence, taking $A_n = \Omega \setminus V_n$ we see that $\mathcal{D}$ has property g.

If additionally Ω is compact and $\mathcal{D}$ has property g, let $\{A_n, n \in \mathbb{N}\}$ be the strictly increasing special sequence in $\mathcal{D}$ guaranteed by property

g. Consider the set $\Omega \setminus \left(\bigcup\{A_n,\ n \in \mathbb{N}\}\right)$. Compactness ensures that it is not empty. Moreover, if t and ω are two different points in $\Omega \setminus \left(\bigcup\{A_n :\ n \in \mathbb{N}\}\right)$ and A is a clopen neighbourhood of ω not containing t, we have that A and $\Omega \setminus A$ are not contained in every A_n, which is not possible by property g. Then $\Omega \setminus \left(\bigcup\{A_n,\ n \in \mathbb{N}\}\right) = \{\omega\}$. It is clear that $\{\Omega\setminus A_n,\ n \in \mathbb{N}\}$ is a strictly decreasing neighbourhood base of ω. ■

Example 6.5.7 If Ω is Alexandroff's compactification $\mathbb{N}^*$ of the positive integers we have that $\mathcal{D}$ has property g, so $\ell_0^\infty(\mathbb{N}^*,\mathcal{D})$ is not a barrelled space.

Corollary 6.5.8 *Assuming Ω is a compact space such that the algebra $\mathcal{D}$ satisfies property g, then Ω admits non-trivial convergent sequences.*

Remark 6.5.9 It is well known that $\beta\mathbb{N}$ has no non-trivial convergent sequence. Thus, in light of Corollary 6.5.8, the algebra $\mathcal{D}$ of clopen subsets of $\beta\mathbb{N}$ cannot satisfy property g.

As we have seen, if Ω is a zero-dimensional Hausdorff topological space that has a point which possesses a strictly decreasing countable neighbourhood base, its algebra $\mathcal{D}$ of clopen subsets satisfies property g and consequently $\ell_0^\infty(\mathcal{D})$ is not barrelled. We can reach this conclusion directly via the next theorem.

Theorem 6.5.10 *Let Ω be a zero-dimensional Hausdorff topological space and let $\mathcal{D}$ denote the algebra of the clopen subsets of Ω. If there is a non-trivial convergent sequence in Ω, then $\ell_0^\infty(\mathcal{D})$ is not barrelled.*

Proof. Suppose that $\{\omega_n,\ n \in \mathbb{N}\}$ is an injective sequence in Ω which converges to some $\omega \in \Omega$. For each $n \in \mathbb{N}$, consider the subspace

$$E_n := \{f \in \ell_0^\infty(\mathcal{D}),\ f(\omega_p) = f(\omega_q) \text{ with } p,\ q \geq n\}.$$

Clearly, $\{E_n,\ n \in \mathbb{N}\}$ is an increasing sequence of closed subspaces of $\ell_0^\infty(\mathcal{D})$. Let us see that this sequence covers $\ell_0^\infty(\mathcal{D})$. In fact, suppose that $f = \sum_{i=1}^{p} a_i e(C_i) \in \ell_0^\infty(\mathcal{D})$, $a_i \in \mathbb{K}$, $1 \leq i \leq p$, with $\{C_i,\ 1 \leq i \leq p\}$ a partition of Ω formed by elements of $\mathcal{D}$. If $j \in \{1, .., p\}$ is such that $\omega \in C_j$, then there is some $m \in \mathbb{N}$ with $\omega_n \in C_j$ for every $n \geq m$. This implies that f belongs to E_m and consequently the sequence $\{E_n, n \in \mathbb{N}\}$ covers $\ell_0^\infty(\mathcal{D})$. If $\ell_0^\infty(\mathcal{D})$ were barrelled, there would be some $q \in \mathbb{N}$ such that $E_q = \ell_0^\infty(\mathcal{D})$. But since Ω is T_1 there must be a clopen neighbourhood U of ω in Ω with $\omega_q \notin U$ and clearly $e(U) \in \ell_0^\infty(\mathcal{D}) \setminus E_q$. Obviously, this is a contradiction. ∎

Corollary 6.5.11 *Let Ω be a set, Σ an algebra of subsets of Ω and Y the Stone space of Σ. If there exists some non-trivial convergent sequence in Y, then the space $\ell_0^\infty(\Omega,\Sigma)$ is not barrelled.*

Proof. If $\mathcal{D}$ denotes the algebra of the clopen subsets of Y, $\ell_0^\infty(Y,\mathcal{D})$ is the subspace of C(Y) formed by those continuous functions taking only a finite number of values. Since Y is zero-dimensional, it follows from the theorem above that $\ell_0^\infty(Y,\mathcal{D})$ is not barrelled. Now the mapping T that carries each function $\sum_{i=1}^{n} a_i e(A_i)$ of $\ell_0^\infty(\Omega,\Sigma)$ to the function $\sum_{i=1}^{n} a_i e(C_i)$ of $\ell_0^\infty(Y,\mathcal{D})$, where $C_i \in \mathcal{D}$ is the image of $A_i \in \Sigma$ by the isomorphism between Σ and $\mathcal{D}$ defined by the Stone representation theorem, is an isometry of $\ell_0^\infty(\Omega,\Sigma)$ onto $\ell_0^\infty(Y,\mathcal{D})$. ∎

Theorem 6.5.12 *Assume Ω is a Hausdorff completely regular topological space and $\beta\Omega$ is the Stone-Čech compactification of Ω. If $\mathcal{B}$ and $\mathcal{D}$ denote respectively the algebras of clopen subsets of Ω and $\beta\Omega$, the space $\ell_0^\infty(\Omega,\mathcal{B})$ is isometric to $\ell_0^\infty(\beta\Omega,\mathcal{D})$.*

Proof. Consider the linear mapping T from $\ell_0^\infty(\beta\Omega,\mathcal{D})$ into the linear subspace $\ell_0^\infty(\Omega,\mathcal{B})$ of $\ell_0^\infty(\beta\Omega,\mathcal{D})$ defined by the restriction to Ω of each $f \in \ell_0^\infty(\beta\Omega,\mathcal{D})$. Since Ω is dense in $\beta\Omega$, each member of $\mathcal{D}$ meets Ω and we have $\sup\{|f(\omega)|, \omega \in \beta\Omega\} = \sup\{|f(\omega)|, \omega \in \Omega\}$ for each $f \in \ell_0^\infty(\beta\Omega,\mathcal{D})$. Now, T is onto, since if $h = \sum_{i=1}^{p} a_i e(C_i) \in \ell_0^\infty(\Omega,\mathcal{B})$ with $C_i \in \mathcal{B}$, $1 \leq i \leq p$, then $f = \sum_{i=1}^{p} a_i e(\overline{C}_i^{\beta X}) \in \ell_0^\infty(\beta\Omega,\mathcal{D})$ and $T(f) = h$. Now, T is injective since Ω is dense in $\beta\Omega$ and $\|T(f)\| = \|f\|$. Thus, T is an isometry. ∎

Examples 6.5.13 If we consider the sets $\mathbb{Q}$ and $\mathbb{P}$ of the rational and irrational numbers respectively as topological subspaces of $\mathbb{R}$, Theorem 6.5.10 asserts that the spaces $\ell_0^\infty(\mathbb{Q},\mathcal{D}_Q)$ and $\ell_0^\infty(\mathbb{P},\mathcal{D}_P)$, $\mathcal{D}_Q$ and $\mathcal{D}_P$ being the algebras of the clopen subsets of $\mathbb{Q}$ and $\mathbb{P}$, respectively, are not barrelled. Now, it follows from the previous theorem that the spaces $\ell_0^\infty(\beta\mathbb{Q},\mathcal{B})$ and $\ell_0^\infty(\beta\mathbb{P},\mathcal{D})$ with $\mathcal{B}$ and $\mathcal{D}$ standing for the respective algebras of clopen subsets, are not barrelled either. On the other hand, if Ω is an infinite zero-dimensional metric space without any isolated point, a non-trivial sequence converges to each $\omega \in \Omega$ and, by Theorem 6.5.10, the space $\ell_0^\infty(\Omega,\mathcal{D})$ generated by the algebra of clopen subsets of Ω is not barrelled. In particular, if Ω is homeomorphic to the Cantor space, then $\ell_0^\infty(\Omega,\mathcal{D})$ is not barrelled.

6.6 $\ell_0^\infty(\Sigma)$ AND TOTALLY BARRELLED SPACES

Next we shall see that $\ell_0^\infty(\Sigma)$ is not totally barrelled whenever Ω is an infinite set and Σ is the σ-algebra of all its subsets. If S_Σ denotes the Stone space corresponding to the algebra Σ and $C \in \Sigma$ we shall write $\check{C}$ for the set $\{\mu \in S_\Sigma : \mu(C) = 1\}$. Is is plain that $\{\check{C} : C \in \Sigma\}$ is a base for the topology of S_Σ and that this space coincides with the Stone-Čech compactification of Ω.

Let $\{E_0,E_1\}$ be a partition of Ω, E_0 and E_1 being two infinite subsets of Ω. Now let $\{E_{00},E_{01}\}$ and $\{E_{10},E_{11}\}$ be a partition of E_0 and E_1, respectively, each E_{ij} being an infinite subset of Ω for i, j $\in \{0, 1\}$. In this way we build up a tree $\mathcal{F} = \{E_{n_1,\ldots,n_p}$, $p \in \mathbb{N}$, $n_1, \ldots, n_p \in \{0,1\}\}$ of infinite subsets of Ω. It is easy to note that the family $\mathfrak{B}$ of all the finite unions of elements of $\mathcal{F}$ is an algebra, since the complement of a finite union of elements of $\mathcal{F}$ is also a finite union of elements of $\mathcal{F}$. Then, if $S_\mathcal{B}$ stands for the Stone space of the subalgebra $\mathcal{B}$ of Σ, we will write $\hat{E} := \{\lambda \in S_\mathcal{B} : \lambda(E) = 1\}$ whenever $E \in \mathcal{B}$. Now it can be easily proven that $\{\hat{E}_{n_1,\ldots,n_p}$, $p \in \mathbb{N}$, $n_1,\ldots,n_p \in \{0,1\}\}$ is a base of the topology of $S_\mathcal{B}$.

Consider the dual spaces $\ell_0^\infty(\Omega,\Sigma)'$ and $\ell_0^\infty(\Omega,\mathcal{B})'$ both provided with its corresponding weak*-topologies, and define the map ψ from $\ell_0^\infty(\Omega,\Sigma)'$ into $\ell_0^\infty(\Omega,\mathcal{B})'$ such that $\psi(\mu)$ designates the restriction of $\mu \in \ell_0^\infty(\Omega,\Sigma)'$ to the (closed) subspace $\ell_0^\infty(\Omega,\mathcal{B})$ of $\ell_0^\infty(\Omega,\Sigma)$. If ϕ is the restriction of ψ to S_Σ, it is obvious that $\phi(S_\Sigma) \subseteq S_{\mathcal{B}}$.

Proposition 6.6.1 *ϕ is a continuous mapping from S_Σ onto $S_{\mathcal{B}}$.*

Proof. Given $\nu \in S_{\mathcal{B}}$, let $\{m_p,\ p \in \mathbb{N}\} \subseteq \{0,1\}^{\mathbb{N}}$ be such that

$$\nu(E_{m_1,\dots,m_p}) = 1 \quad \forall p \in \mathbb{N}.$$

As $\{\check{E}_{m_1,\dots,m_p},\ p \in \mathbb{N}\}$ has the finite intersection property and S_Σ is compact, there exists some $\mu \in \bigcap\{\check{E}_{m_1,\dots,m_p},\ p \in \mathbb{N}\}$. Then, $\phi(\mu) = \nu$ because of the choice of $\{m_p,\ p \in \mathbb{N}\}$ and consequently ϕ is onto. On the other hand, since ψ is a weak* to weak* continuos mapping, then its restriction ϕ to S_Σ is continuous too. We give another proof of this. As $\{\hat{E}_{n_1,\dots,n_p},\ p \in \mathbb{N},\ n_1,\ \dots,\ n_p \in \{0,1\}\}$ is a base formed by clopen subsets of the topology of $S_{\mathcal{B}}$, then given an open set $U \subseteq S_{\mathcal{B}}$ such that $\phi(\mu) = \nu \in U$, there is some $\hat{E}_{h_1,\dots,h_p}$ such that $\nu \in \hat{E}_{h_1,\dots,h_p} \subseteq U$. From $\phi(\check{E}_{h_1,\dots,h_p}) \subseteq \hat{E}_{h_1,\dots,h_p} \subseteq U$, it follows that ϕ is continuous at ν. Hence ϕ is continuous. ∎

Let us now recall that given two Hausdorff topological spaces X and Y, a continuous mapping of X onto Y is called *irreducible* if, for each closed subset A of X, f(A) = Y implies A = X.

Proposition 6.6.2 *Let X and Y be two Hausdorff topological spaces. If f is a continuous mapping of X onto Y and X is compact then there exists some closed subset A of X such that f(A) = Y and $f|_A : A \longrightarrow Y$ is irreducible.*

Proof. Let $\mathcal{A} := \{M \subseteq X,\ M \text{ is closed and } f(M) = Y\}$, and let us show that if $\{M_\alpha,\ \alpha \in \Lambda\}$ is a decreasing family of elements of $\mathcal{A}$, then $M := \bigcap\{M_\alpha,\ \alpha \in \Lambda\} \in \mathcal{A}$. So, assume by contradiction that there is some $y_0 \in Y \setminus f(M)$.

Then $B := f^{-1}(y_0)$ is a closed subset of X with $M \cap B = \emptyset$. On the other hand, from $y_0 \in f(M_\alpha)$ $\forall\alpha \in \Lambda$, it follows that for each $\alpha \in \Lambda$ there is some $x_\alpha \in M_\alpha$ such that $f(x_\alpha) = y_0$, i.e. $x_\alpha \in f^{-1}(y_0) = B$, and therefore $B \cap M_\alpha \neq \emptyset$ $\forall\alpha \in \Lambda$. Hence the family of closed subsets of B $\{B \cap M_\alpha, \alpha \in \Lambda\}$ has the finite intersection property and as X is a compact Hausdorff topological space, we deduce that $B \cap M = \bigcap\{B \cap M_\alpha, \alpha \in \Lambda\} \neq \emptyset$, which is a contradiction.

According to Zorn's lemma there is a minimal set $A \in \mathcal{A}$ and clearly $f_{|A} : A \longrightarrow Y$ is irreducible. Indeed $f_{|A} : A \to Y$ is continuous, and if there were a closed subset P of A, with $f(P) = Y$, this would imply that P would coincide with A, since A is just a minimal set with respect to the inclusion relationship. ∎

Proposition 6.6.3 *Let* X *be a compact Hausdorff topological space and let* f *be a continuous mapping from* X *onto a Hausdorff topological space* Y. *If* f *is irreducible, given any non-void open subset* U *of* X, *there exists some non-empty open subset* V *of* Y *such that* $\emptyset \neq f^{-1}(V) \subseteq U$.

Proof. Given a non-empty open subset U of X, $X \setminus U$ is a compact subset of X and hence $f(X \setminus U)$ is a closed subset of Y. If $f(X \setminus U)$ were dense in Y, $f(X \setminus U)$ would coincide with Y and, f being irreducible, $X \setminus U = X$ which is not possible since $U \neq \emptyset$. Therefore $f(X \setminus U)$ is not dense in Y and there must be some non-empty subset V of Y such that $V \cap f(X \setminus U) = \emptyset$. In fact, we could have taken $V = Y \setminus f(X \setminus U)$.

Finally it remains to see that $f^{-1}(V) \subseteq U$. But if $x \in f^{-1}(V)$ then $f(x) \in V$ and therefore $f(x) \notin f(X \setminus U)$. Hence $x \notin X \setminus U$, i.e. $x \in U$. ∎

Theorem 6.6.4 $\ell_0^\infty(\Sigma)$ *is not totally barrelled.*

Proof. Let $\mathfrak{B}$, ϕ and Σ be the same as in Proposition 6.6.1, and let us denote by $\{\Omega_n, n \in \mathbb{N}\}$ the tree $\mathcal{F} = \{E_{n_1,\dots,n_p}, p \in \mathbb{N}, n_1, \dots, n_p \in \{0,1\}\}$ of infinite subsets of Ω defined above. Given that S_Σ and $S_\mathfrak{B}$ are compact Hausdorff spaces and $\phi : S_\Sigma \longrightarrow S_\mathfrak{B}$ is continuous and onto, then by Proposition 6.6.2, there exists a closed subset Z of S_Σ such that $\phi_{|Z} : Z \longrightarrow S_\mathfrak{B}$ is irreducible, and, hence, onto.

If $G \in \Sigma$ and $\check{G} \cap Z \neq \emptyset$, by Proposition 6.6.3 there exists an open subset U of $S_{\mathfrak{B}}$ such that $\emptyset \neq \phi^{-1}(U) \cap Z \subseteq \check{G} \cap Z \subseteq \check{G}$, and since $\{\hat{\Omega}_n, n \in \mathbb{N}\}$ is a base of the topology of $S_{\mathfrak{B}}$, there must be some $p \in \mathbb{N}$ such that $\emptyset \neq \phi^{-1}(\hat{\Omega}_p) \cap Z \subseteq \check{G}$. For each $n \in \mathbb{N}$ now set

$$F_n := \{f \in \ell_0^\infty(\Omega,\Sigma): \mu(f) = \lambda(f), \forall \mu, \lambda \in \phi^{-1}(\hat{\Omega}_n) \cap Z\}.$$

It is worth noticing that $\phi^{-1}(\hat{\Omega}_n) \cap Z \neq \emptyset$ for each $n \in \mathbb{N}$. In fact, $\phi(Z) = S_{\mathfrak{B}}$ since $\phi_{|Z}$ is onto and, on the other hand, $\hat{\Omega}_n \subseteq S_{\mathfrak{B}}$. Therefore, there must be some $\mu_n \in Z$ such that $\phi(\mu_n) \in \hat{\Omega}_n$ and, consequently, $\mu_n \in Z \cap \phi^{-1}(\hat{\Omega}_n)$.

Clearly each F_n is a closed vector subspace of $\ell_0^\infty(\Omega,\Sigma)$. Let us now see that $\ell_0^\infty(\Omega,\Sigma) = \bigcup\{F_n, n \in \mathbb{N}\}$. Given $f = \sum_{1\le k\le n} a_k e(G_k) \in \ell_0^\infty(\Omega,\Sigma)$, as $\bigcup\{\check{G}_k, 1 \le k \le n\} = S_\Sigma$, it follows that there is some $j \in \{1,\dots,n\}$ such that $\check{G}_j \cap Z \neq \emptyset$. Consequently there exists some positive integer q such that $\phi^{-1}(\hat{\Omega}_q) \cap Z \subseteq \check{G}_j$. Therefore, $f \in F_q$.

Let us now see that each F_n is infinite-codimensional in $\ell_0^\infty(\Omega,\Sigma)$. Given $n \in \mathbb{N}$, let $\{\Omega_{n_i}, i \in \mathbb{N}\}$ be a sequence of pairwise disjoint members of $\mathcal{F}$, all of them contained in Ω_n.

As $\phi_{|Z}: Z \longrightarrow S_{\mathfrak{B}}$ is onto, for each $i \in \mathbb{N}$ there exists some $\mu_i \in Z$ with $\phi(\mu_i) \in \hat{\Omega}_{n_i} \subseteq S_{\mathfrak{B}}$. Hence, $\{\mu_i, i \in \mathbb{N}\}$ is an injective sequence of elements of Z with $\phi(\mu_i) \in \hat{\Omega}_n$ for each $i \in \mathbb{N}$. Fix $k \in \mathbb{N}$ and for $1 \le i \le k$, let $A_i \subseteq \Omega$ be such that $\mu_i \in \check{A}_i$ but $\mu_j \notin \check{A}_i$ for $j \in \{1,\dots,k+1\}\setminus\{i\}$. Owing to the fact that $\mu_{k+1} \in \phi^{-1}(\hat{\Omega}_n) \cap Z$, if $a_1 e(A_1) + \dots + a_k e(A_k) \in F_n$, then $\mu_{k+1}\left(a_1 e(A_1) + \dots + a_k e(A_k)\right) = 0$, and since $\mu_i \in \phi^{-1}(\hat{\Omega}_n) \cap Z$ for $1 \le i \le k$, it follows that $a_i = 0$ for $1 \le i \le k$. This proves that $\operatorname{codim} F_n \ge k$ for each $k \in \mathbb{N}$. Hence each F_n is an infinite-codimensional closed subspace of $\ell_0^\infty(\Omega,\Sigma)$, which is not totally barrelled. ∎

6.7 EXERCISES

1. *Dimension of* $\ell_0^\infty(\Sigma)$. If Σ is an infinite σ-algebra of subsets of a set Ω, then $\dim \ell_0^\infty(\Sigma) = |\Sigma|$. Furthermore, if $|\Sigma| \ge c$, then we have that $|\ell_0^\infty(\Sigma)| = \dim \ell_0^\infty(\Sigma)$.

2. *Non-barrelled* $\ell_0^\infty(\Sigma)$ *spaces.* Assuming Ω is an infinite set and Σ is the algebra of all finite and cofinite subsets of Ω, then $\ell_0^\infty(\Sigma)$ is not barrelled and Σ does not satisfy condition g.

3. *Ultrabornological subspaces of* $\ell^\infty(I)$ *containing* $\ell_0^\infty(I)$. Let I be an infinite index set and let $\ell_0^\infty(I)$ denote here the subspace of the Banach space $\ell^\infty(I)$ generated by the indicator functions of all the subsets of I. We denote by $\mathfrak{F}$ the family of all ultrafilters $\mathcal{U}$ on I containing the filter of the cofinite subsets of I.

(i) If $\mathcal{U} \in \mathfrak{F}$ and $U \in \mathcal{U}$, define

$$L(U) := \bigcup\{x \in \ell^\infty(I) : x \textit{ takes a constant value on } U\}$$

and let $L(\mathcal{U}) = \bigcup\{L(U) : U \in \mathcal{U}\}$. Then each L(U) is a closed linear subspace of $\ell^\infty(I)$ and $L(\mathcal{U})$ is a proper linear subspace of $\ell^\infty(I)$ containing $\ell_0^\infty(I)$.

(ii) For each $\mathcal{U} \in \mathfrak{F}$, $L(\mathcal{U})$ is ultrabornological. *[Hints: If* V *is an absolutely convex subset of* $L(\mathcal{U})$ *such that* $V \cap L(U)$ *is a neighbourhood of the origin in* L(U) *for each* $U \in \mathcal{U}$, *find first a* $\lambda > 0$ *with* $\lambda e(A) \in V$ *for each* $A \subseteq I$; *then prove that if B denotes the closed unit ball of* $\ell^\infty(I)$, *we have that* $\lambda B \cap L(\mathcal{U}) \subseteq 5V$.*]*

(iii) If $I = \mathbb{N}$, there is no $L(\mathcal{U})$ which is unordered Baire-like.

4. *A projective description of the space* $\ell_0^\infty(I)$. Using the same notation as in the previous exercise, we state the following.

(i) $\ell_0^\infty(I) = \bigcap\{L(\mathcal{U}),\ \mathcal{U} \in \mathfrak{F}\}$

(ii) $\ell_0^\infty(I)$ is the projective limit of the ultrabornological subspaces $\{L(\mathcal{U}),\ \mathcal{U} \in \mathfrak{F}\}$ of $\ell^\infty(I)$ with respect to the canonical inclusion maps.

(iii) If $|\mathfrak{F}|$ is smaller than the first strongly inaccessible cardinal, $\ell_0^\infty(I)$ is isomorphic to a closed subspace of an ultrabornological space.

5. *A barrelled* $\ell_0^\infty(\Sigma)$ *space generated by an algebra which is not a* σ*-algebra.* Let Ω be an infinite set provided with its discrete topology and denote by $\mathcal{D}$ the algebra of all clopen subsets of $\beta\Omega$.

(i) $\mathcal{D}$ is not a σ-algebra.

(ii) $\ell_0^\infty(\beta\Omega,\mathcal{D})$ is barrelled of class $\aleph_0$.

6. *The spaces* $\ell_0^\infty(\beta\Omega\backslash\Omega,\mathcal{D})$. Suppose that Ω is a completely regular Hausdorff topological space and denote by $\mathcal{D}$ and $\mathcal{D}'$ respectively the algebras of clopen subsets of $\beta\Omega$ and $\beta\Omega \setminus \Omega$.

(i) If for each clopen set A in $\beta\Omega\backslash\Omega$ there exists some clopen subset B in $\beta\Omega$ such that $A = B \setminus \Omega$, then $\ell_0^\infty(\beta\Omega\backslash\Omega,\mathcal{D}')$ is isomorphic to a quotient of $\ell_0^\infty(\beta\Omega,\mathcal{D})$.

(ii) $\ell_0^\infty(\beta\Omega\backslash\Omega,\mathcal{D}')$ is isomorphic to a quotient of ℓ_0^∞, hence barrelled of class $\aleph_0$.

7. $\ell_0^\infty(\mathcal{A})$ *spaces generated by rings of subsets.* If Ω is a set and $\mathcal{A}$ is an infinite ring of subsets of Ω, $\ell_0^\infty(\mathcal{A})$ is not unordered Baire-like.

8. *A barrelled locally convex space which is not ultrabarrelled.* Let E be a linear space of uncountable dimension and $\{e_i,\ i \in I\}$ be a Hamel basis of E. Suppose that p is a real number with $0 < p < 1$ and consider the string $\mathcal{U} = \{U_n,\ n \in \mathbb{N}\}$ where

$$U_n = \left\{\Sigma\{\lambda_i e_i,\ i \in J\}: J \subseteq I,\ J \text{ finite and } \Sigma\{|\lambda_i|^p,\ i \in J\} \leq 1/2^{n-1}\right\}$$

(i) U_n is not an absolutely convex set for any $n \in \mathbb{N}$.

(ii) U_1 does not contain any absolutely convex absorbing set U of E. *[Hint: For each* $i \in I$ *let* $\varepsilon_i > 0$ *be such that* $\varepsilon_i e_i \in U$. *Since* I *is uncountable, there exists some* $\varepsilon > 0$ *with the property that the set* $I_\varepsilon = \{i \in I,\ \varepsilon_i > \varepsilon\}$ *is infinite. For any finite sequence* $\{i_1, i_2, \ldots, i_n\}$ *of elements of* I_ε, $\sum_{j=1}^{n}(\varepsilon/n)e_{i_j} \in U_1$, *against the arbitrarity of* n.*]*

(iii) If we consider on E the locally convex topology ρ induced by the product topology of $\Pi\{\langle e_i\rangle,\ i \in I\}$, then the string $\mathcal{U}$ is ρ-closed.

(iv) If τ^c denotes the strongest locally convex topology on E, $E(\tau^c)$ is a barrelled Hausdorff locally convex space which is not ultrabarrelled.

9. *Non-totally barrelledness of* $\ell_0^\infty(\Sigma)$. If Ω is a set, assuming that Σ is any infinite σ-algebra of subsets of Ω then $\ell_0^\infty(\Omega,\Sigma)$ is not a totally barrelled space.

6.8 NOTES AND REMARKS

G. Bennet and J. Kalton observed in [BK3] that the space ℓ_0^∞ is of the first category, since it can be covered by the sequence (X_n) of the closed sets X_n formed by all sequences taking at most n values. Later, S.A. Saxon notices in [S3] that this space cannot be even unordered Baire-like. This prompted M. Valdivia and P. Pérez Carreras to ask in [VP1] whether or not ℓ_0^∞ is totally barrelled. This is really a deep question which was answered in the negative by J. Arias de Reyna [Ar2]. On the other hand, J. Batt, P. Dierolf and J. Voigt [BDV] show that if I is an infinite index set and $\mathcal{P}$ is the σ-algebra of all subsets of I, then the space $\ell_0^\infty(\mathcal{P})$ is not ultrabornological. Actually, Valdivia [V2] was the first to discover the existence of bornological barrelled spaces which are not ultrabornological, solving a problem raised in Bourbaki [B]. One of the results obtained by Valdivia can be stated as follows.

Theorem 6.8.1 *Let* I *be an infinite index set. If* (α_n) *is a sequence of cardinal numbers such that* $0 < \alpha_n \le \aleph_0$, *there exists a sequence* (E_n) *of dense linear subspaces of* $\ell^1(I)$ *such that for each* $n \in \mathbb{N}$ *one has*

(i) $E_{n+1} \subseteq E_n$.

(ii) E_{2n-1} *is an ultrabornological space.*

(iii) E_{2n} *is bornological and barrelled but not ultrabornological.*

(iv) The codimension of E_{2n+1} *in* E_{2n} *is* α_n.

According to Theorem 6.2.2 $\ell_0^\infty(\Sigma)$ is not ultrabornological whenever Σ is an infinite σ-algebra, despite being bornological and barrelled of class $\aleph_0$.

As P. Dierolf, S. Dierolf and L. Drewnowski [DDD] pointed out, N.J. Kalton was the first to notice that with I and $\mathcal{P}$ as above, $\ell_0^\infty(\mathcal{P})$ is not ultrabarrelled, and A. Pelczyski was the first to notice that $\ell_0^\infty(\mathcal{P})$ contains no infinite-dimensional separable barrelled subspace. By the way, ultrabarrelled spaces were introduced by W. Robertson in [R1] and we refer the reader to Ferrer et al. [FLS2], Pérez Carreras [P] and Sánchez Ruiz [Sa1] and [Sa8] for research on some properties of this and other related classes of topological vector spaces. The extension of the two aforementioned results to any σ-algebra of subsets of I has been given in the quoted paper [DDD]. In our Theorem 6.4.4 we have proved that if E is an infinite-dimensional subspace of $\ell_0^\infty(\Sigma)$, Σ being an infinite σ-algebra of subsets of Ω, then E is not even countably $\mathcal{L}$-barrelled.

Our Lemma 6.2.1 and Theorem 6.2.2 are taken from Dierolf, Dierolf and Drewnowski [DDD]. Proposition 6.2.4 comes from Batt, Dierolf and Voigt [BDV], although the proof given is taken half from this last paper and half from [DDD]. Results 6.2.5, 6.2.6, 6.3.1 and 6.3.2 can be found in Dierolf, Dierolf and Drewnowski [DDD]. Lemmas 6.4.1, 6.4.2 and 6.4.3 are from Adasch, Ernst and Keim [AEK]. Definition 6.5.3 and Propositions 6.5.5, 6.5.6 and Corollary 6.5.8 come from J. Ferrer and M. López Pellicer [FL]. Theorem 6.5.10, Corollary 6.5.11, Theorem 6.5.12 and Examples 6.5.13 have been chosen from Ferrando and Ferrer [FF]. Other examples concerning Boolean algebras $\mathcal{A}$ such that $\ell_0^\infty(\mathcal{A})$ is not barrelled have been considered in [Sc].

Our exposition of Arias de Reyna's result in case Ω is infinite and Σ is the σ-algebra of all subsets of Ω is simpler than his original proof, [Ar2]. Propositions 6.6.2 and 6.6.3 have been taken from Semadeni [Sd]. The case when Σ is an arbitrary infinite σ-algebra is left to Exercise 9, and can be deduced from our Theorem 6.6.4. Exercises 3 and 4 are selected from Ferrando [Fe2]. Exercises 5, 6 and 7 can be seen in Ferrando and Ferrer [FF]. Exercise 8 is taken from Adasch, Ernst and Keim [AEK]. Exercise 9 comes naturally from Arias de Reyna [Ar2].

7. Applications to measure theory

7.1 PRELIMINARIES AND VECTOR MEASURES

In this chapter we shall obtain some results in measure theory, mainly taking advantage of the barrelledness of class $\aleph_0$ of the space $\ell_0^\infty(\Sigma)$ shown in Chapter 5 when Σ is a σ-algebra of subsets of a set Ω. As in that chapter, we shall identify the space $ba(\Sigma)$ of the bounded finitely additive scalar measures on the σ-algebra Σ on Ω with the topological dual of the space $\ell_0^\infty(\Sigma)$. The subspace of all the countably additive scalar measures will be denoted by $ca(\Sigma)$.

First, let us recall that a function F defined in the σ-algebra Σ with values in a Banach space X is called a *finitely additive measure*, or simply a *vector measure*, if whenever A_1 and A_2 are disjoint members of Σ, then $F(A_1 \cup A_2) = F(A_1) + F(A_2)$. If, in addition, $F(\bigcup\{A_n, n \in \mathbb{N}\}) = \sum_{n=1}^{\infty} F(A_n)$ in the norm topology of X for all sequences (A_n) of pairwise disjoint members of Σ, then F is said to be a *countably additive vector measure* or, simply, that F is countably additive.

Let us consider the linear space C[0,1] of all continuous real functions defined in [0,1] as a subspace of $L^\infty[0,1]$. Given $a \in [0,1]$, we consider the point mass functional δ_a as a continuous linear form on the subspace C[0,1] of $L^\infty[0,1]$ such that if $\hat{f} \in L^\infty[0,1]$, then $\delta_a(\hat{f}) = f(a)$, where $f \in \hat{f} \cap C[0,1]$. Let T be a Hahn-Banach extension of $\delta_{1/2}$ to the whole of $L^\infty[0,1]$.

Now we define the scalar measure F by $F(A) = T(e(A))$ for each Lebesgue measurable set $A \subseteq [0,1]$. The linearity of T implies that F is a finitely additive vector measure which, as we are going to see, may fail to be countably additive. In fact, if F were countably additive we would have

$$\lim_n F\left\{[0,(n-1)/2n] \cup [(n+1)/2n,1]\right\} = 1$$

since it verifies that

$$F([0,1]\setminus\{1/2\}) = T(e([0,1]\setminus\{1/2\})) = \delta_{1/2}(e([0,1]) = 1.$$

But if $u_n = e([0,(n-1)/2n] \cup [(n+1)/2n,1])$, it is easy to determine a continuous function v_n such that $v_n(1/2) = 0$ and $\|u_n - v_n\| = 1/2$. Then, as $Tv_n = 0$ and $\|T\| = 1$, one has that $|F([0,(n-1)/2n] \cup [(n+1)/2n,1])| = |Tu_n| = |Tu_n - Tv_n| \le \|u_n - v_n\| = 1/2$, and taking limits we obtain $1 \le 1/2$, a contradiction.

However, if $T : L^1[0,1] \longrightarrow X$ is a continuous linear operator and, for every Lebesgue measurable set $A \subseteq [0,1]$, we define as above $F(A) = T(e(A))$, then the measure F happens to be countably additive, since $\|F(\bigcup\{A_n,\ n \in \mathbb{N}\})\| \le \|T\|\cdot\|e(\cup_n A_n)\|_1 = \|T\|\cdot\lambda(\bigcup\{A_n,\ n \in \mathbb{N}\})$, λ being the Lebesgue measure.

These two examples are related to the following two definitions.

1. The *variation* of a measure $F : \Sigma \longrightarrow X$ is the function $|F| : \Sigma \longrightarrow [0, +\infty]$ defined by

$$|F|(A) = \sup\left\{\textstyle\sum_{B\in\pi}\|F(B)\| : \pi \in \Pi(A)\right\}$$

for each $A \in \Sigma$ where $\Pi(A)$ denotes the family of all finite partitions of A by elements of Σ. If $|F|(\Omega) < \infty$, F is called a *measure of bounded variation.* If F represents the measure corresponding to the linear continuous operator $T : L^1[0,1] \longrightarrow X$, then F is a measure of bounded variation since $|F|(\Omega) \le \|T\|$. Nevertheless, if F denotes the measure corresponding to the identity mapping $I : L^\infty[0,1] \longrightarrow L^\infty[0,1]$, then F is not of bounded variation since if $\{A_n,\ 1 \le n \le m\}$ is a finite partition of the interval [0,1] by *m* Lebesgue measurable sets of positive measure one has that $\sum_{n=1}^{m} \|F(A_n)\|_\infty = m$.

2. The *semivariation* of the measure $F : \Sigma \longrightarrow X$ is the extended non-negative function $\|F\| : \Sigma \longrightarrow [0, +\infty]$ defined on each $A \in \Sigma$ as $\|F\|(A) = \sup\{|x^*F|(A) :\ x^* \in X',\ \|x^*\| \le 1\}$,
where $|x^*F|$ is the variation of the scalar measure x^*F. If $\|F\|(\Omega) < \infty$, F will be called a measure of bounded semivariation. If F is the measure corresponding to the linear continuous operator $T : L^\infty[0,1] \longrightarrow X$, then F is a *measure of bounded semivariation.* In fact, if π is a

finite partition of the interval [0,1] into Lebesgue measurable subsets, then $\sum_{A\in\pi}|x^*F(A)| = \sum_{A\in\pi}|x^*T(e(A))|$. Thus if ε_A is such that

$$|x^*T(e(A))| = \varepsilon_A x^*T(e(A)),$$

we have

$$\sum_{A\in\pi}|x^*F(A)| = x^*T(\sum_{A\in\pi}\varepsilon_A e(A)) \le \|T\|.$$

Direct verifications show that the variation of F is a monotone finitely additive function on Σ, while the semivariation of F is a monotone subadditive function on Σ. Also it is easy to check that for each $A \in \Sigma$ one has $\|F\|(A) \le |F|(A)$.

It is easily proven that the linear space over $\mathbb{K}$ of all finitely additive measures of bounded variation defined in Σ with values in the Banach space X, a space that we will denote by $bv(\Sigma,X)$, equipped with the variation norm, is a Banach space. The linear subspace $bvca(\Sigma,X)$ of all countably additive measures is also a Banach space, and $F \in bv(\Sigma,X)$ is countably additive if and only if its variation $|F|$ is a countably additive scalar measure. It is natural to ask if the space $ba(\Sigma,X)$ of all finitely additive measures with bounded semivariation defined in Σ with values in the Banach space X, and equipped with the semivariation norm, is a Banach space. This question can be easily answered after the next results concerning the semivariation.

Proposition 7.1.1 *Given a vector measure* $F : \Sigma \longrightarrow X$, *the semivariation of* F *on* $A \in \Sigma$ *is also given by*

$$\|F\|(A) = \sup\left\{\|\sum_{B\in\pi}\varepsilon_B F(B)\|,\ |\varepsilon_B| \le 1,\ \pi \in \Pi(A)\right\}.$$

Proof. Easy computations yield

$$\sup_{\pi}\left\{\|\sum_{B\in\pi}\varepsilon_B F(B)\|\right\} = \sup_{\pi}\left\{|x^*\{\sum_{B\in\pi}\varepsilon_B F(B)\}| : x^* \in X',\ \|x^*\| \le 1\right\} \le$$

$$\sup_{\pi}\left\{\sum_{B\in\pi}|\varepsilon_B x^*F(B)| : x^* \in X',\ \|x^*\| \le 1\right\} \le$$

$$\sup_{\pi}\left\{\sum_{B\in\pi}|x^*F(B)| : x^* \in X',\ \|x^*\| \le 1\right\} \le \|F\|(A).$$

For the reverse inequality, let ε_B be a scalar such that $\varepsilon_B x^*F(B) = |x^*F(B)|$, then

$$\sum_{B\in\pi}|x^*F(B)| = \sum_{B\in\pi}\varepsilon_B x^*F(B) = x^*\left\{\sum_{B\in\pi}\varepsilon_B F(B)\right\} \le \|\sum_{B\in\pi}\varepsilon_B F(B)\|. \blacksquare$$

Proposition 7.1.2 *Given a vector measure* $F : \Sigma \longrightarrow X$, *the semivariation of* F *on* $A \in \Sigma$ *satisfies*

$$\sup\{\|F(B)\|,\ B \subseteq A,\ B \in \Sigma\} \le \|F\|(A) \le 4\sup\{\|F(B)\|,\ B \subseteq A,\ B \in \Sigma\}$$

Proof. From $\|F\|(A) = \sup_{\pi}\{\|\sum_{B\in\pi}\varepsilon_B F(B)\|\}$, it follows that

$$\sup\{\|F(B)\|,\ B \subseteq A,\ B \in \Sigma\} \le \|F\|(A).$$

On the other hand, if the Banach space X is real, $\pi = \{B_1, \dots, B_m\}$ is a partition of the member A of Σ by elements of Σ, $x^* \in X^*$ satisfies $\|x^*\| \le 1$, $\pi^+ = \{n \in \mathbb{N} : 1 \le n \le m,\ x^*F(B_n) \ge 0\}$ and $\pi^- = \{n \in \mathbb{N} : 1 \le n \le m,\ x^*F(B_n) \ge 0\}$, we have $\sum_{B\in\pi}|x^*F(B)| = \sum_{n\in\pi^+}|x^*F(B_n)| - \sum_{n\in\pi^-}|x^*F(B_n)| = x^*\{\sum_{n\in\pi^+}F(B_n)\} - x^*\{\sum_{n\in\pi^-}F(B_n)\} \le 2\sup\{\|F(B)\|,\ B \subseteq A,\ B \in \Sigma\}$. This fact, along with the definition of semivariation, shows that in the real case $\|F\|(A) \le 2\cdot\sup\{\|F(B)\|,\ B \subseteq A,\ B \in \Sigma\}$. If X is a complex Banach space, a similar argument gives an analogous estimate with the number 2 replaced by 4. ∎

Remark 7.1.3 By Proposition 7.1.2, *a vector measure* F *is of bounded semivariation on* Σ *if and only if its range is bounded in* X. For this reason, in this case we also say that F is a bounded measure.

Owing to the previous remark, we have two equivalent norms in $ba(\Sigma,X)$, the semivariation and the norm defined by $\sup\{\|F(B)\|,\ B \in \Sigma\}$, it being plain that $ba(\Sigma,X)$ is a Banach space with the second norm, as well as the subspace $ca(\Sigma,X)$ of the countably additive vector measures.

Using the word ‘space’ to mean ‘Hausdorff locally convex space’, as usual, in order to consider measures with values in a locally convex space some of the definitions are formally the same as in the case of measures with values in a Banach space, taking into account every one of the seminorms belonging to a fundamental system defining the topology of the locally convex space. For example a vector measure F on a σ-algebra Σ of subsets of a set Ω with values in a space E is called

- countably additive, if $F(\cup\{A_n,\ n \in \mathbb{N}\}) = \sum_{n=1}^{\infty} F(A_n)$ whenever $\{A_n,\ n \in \mathbb{N}\}$ is a sequence of pairwise disjoint elements of Σ;

• strongly additive, whenever given a sequence $\{A_n, n \in \mathbb{N}\}$ of pairwise disjoint elements of Σ, the series $\sum_{n=1}^{\infty} F(A_n)$ converges in E;

• exhaustive, whenever given a sequence $\{A_n, n \in \mathbb{N}\}$ of pairwise disjoint elements of Σ, $\lim_{n\to\infty} F(A_n) = 0$ in E.

• Bounded, if $\{F(A), A \in \Sigma\}$ is a bounded subset of E.

The relationship among these types of measures is

countably additive ⇒ strongly additive ⇒ exhaustive ⇒ bounded.

The first two implications are immediate. The third implication is due to the fact that if F is not bounded, then there is some absolutely convex neighbourhood of the origin U such that for each $n \in \mathbb{N}$ there is $E_n \in \Sigma$, with $F(E_n) \notin nU$. Now consider some $B \in \Sigma$ with $F(B) \notin (\lambda+1)U$ where $\lambda > 0$ is such that $F(\Omega) \in \lambda U$. Then, since $F(B) = F(\Omega) - F(\Omega \setminus B)$, it follows that $F(\Omega \setminus B) \notin U$. Now, F being not bounded, at least one of the sets $\{F(A \cap B), A \in \Sigma\}$ or $\{F(A \cap (\Omega\setminus B)), A \in \Sigma\}$ is not bounded in E. In the first case set $A_1 := \Omega\setminus B$, in the second set $A_1 := B$. In any case $A_1 \in \Sigma$, $F(A_1) \notin U$ and $\{F(A \cap (\Omega\setminus A_1)), A \in \Sigma\}$ is not bounded in E.

Replacing Ω by $\Omega\setminus A_1$ in the above we find $A_2 \in \Sigma$, $A_2 \subseteq \Omega\setminus A_1$, with $F(A_2) \notin U$ and $\{F(A \cap (\Omega\setminus(A_1 \cup A_2))), A \in \Sigma\}$ not bounded in E. Iterating this, we find a pairwise disjoint sequence $\{A_n, n \in \mathbb{N}\}$ of members of Σ such that $F(A_n) \notin U$ for each $n \in \mathbb{N}$ and hence F is not exhaustive.

On the other hand, let us recall that a space E is said to be

• locally complete, if each bounded closed absolutely convex subset B of E is a Banach disk, i.e., if E_B is a Banach space;

• B_r-complete, if, given any dense subspace G of $E'(\sigma(E',E))$ which meets every equicontinuous closed subset of $E'(\sigma(E',E))$ in a closed subset, then G coincides with E';

• dual locally complete, if $E'(\sigma(E',E))$ is locally complete;

• $\Gamma_r(\Lambda_r)$ if, given any quasicomplete (locally complete) subspace G of $E^*(\sigma(E^*,E))$ which meets E' in a dense subspace of $E'(\sigma(E',E))$, then G contains E'.

As we have said before, Γ_r-spaces are the maximal class of spaces satisfying the closed graph theorem when barrelled spaces are considered in the domain, while it is well known that B_r-complete spaces are the maximal class of spaces satisfying the closed graph theorem for nearly continuous linear mappings when a locally convex space is considered in the domain. On the other hand, we need to recall that Λ_r-spaces are the maximal class of spaces satisfying the closed graph theorem for weakly continuous linear mappings whenever dual locally complete spaces are located in the domain.

7.2 APPLICATIONS TO VECTOR MEASURES

In what follows we are going to obtain several conditions concerning countable additivity of a bounded vector measure.

Theorem 7.2.1 *Let* F *be a bounded additive measure on* Σ *with values in a space* E. *Suppose that* X *is a space with a* p*-net* W *such that each* $L \in W_p$ *has a locally convex topology* $\mathcal{T}_L$, *finer than that induced by* X, *under which* $L(\mathcal{T}_L)$ *is a* Γ_r*-space. If* T *is a linear mapping from* E *into* X *with closed graph, then there exists some* $G \in W_p$ *such that* $T \circ F$ *is a bounded* $G(\mathcal{T}_G)$*-valued finitely additive measure.*

Proof. Since F is bounded, the map $S : \ell_0^\infty(\Sigma) \longrightarrow E$ such that $S(e(A)) = F(A)$ for every $A \in \Sigma$ is continuous and hence the linear mapping $K = T \circ S$ has a closed graph. On the other hand by Theorem 5.3.13 there exists some $G \in W_p$ such that $K^{-1}(G)$ is barrelled and dense in $\ell_0^\infty(\Sigma)$. According to Exercise 1 of Chapter 5, the restriction of K to $K^{-1}(G)$ admits a linear continuous extension J in $\ell_0^\infty(\Sigma)$ with values in G and, since K has a closed graph, $K = J$. ∎

Theorem 7.2.2 *Let* T *be a continuous linear mapping of* ℓ^∞ *into a space* E. *If there exists a neighbourhood of the origin* U *in* E *such that the set* $\{n \in \mathbb{N},\ Te_n \notin U\}$ *is infinite, then* E *contains a copy of* ℓ^∞.

Proof. If we show that there exists an infinite subset $\mathbb{M}$ of $\mathbb{N}$ such that the restriction of T to $\ell^\infty(\mathbb{M})$ is an isomorphism in E, then E will contain a copy of ℓ^∞ and we will be done.

Without loss of generality we may assume that $Te_n \notin U\ \forall n \in \mathbb{N}$. Let W be an absolutely convex neighbourhood of the origin with $2W \subseteq U$ and denote by B the closed unit ball of ℓ^∞, $B = \{x \in \ell^\infty, \|x\|_\infty \le 1\}$. As T is continuous there is some positive integer p such that $TB \subseteq pW$. Let us show that there is an infinite subset $\mathbb{M}$ of $\mathbb{N}$ such that, if $x \in \ell^\infty(\mathbb{M})$ and $\|x\|_\infty = 1$, then $Tx \notin W$. This will imply that T is an isomorphism and the proof will be finished.

In order to see this, it is enough to show that, if for each $A \subseteq \mathbb{N}$ we set $B(A) := \{x \in \ell^\infty, x(n) = 0 \; \forall n \notin A\}$, then

(a) There is an infinite $\mathbb{M} \subseteq \mathbb{N}$ with $Tx \notin W$ for each $x \in B(\mathbb{M})$ such that $x(m) = 1$ for some $m \in \mathbb{M}$.

<u>Claim.</u> Condition (a) is equivalent to:

(b) There exists an infinite subset A of $\mathbb{N}$ such that, for each infinite subset A' of A, there is some $n \in A'$ such that $T(x + e_n) \notin W$ for each x in $B(A \setminus A')$.

<u>Proof</u> <u>of</u> <u>the</u> <u>claim.</u> (a) $\Rightarrow$ (b). Considering $A = \mathbb{M}$, then for any infinite subset A' of A and each $n \in A'$, $(x + e_n)(n) = 1 \; \forall x \in B(A \setminus A')$. Thus, by (a), $T(x + e_n) \notin W$.

(b) $\Rightarrow$ (a). Let $\{A_i, i \in \mathbb{N}\}$ be an infinite partition of A such that $|A_i| = \aleph_0$ for each $i \in \mathbb{N}$. By (b), for each $r \in \mathbb{N}$ there exists some $n_r \in A_r$ such that $T(x + e_{n_r}) \notin W$ if $x \in B(A \setminus A_r)$. Clearly $\mathbb{M} = \{n_r, r \in \mathbb{N}\}$ verifies (a).

Let us continue with the proof of the theorem. If the conclusion were false, that is, if E did not contain any copy of ℓ^∞ then (b) would be false and, applying p times the fact that (b) is false, we would be able to find a decreasing sequence of infinite subsets of $\mathbb{N} = A_0 \supseteq A_1 \supseteq \ldots \supseteq A_p$ such that for each $j \in A_i$ there exists some $x_j \in B(A_{i-1} \setminus A_i)$ such that $T(x_j + e_j) \in W$ for $1 \le i \le p$. Fixing now any $k \in A_p$, given that $k \in A_i$ for $1 \le i \le p$, there must be some $y_i \in B(A_{i-1} \setminus A_i)$, for $i = 1, \ldots, p$ such that $T(y_i + e_k) \in W$ for $1 \le i \le p$.

Taking into account that the supports of the functions y_i for $1 \le i \le p$ are pairwise disjoint and that $y_i \in B$ for $1 \le i \le p$, then we have that $pT(e_k) + T(\sum_{i=1}^{p} y_i) \in pW$. Hence $pT(e_k) \in -T(\sum_{i=1}^{p} y_i) + pW \subseteq pW + pW \subseteq pU$ and thus $Te_k \in U$, a contradiction. ■

Corollary 7.2.3 *Let* T *be a continuous linear mapping of* ℓ^∞ *into a space* E. *If* E *does not contain a copy of* ℓ^∞, *then* $\{Te_n, n \in \mathbb{N}\}$ *is a null sequence.*

Proof. If $\{Te_n, n \in \mathbb{N}\}$ does not converge to the origin, then there is a neighbourhood of the origin U in E such that the set $\{n \in \mathbb{N}, Te_n \notin U\}$ is infinite. Hence, by the previous theorem, E must contain a copy of ℓ^∞, which is a contradiction. ■

Theorem 7.2.4 *Let* F *be a finitely additive measure on* Σ *with values in a space* E *and let* H *be a* $\sigma(E',E)$*-total subset of* E'. *Suppose that* E *has a p-net* W *such that in each* $L \in W_p$ *there is a locally convex topology* $\mathcal{T}_L$ *finer than the relative topology, under which either* $L(\mathcal{T}_L)$ *is a* B_r*-complete space or a sequentially complete* Γ_r*-space not containing a copy of* ℓ^∞. *If* $u \circ F \in ca(\Sigma)$ *for each* $u \in H$, *then there is some* $G \in W_p$ *such that* F *is a* $G(\mathcal{T}_G)$*-valued countably additive vector measure.*

Proof. The mapping S from $\ell_0^\infty(\Sigma)$ into E defined by $S(e(A)) = F(A)$ for every $A \in \Sigma$ has a closed graph since by hypothesis $u \circ F \in ca(\Sigma)$ for each $u \in \langle H \rangle$. According to Theorem 5.3.13 there exists some $G \in W_p$ such that $S^{-1}(G)$ is barrelled and dense in $\ell_0^\infty(\Sigma)$. Since each B_r-complete space is a Γ_r-space, then the restriction of S to $S^{-1}(G)$ admits a continuous extension U in $\ell^\infty(\Sigma)$ with values in $G(\mathcal{T}_G)$.

We are going to show now that given a pairwise disjoint sequence $\{A_n, n \in \mathbb{N}\}$ of elements of Σ, then $\sum_n F(A_n)$ converges to $F(\bigcup\{A_n, n \in \mathbb{N}\})$ in E. We use Theorem 7.2.2 to show first that F is exhaustive. To this end, let L be the subspace of $\ell^\infty(\Sigma)$ of all those $f \in \ell^\infty(\Sigma)$ which are null on $\Omega \setminus \bigcup\{A_i, i \in \mathbb{N}\}$ and take a constant value in each A_i for every $i \in \mathbb{N}$. Clearly, L is isometric to ℓ^∞, hence if K denotes the isometry from ℓ^∞ onto L and T stands for the restriction of U on L, the mapping $T \circ K : \ell^\infty \longrightarrow G(\tau_G)$ is linear and continuous. As $G(\tau_G)$ does not contain a copy of ℓ^∞, according to Theorem 7.2.2 if follows that $\{T \circ Ke_n, n \in \mathbb{N}\}$ converges to the origin in E, that is $\lim_n F(A_n) = 0$.

Next we will show that F is strongly additive. This will establish that $\sum_n F(A_n)$ converges to $F(\bigcup\{A_n, n \in \mathbb{N}\})$ since $u \circ F \in ca(\Sigma)$ for $u \in H$

and H is weak* total in E'. But since $G(\tau_G)$ is sequentially complete it suffices to show that $\sum_n F(A_n)$ is a Cauchy series in E. Let us assume it were not. Then there would be a neighbourhood of the origin V in E such that for each $N \in \mathbb{N}$, we would be able to find some positive integers $m_N \geq n_N \geq N$ with $F(\bigcup\{A_i, n_N \leq i \leq m_N\}) \notin V$.

Considering N = 1, choose $m_1 \geq n_1 \geq 1$ and define $B_1 := \bigcup\{A_i, n_1 \leq i \leq m_1\}$. Then we consider $N = m_1+1$, choose $m_2 \geq n_2 \geq N$, and define $B_2 := \bigcup\{A_i, n_2 \leq i \leq m_2\}$. By recurrence we find a pairwise disjoint sequence of elements $\{B_n, n \in \mathbb{N}\}$ of Σ such that $F(B_n) \notin V$ for each $n \in \mathbb{N}$.

But this contradicts the assumption that F is exhaustive. Hence $\sum_n F(A_n)$ is a Cauchy series in $G(\tau_G)$ and the proof is complete. ■

Remark 7.2.5 In the last theorem the subspaces L belonging to W_p do not contain a copy of ℓ^∞ and all of them are Γ_r under a topology finer than the relative one. Now we are going to prove that this theorem can also be established if these subspaces L are Λ_r under a topology finer than the relative one. We recall the following well-known result of measure theory, which will be used later: (a) *Assume* $\{\lambda_n, n \in \mathbb{N}\}$ *is a sequence of elements of* $ca(\Sigma)$. *If* $\lim_n \lambda_n(A) = \lambda(A)$ *for each* $A \in \Sigma$, *then* $\lambda \in ca(\Sigma)$. This result states that $ca(\Sigma)(\sigma(ca(\Sigma),\ell_0^\infty(\Sigma)))$ is sequentially complete. The next proposition shows that if E is a dense barrelled subspace of $\ell_0^\infty(\Sigma)$ then $ca(\Sigma)(\sigma(ca(\Sigma),E))$ is also sequentially complete. Notice that if Σ is infinite there are bounded sequences in $ca(\Sigma)(\sigma(ca(\Sigma),\ell_0^\infty(\Sigma))$ without any adherent point. In fact, let $\{A_n, n \in \mathbb{N}\}$ be a sequence of non-empty pairwise disjoint elements of Σ, let t_n be a point of A_n and let δ_n be the Dirac measure on t_n. If $\lambda \in ca(\Sigma)$ we can find some $p \in \mathbb{N}$ such that if $M := \bigcup\{A_n, n \geq p\}$ then $|\lambda(M)| < 1/2$ and, therefore, for $n \geq p$ we have $|(\delta_n - \lambda)| \geq \delta_n(M) - |\lambda(M)| > 1/2$. By the way, it is well known that there are bounded sequences in $\ell^1(\sigma(\ell^1,\ell^\infty))$ without any adherent point.

Proposition 7.2.6 *If* E *is a dense barrelled subspace of* $\ell_0^\infty(\Sigma)$, *then the space* $ca(\Sigma)(\sigma(ca(\Sigma),E))$ *is sequentially complete and* $\ell_0^\infty(\Sigma)$ *is contained in the bounded closure of* E *with respect to the dual pair* $(E,ca(\Sigma))$.

Proof. The E-bounded subsets of $ca(\Sigma)$ are E-equicontinuous and, since E is dense in $\ell_0^\infty(\Sigma)$, they are $\ell_0^\infty(\Sigma)$-equicontinuous too. So, the E-bounded subsets of $ca(\Sigma)$ are $\ell_0^\infty(\Sigma)$-bounded and the second affirmation follows.

Let now D be an absolutely convex subset of $ca(\Sigma)$ which is bounded and closed under $\sigma(ca(\Sigma),E)$. Since D is $\sigma(ca(\Sigma),\ell_0^\infty(\Sigma))$-compact we have that the topologies and the translation invariant uniformities induced by both $\sigma(ca(\Sigma),\ell_0^\infty(\Sigma))$ and $\sigma(ca(\Sigma),E)$ coincide on D. Consequently, any $\sigma(ca(\Sigma),E)$-Cauchy sequence (λ_n) in $ca(\Sigma)$ is also $\sigma(ca(\Sigma),\ell_0^\infty(\Sigma))$-Cauchy and by result (a) above there is some $\lambda \in ca(\Sigma)$ such that $\lim_n \lambda_n = \lambda$ under $\sigma(ca(\Sigma),\ell_0^\infty(\Sigma))$. ■

Remark 7.2.7 In particular, when $\Omega = \mathbb{N}$ and $\Sigma = 2^{\mathbb{N}}$ we have that if E is any dense barrelled subspace of ℓ_0^∞ then the space $\ell^1(\sigma(\ell^1,E))$ is weakly sequentially complete.

Proposition 7.2.8 *Let* W *be a* p*-net contained in a space* F *and let* T *be a linear mapping from* $\ell_0^\infty(\Sigma)$ *into* F *with closed graph in the product* $\ell_0^\infty(\Sigma)(\sigma(\ell_0^\infty(\Sigma),ca(\Sigma))) \times F$. *If each* $L \in W_p$ *has a locally convex topology* $\mathcal{T}_L$ *finer than the relative topology such that* $L(\mathcal{T}_L)$ *is a* Λ_r*-space, then there is some* $G \in W_p$ *containing the range of* T *such that* T *is weakly continuous with respect to the dual pairs* $(\ell_0^\infty(\Sigma),ca(\Sigma))$ *and* $(G,G(\mathcal{T}_G)')$.

Proof. By Theorem 5.3.13 and Proposition 7.2.6, there is some G such that $E := T^{-1}(G)$ is dual locally complete with respect to the dual pair $(E,ca(\Sigma))$ and its bounded closure $\check{E}$ contains $\ell_0^\infty(\Sigma)$. Let us denote by S the restriction of T to E. Since S has a closed graph in the product $E(\sigma(E,ca(\Sigma)) \times (G,\mathcal{T}_G))$ then, by Exercise 2 of Chapter 5, the mapping S has a continuous extension K from $\ell_0^\infty(\Sigma)(\sigma(\ell_0^\infty(\Sigma),ca(\Sigma)))$ with values in $G(\sigma(G,G(\mathcal{T}_G)'))$. Now, the mappings T and K are continuous when a locally convex topology coarser than the initial one is considered in F and both coincide in E. This fact concludes the proof. ■

In the following result we will bear in mind that $\sigma(\ell^1,\ell_0^\infty)$ and $\|\cdot\|_1$ define on ℓ^1 the same converging sequences and the same compact subsets, and that a subset K of ℓ^1 is relatively compact if and only if

$$\lim_{n\to\infty} \left\{\sup\left(\sum_{j=n}^{\infty} |x_j|, \; x \in K\right)\right\} = 0.$$

Theorem 7.2.9 *If the series* $\sum_n x_n$ *of* E *is* $\sigma(E,E')$*-subseries convergent in* E, $\sum_n x_n$ *converges under* $\lambda(E,E')$, *where* $\lambda(E,E')$ *denotes the topology of the uniform convergence on the* $\sigma(E',E)$*-compact subsets of* E'.

Proof. By hypothesis, for every $u \in E'$ and every subseries $\sum_{k=1}^{\infty} x_{n_k}$ of $\sum_{n=1}^{\infty} x_n$, the series $\sum_{k=1}^{\infty} \langle x_{n_k}, u\rangle$ converges in $\mathbb{K}$. Consequently, the series $\sum_{n=1}^{\infty} \langle x_n, u\rangle$ is absolutely convergent in $\mathbb{K}$, that is $\sum_{n=1}^{\infty} |\langle x_n, u\rangle| < \infty$.

Denote by T the linear mapping from E' into ℓ^1 defined by $T(u) = \{\langle x_n, u\rangle, \; n \in \mathbb{N}\}$, which is well defined because of what we have seen above, and let $\theta = \{\theta_n, \; n \in \mathbb{N}\}$ be a sequence of zeros and ones. One has

$$\sum_{n=1}^{\infty} \theta_n (Tu)_n = \sum_{n=1}^{\infty} \theta_n \langle x_n, u\rangle = \langle \sum_{n=1}^{\infty} \theta_n x_n, u\rangle$$

since $\sum_n \theta_n \langle x_n, u\rangle$ is a subseries of $\sum_n \langle x_n, u\rangle$, $\theta \in \ell^{\infty}$ and the series $\sum_n x_n$ is (BM)-convergent in $E(\sigma(E,E'))$. Taking into account the fact that the elements of ℓ_0^{∞} are finite linear combinations of sequences of zeros and ones (the characteristic functions of the subsets of $\mathbb{N}$), it follows that T is $\sigma(E',E)$-$\sigma(\ell^1, \ell_0^{\infty})$ continuous. In fact, let A be a subset of $\mathbb{N}$ and let e(A) denote its indicator function. Assume $\theta_n = e(A)_n$ for each $n \in \mathbb{N}$, denote $\sum_n \theta_n x_n$ (weak convergence) by x and take a sequence $\{u^m, \; m \in \mathbb{N}\}$ in E' converging weakly* to some element u in E'. Then we have

$$\langle T(u^m - u), e(A)\rangle = \sum_{n=1}^{\infty} \theta_n \langle x_n, u^m - u\rangle = \langle \sum_{n=1}^{\infty} \theta_n x_n, u^m - u\rangle = \langle x, u^m - u\rangle \longrightarrow 0$$

Having established that T is $\sigma(E',E)$-$\sigma(\ell^1, \ell_0^{\infty})$ continuous, if K is a $\sigma(E',E)$-compact subset of E', then T(K) is $\sigma(\ell^1, \ell_0^{\infty})$-compact and hence $\|\cdot\|_1$-compact. Thus,

$$\lim_{n\to\infty} \sup \left\{ \sum_{j=n}^{\infty} |\langle x_j, \xi\rangle|, \; \xi \in K\right\} = 0.$$

Therefore,

$$\lim_{n\to\infty} \sup \left\{\sup_{m\geq n} \left|\sum_{j=n}^{m} \langle x_j, \xi\rangle\right|, \; \xi \in K\right\} \leq \lim_{n\to\infty} \sup \left\{\sup_{m\geq n} \sum_{j=n}^{m} |\langle x_j, \xi\rangle|, \; \xi \in K\right\} \leq$$

$$\lim_{n\to\infty} \sup\left\{\sum_{j=n}^{\infty} |<x_j,\xi|, \ \xi \in K\right\} = 0.$$

This proves that $\sum_n x_n$ is a $\lambda(E,E')$-Cauchy series in E, and hence $\{\sum_{j=1}^{n} x_j, \ n \in \mathbb{N}\}$ is a $\lambda(E,E')$-precompact subset of E. Thus, if x denotes the weak limit $\sum_n x_n$, then

$$M := \{\sum_{j=1}^{n} x_j, \ n \in \mathbb{N}\}\cup\{x\}$$

is a $\lambda(E,E')$-precompact subset of E. Finally, denoting by $\mathfrak{M}$ the family of all the $\lambda(E,E')$-precompact subsets of E and by $\mathfrak{N}$ the family of all the $\sigma(E',E)$-compact subsets of E', then each $A \in \mathfrak{M}$ is $\mathfrak{T}_{\mathfrak{N}}$-precompact. By Grothendieck's theorem $\mathfrak{T}_{\mathfrak{N}}$ and $\sigma(E,E')$ coincide on each $A \in \mathfrak{M}$. Thus, in particular $\lambda(E,E')$ and $\sigma(E,E')$ coincide on M. Therefore the series $\sum_{n=1}^{\infty} x_n$ converges to x in $\lambda(E,E')$. ■

Corollary 7.2.10 *Let Ω be a set and let Σ be a σ-algebra of subsets of Ω. If F is a weakly countably additive measure on Σ with values in a space E, then F is countably additive.*

Theorem 7.2.11 *Let F be a countably additive measure on Σ with values in a space E, and let W be a p-net in a space H such that each $L \in W_p$ has a locally convex topology $\mathcal{T}_L$, finer than the relative one, under which $L(\mathcal{T}_L)$ is a Λ_r-space. If T is a linear mapping from E into H with closed graph, there is some $G \in W_p$ such that $T\circ F$ is a $G(\mathcal{T}_G)$-valued countably additive measure.*

Proof. The mapping $S : \ell_0^{\infty}(\Sigma)(\sigma(\ell_0^{\infty}(\Sigma),ca(\Sigma))) \longrightarrow E(\sigma(E,E'))$ defined by $S(e(A)) = F(A)$ for every $A \in \Sigma$ is continuous, since $uF \in ca(\Sigma)$ for each $u \in E'$. Hence $J = T\circ S$ has a closed graph in $\ell_0^{\infty}(\Sigma)((\sigma(\ell_0^{\infty}(\Sigma),ca(\Sigma))) \times H$.

By Proposition 7.2.8 there is some $G \in W_p$ such that $J(\ell_0^{\infty}(\Sigma)) \subseteq G$ and $J : \ell_0^{\infty}(\Sigma)(\sigma(\ell_0^{\infty}(\Sigma),ca(\Sigma))) \rightarrow G(\sigma(G,G(\mathcal{T}_G)'))$ is continuous.

If $v \in G(\mathcal{T}_G)'$, the continuity of J implies that $v\circ T\circ F \in ca(\Sigma)$, and from Theorem 7.2.9 it follows that $T\circ F$ is $\mathcal{T}_G$-countably additive. ■

Corollary 7.2.12 *Let* F *be an additive measure on* Σ *with values in a space* W *and let* H *be a* $\sigma(W',W)$*-total subset of* W'. *Suppose that* W *has a p-net* $\mathcal{W}$ *such that each* $L \in \mathcal{W}_p$ *has a locally convex topology* $\mathcal{T}_L$, *finer than that induced by* W, *under which* $L(\mathcal{T}_L)$ *is a* Λ_r*-space. If* $u \circ F \in ca(\Sigma)$ *for each* $u \in H$, *there exists some* $G \in \mathcal{W}_p$ *such that* F *is a* $G(\mathcal{T}_G)$*-valued countably additive measure.*

Proof. The corollary follows directly from Theorems 7.2.4 and 7.2.9 by taking $E = W(\sigma(W,\langle H\rangle))$ and T as the identity map on E. ∎

7.3 APPLICATIONS TO SCALAR MEASURES

We end this chapter with some results concerning scalar measures. By analogy with the definition of a p-net in a space E, and for the sake of simplicity, we introduce the following definition.

Definition 7.3.1 *Given a positive integer* s, *we shall say that a countable family of subsets* $\mathcal{A} = \{\Sigma_{m_1 m_2 \ldots m_p},\ m_r \in \mathbb{N},\ 1 \le r \le p \le s\}$ *of the* σ*-algebra* Σ *is an s-net if* $\{\Sigma_m,\ m \in \mathbb{N}\}$ *is an increasing covering of* Σ *and* $\{\Sigma_{m_1 \ldots m_{j-1} m},\ m \in \mathbb{N}\}$ *is an increasing covering of* $\Sigma_{m_1 \ldots m_{j-1}}$ *for each* $m_j \in \mathbb{N}$, $2 \le j \le s$.

A subset M of *ba*(Σ) is said to be simply bounded in a subset $\mathcal{B}$ of Σ if, for every $A \in \mathcal{B}$, $\sup\{|\mu(A)|,\ \mu \in M\} < \infty$.

Theorem 7.3.2 *Let* $\mathcal{A} = \{\Sigma_{m_1 m_2 \ldots m_p},\ m_r \in \mathbb{N},\ 1 \le r \le p \le s\}$ *be an s-net of the* σ*-algebra* Σ. *Then, there are* s *positive integers* $p_1, p_2, \ldots, p_s$ *such that:*

(i) Every subset M *of* $ba(\Sigma)$ *which is simply bounded in* $\Sigma_{p_1 \ldots p_s}$ *is bounded in* $ba(\Sigma)$.

(ii) If $\mu_n \in ba(\Sigma)$ *for each* $n \in \mathbb{N}$ *and* $\{\mu_n(A),\ n \in \mathbb{N}\}$ *is a Cauchy sequence for every* $A \in \Sigma_{p_1 \ldots p_s}$, *then there is some* $\mu \in ba(\Sigma)$ *such that* $\lim_n \mu_n(A) = \mu(A)$, *for every* $A \in \Sigma$.

(iii) If, additionally, the subfamily $\Sigma_{p_1p_2\ldots p_s}$ is a σ-algebra on $A_{p_1p_2\ldots p_s} := \bigcup\{A,\ A \in \Sigma_{p_1p_2\ldots p_s}\}$, then the linear mapping from $ba(\Sigma)$ into $ba(\Sigma_{p_1\ldots p_s})$ defined by its restriction is a surjective isometry.

Proof. (i) For every $p \in \{1, \ldots, s\}$ and every $m_1, \ldots, m_p \in \mathbb{N}$, let $E_{m_1m_2\ldots m_{j-1}m_j}$ denote the subspace of $\ell_0^\infty(\Sigma)$ generated by $\{e(A),\ A \in \Sigma_{m_1m_2\ldots m_j}\}$. According to Theorem 5.3.13 there are s positive integers $p_1, p_2, \ldots, p_s$ such that $E_{p_1p_2\ldots p_s}$ is a dense barrelled subspace of $\ell_0^\infty(\Sigma)$. If M is simply bounded in $\Sigma_{p_1p_2\ldots p_s}$ then M is a bounded subset of $ba(\Sigma)(\sigma(ba(\Sigma),E_{p_1p_2\ldots p_s}))$ and, $E_{p_1p_2\ldots p_s}$ being barrelled and dense in $\ell_0^\infty(\Sigma)$, M is $\ell_0^\infty(\Sigma)$-equicontinuous. Hence $\sup\{|\mu|,\ \mu \in M\} < \infty$, i.e., M is a bounded subset of $ba(\Sigma)$.

(ii) By (i), $\{\mu_n,\ n \in \mathbb{N}\}$ is contained in a $\sigma(ba(\Sigma),\ell_0^\infty(\Sigma))$-compact absolutely convex subset N of $ba(\Sigma)$. This implies that the translation-invariant uniformities defined by the weak topologies $\sigma(ba(\Sigma),\ell_0^\infty(\Sigma))$ and $\sigma(ba(\Sigma),E_{p_1p_2\ldots p_s})$ are the same on N. So (ii) follows immediately.

(iii) Since $E_{p_1p_2\ldots p_s}$ and $\ell_0^\infty(A_{p_1\ldots p_s},\Sigma_{p_1\ldots p_s})$ are isometric and $\ell_0^\infty(A_{p_1p_2\ldots p_s},\Sigma_{p_1p_2\ldots p_s})$ is dense in $\ell_0^\infty(\Sigma)$, then (iii) is true.

This completes the proof. ■

Remark 7.3.3 Theorem 7.3.2 (i) has been deduced from Theorem 5.3.13 and we are going to show now that both theorems are indeed equivalent. In fact, we are going to conclude from Theorem 7.3.2 (i) that assuming that $W = \{E_{m_1m_2\ldots m_p},\ m_r \in \mathbb{N},\ 1 \le r \le p \le s\}$ is an s-net in $\ell_0^\infty(\Sigma)$, then there is some $L \in W_s$ that is barrelled and dense in $\ell_0^\infty(\Sigma)$.

For each $p \in \{1, \ldots, s\}$ and each $m_1, \ldots, m_p \in \mathbb{N}$, let us denote by $\Sigma_{m_1m_2\ldots m_p}$ the subset of Σ whose characteristic functions belong to $E_{m_1m_2\ldots m_p}$. Then $\mathcal{A} := \{\Sigma_{m_1m_2\ldots m_p},\ m_r \in \mathbb{N},\ 1 \le r \le p \le s\}$ is an s-net in Σ, and if we suppose that there is $p_1, \ldots, p_s \in \mathbb{N}$ such that if a set M in $ba(\Sigma)$ is simply bounded in $\Sigma_{p_1p_2\cdots p_s}$ then $\sup\{|\mu|,\ \mu \in M\} < \infty$, it follows that the subspace $F_{p_1p_2\cdots p_s}$ of $\ell_0^{\infty}(\Sigma)$ generated by $\{e(A),\ A \in \Sigma_{p_1p_2\cdots p_s}\}$ is barrelled and according to the Hahn-Banach theorem dense in $\ell_0^{\infty}(\Sigma)$. Hence, since $F_{p_1p_2\cdots p_s} \subseteq E_{p_1p_2\cdots p_s}$, $E_{p_1p_2\cdots p_s}$ is barrelled and dense in $\ell_0^{\infty}(\Sigma)$ too.

7.4 EXERCISES

1. *Strongly additive measures.* Let E be a sequentially complete space and let $F : \Sigma \to E$ be an E-valued bounded measure defined on a σ-algebra Σ of sets.

(i) F is exhaustive if and only if F is strongly additive.

(ii) If F is bounded and E does not contain a copy of ℓ^{∞}, then F is strongly additive.

2. *Exhaustive measures.* Let E be a Γ_r-space. The following facts are equivalent.

(a) Every E-valued bounded measure on a σ-algebra Σ of sets is exhaustive.

(b) E does not contain a copy of ℓ^{∞}.

3. *A property of strongly additive vector measures.* If Ω is a set, Σ a σ-algebra of subsets of Ω and X is a Banach space, let $F : \Omega \to X$ be a vector measure. If $F(\Sigma)$ is a relatively weakly compact subset of X, then F is strongly additive.

(i) The closed linear span Y of the set $F(\Sigma)$ is a weakly compactly generated Banach space.

(ii) Each subspace of Y has a weak* sequentially compact dual ball. *[Hint: This is a consequence of a well-known theorem of D. Amir and J. Lindenstrauss.]*

(iii) Y does not contain any copy of ℓ^{∞}. *[Hint: Use Exercise 1 of Chapter 4.]*

(iv) F is strongly additive.

4. *An application of De Wilde's closed graph theorem.* Let Σ be a σ-algebra of sets and X be a space. Assume that $F : \Sigma \longrightarrow X$ is a bounded measure, let Y be a webbed space which does not contain any copy of ℓ^∞ and suppose that $T : X \longrightarrow Y$ is a linear mapping with closed graph.

(i) If X is sequentially complete, then $T\circ F$ is exhaustive.

(ii) If X contains an s-net W such that every $H \in W_s$ is a Γ_r-space, $T\circ F$ is exhaustive.

5. *Another property concerning exhaustive measures.* Suppose that Σ is a σ-algebra of sets and X is a space. Now, let $F : \Sigma \longrightarrow X$ be a countably additive measure with bounded variation and suppose that the space X contains an s-net W such that each $L \in W_s$ is a Λ_r-space. Assume that Y is a webbed space that does not contain a copy of ℓ^∞. If T is a linear mapping from X into Y with closed graph, then $T\circ F$ is exhaustive.

(i) The space $K(\Sigma)$ of all bounded Σ-measurable functions with countable range provided with the supremum norm (see Exercise 7 of Chapter 5) is ultrabornological.

(ii) Each series in the space X which is subseries convergent is BM-convergent.

(iii) $T\circ F$ is an exhaustive vector measure.

6. *Strongly additive measures with values in the dual of a Banach or a Fréchet space.* Let Σ be a σ-algebra of subsets of a set Ω.

(i) Assume E is a Banach space with a separable strong dual E'_β and let $F : \Sigma \longrightarrow E'(\sigma(E',E)$ be a bounded measure. Then $F : \Sigma \longrightarrow E'(\beta(E',E))$ is strongly additive.

(ii) Let E be a Fréchet space such that $E'(\beta(E',E))$ does not contain a copy of ℓ_∞. If $F : \Sigma \longrightarrow E'(\sigma(E',E))$ is a bounded vector measure, then $F : \Sigma \longrightarrow E'(\beta(E',E))$ is strongly additive.

7. *Subcontinuous mappings.* A linear map T between two spaces E and F is said to be subcontinuous if $\sum_{n=1}^{\infty} Tx_n$ converges to $T\left\{\sum_{n=1}^{\infty} x_n\right\}$ when $\sum_{n=1}^{\infty} x_n$ is subseries convergent in E. In what follows let us suppose that T is a linear mapping with closed graph from a space E into a space F containing a p-net W of subspaces such that each $L \in W_p$ is a sequentially complete Γ_r-space which does not contain a copy of ℓ^∞.

(i) There is some $G \in W_p$ such that the map $\lambda : 2^{\mathbb{N}} \to F$ defined by $\lambda(A) = T\left(\sum \{x_n, n \in A\}\right)$ for each subset A of $\mathbb{N}$ is exhaustive and G-valued.

(ii) T is subcontinuous.

7.5 NOTES AND REMARKS

All properties concerning vector measures that have been collected in section 7.1 come from Chapter I of Diestel and Uhl [DU]. Theorem 7.2.1 strengthens the following result obtained by Valdivia [V5].

Theorem 7.5.1 *Let* F *be a bounded additive measure from a* σ*-algebra* Σ *on* Ω *into a space* E. *Let* $\{F_n, n \in \mathbb{N}\}$ *be an increasing sequence of* Γ_r*-spaces covering a space* F. *If* $T : E \longrightarrow F$ *is a linear mapping with closed graph, there is a positive integer* q *such that* $T \circ F$ *is a bounded additive* F_q*-valued measure on* Σ.

Theorem 7.2.4 provides a pleasant extension to the locally convex case of the implication (i) $\Rightarrow$ (iii) of the following well-known theorem of J. Diestel and B. Faires [DF].

Theorem 7.5.2 *Assume that* X *is a Banach space. The following statements concerning* X *are equivalent:*
(i) X *does not contain any copy of* ℓ^∞;
(ii) given any σ*-algebra* Σ *of subsets of a set* Ω *and any bounded additive map* $F : \Sigma \to X$, *then* F *is strongly bounded;*
(iii) given any σ*-algebra* Σ *of subsets of a set* Ω *and any additive map* $F : \Sigma \longrightarrow X$ *and a total family* $H \subseteq X'$, *if* x^*F *is countably additive for each* $x^* \in H$, *then* F *is countably additive.*

Another generalization of this theorem is given in 7.2.12, which does not use Rosenthal's theorem in the locally convex case. This avoids imposing over the Λ_r-spaces which defines the range measure space the condition that they do not contain any copy of ℓ^∞.

Theorem 7.3.2 (i) provides the previously announced generalization of the Nikodym-Grothendieck boundedness theorem which, as we know, asserts that each subset of $ba(\Sigma)$ which is simply bounded in Σ, is uniformly bounded. Sections 7.2 and 7.3 essentially follow Ferrando and López-Pellicer [FL8]. However we have included some basic results of Banach space theory, extended to the locally convex or even topological vector space case, as Rosenthal's ℓ^∞-theorem or Orlicz-Pettis' theorem. Theorem 7.2.1 comes from [FL8]. Theorem 7.2.2 and its corollary constitute Drewnowski's extension to topological vector space theory of the quoted result of Rosenthal. Rosenthal's ℓ^∞-theorem can be found in [Rs2], while Drewnowski's results can be seen either in [Dr1] or [Dr2]. Result (a) before Proposition 7.2.6 is part of Nikodym's convergence theorem and it can be found in Diestel [D]. Propositions 7.2.6 and 7.2.8 are taken again from [FL8]. Theorem 7.2.9 and its corollary constitute the locally convex space version of the Orlicz-Pettis theorem. Both have been chosen from G. Benett and N.J. Kalton [BK1]. Other extensions of the Orlicz-Pettis theorem to the locally convex case have been given by A. Grothendieck [G3] and C.W. McArthur [Mc]. Theorem 7.2.11 and Corollary 7.2.12 are from [FL8], and Theorem 7.3.2

has been taken from Ferrando and López-Pellicer [FL3].

De Wilde's closed graph theorem used in Exercise 3 can be found in Chapter 1, in [DW] or in [K2]. Part (i) of Exercise 4 is hard but interesting. Subcontinuous mappings were defined for the first time in Bennett and Kalton [BK3].

Obtaining properties of vector measures in terms of strong barrelledness properties of the space $\ell_0^\infty(\Sigma)$ by means of closed graph theorems was started by M. Valdivia [V5] and continued by B. Rodríguez-Salinas [Ro1] as well as by Pérez Carreras and Bonet [PB2]. All these results use either Valdivia's Γ_r-spaces or Pták's B_r-spaces in the range for the action of closed graph theorems. The usufruct of Valdivia's Λ_r-spaces in the range of our closed graph Theorem 7.2.11 and its corollary, or even the use of De Wilde's webbed spaces, as in Exercise 4, may be new at this juncture.

Finally, we also want to point out that results closely related to our Exercise 2 have been obtained by I. Labuda in [La].

8. The spaces $L^p(\mu, X)$, $1 \leq p \leq \infty$

8.1 MEASURABLE FUNCTIONS

In the following (Ω, Σ, μ) will stand for a complete finite measure space, and X will denote a normed space over the real or complex field $\mathbb{K}$. A function $h : \Omega \longrightarrow X$ is said to be Σ-simple if there is a finite partition $\{E_1, E_2, \ldots, E_n\}$ of Ω by n elements of Σ and n different vectors of X such that $h(\omega) = \sum_{1 \leq j \leq n} e(E_j)(\omega) \cdot x_j$ for each $\omega \in \Omega$. A mapping $f : \Omega \rightarrow X$ is called *μ-measurable* or *strongly measurable* if there is a sequence of X-valued simple functions (f_n) defined on Ω such that $\lim_n f_n(\omega) = f(\omega)$ almost everywhere in Ω, i.e. there is an $E \in \Sigma$ with $\mu(E) = 0$ such that $\lim_n \|f_n(\omega) - f(\omega)\| = 0$ for each $\omega \in \Omega \setminus E$. On the other hand, a function $f : \Omega \rightarrow X$ is said to be *weakly μ-measurable* or *scalarly μ-measurable* if the function $x^*f : \Omega \longrightarrow \mathbb{K}$ is μ-measurable for each $x^* \in X'$.

Theorem 8.1.1 (Pettis) *A function* $f : \Omega \longrightarrow X$ *is strongly measurable if and only if* f *is weakly measurable and* f *has an essentially separable range; i.e. there is some* $E \in \Sigma$, *with* $\mu(E) = 0$, *such that* $f(\Omega \setminus E)$ *is a norm separable set.*

Proof. Obviously if f is strongly measurable, f is weakly μ-measurable. Moreover, if (f_n) is a sequence of X-valued simple functions on Ω which converges to f in each point $\omega \in \Omega \setminus E$, with $E \in \Sigma$ such that $\mu(E) = 0$, and $\{x_{nm} : 1 \leq m \leq k(n)\}$ is the range of f_n, it is clear that given $\varepsilon > 0$ and $\omega \in \Omega \setminus E$, there is an $i(\omega) \in \mathbb{N}$ such that $\|f(\omega) - f_n(\omega)\| < \varepsilon$ for each $n \geq i(\omega)$. Assuming that $f_i(\omega) = x_{ij}$, then $\|f(\omega) - x_{ij}\| < \varepsilon$, which shows that $\{x_{nm} : 1 \leq m \leq k(n), n \in \mathbb{N}\}$ is a dense subset of $f(\Omega \setminus E)$.

Assume conversely that $f : \Omega \longrightarrow X$ is weakly measurable and there is some measure-null set $E \in \Sigma$ such that $f(\Omega \setminus E)$ is a separable subset of X. If (x_n) is a dense sequence in $f(\Omega \setminus E)$, according to the Hahn-

Banach theorem for normed spaces, for each x_n there exists some $x^*_n \in X'$ with $\|x^*_n\| = 1$ such that $x^*_n x_n = \|x_n\|$. Since $\|f(\omega)\| = \sup \{|y^*f(\omega)|, y^* \in X', \|y^*\| = 1\}$, then $|x^*_n f(\omega)| \le \|f(\omega)\|$ for each $\omega \in \Omega$ and each $n \in \mathbb{N}$; in particular for each $\omega \in \Omega \setminus E$ and each $n \in \mathbb{N}$. Consequently given $\varepsilon > 0$, if $\omega \in \Omega \setminus E$ there is some $m \in \mathbb{N}$ such that $\|f(\omega) - x_m\| < \varepsilon/2$. Therefore $\|f(\omega)\| \le \|f(\omega) - x_m\| + \|x_m\| \le \varepsilon/2 + |x^*_m x_m| = \varepsilon/2 + |x^*_m(f(\omega) - x_m)| + |x^*_m f(\omega)| \le \varepsilon/2 + \|f(\omega) - x_m\| + |x^*_m f(\omega)| \le \varepsilon + |x^*_m f(\omega)|$. This shows that $|x^*_m f(\omega)| \le \|f(\omega)\| \le \varepsilon + |x^*_m f(\omega)|$, hence $\|f(\omega)\| = \sup \{|x^*_m f(\omega)|, m \in \mathbb{N}\}$. Since each $x^*_m f(\omega)$ is μ-measurable, it follows that the scalar function $\omega \longrightarrow \|f(\omega)\|$ is μ-measurable. Now, choosing a fixed $n \in \mathbb{N}$ and reasoning with the function $h(\omega) = f(\omega) - x_n$, noticing that $\{x_m - x_n, m \in \mathbb{N}\}$ is dense in $h(\Omega \setminus E)$ and h is μ-measurable, we conclude that each function $\omega \longrightarrow \|f(\omega) - x_n\|$ is μ-measurable. Consequently, given some $\varepsilon > 0$, $E_n := \{\omega \in \Omega \setminus E : \|f(\omega) - x_n\| < \varepsilon\} \in \Sigma$ for each $n \in \mathbb{N}$. Furthermore, $\Omega \setminus E = \bigcup\{E_n, n \in \mathbb{N}\}$, since (x_n) is dense in $f(\Omega \setminus E)$. Set $E_0 = \emptyset$.

Define $g(\omega) = x_n$ if $\omega \in E_n \setminus \bigcup\{E_j : 0 \le j \le n-1\}$ while $g(\omega) = 0$ if $\omega \in E$, i.e. if $\omega \notin \bigcup\{E_n : n \in \mathbb{N}\}$. Thus if $\omega \in \Omega \setminus E$, there will exist a first $m \in \mathbb{N}$ such that $\omega \in E_m \setminus \bigcup\{E_j, 0 \le j \le m-1\}$, hence $g(\omega) = x_m$ and therefore $\|f(\omega) - g(\omega)\| = \|f(\omega) - x_m\| < \varepsilon$. Setting for each $n \in \mathbb{N}$

$$A_n(\varepsilon) := E_n \setminus \bigcup\{E_j : 0 \le j \le n-1\}$$

we have shown that for each $\varepsilon > 0$ there is a partition $\{A_n(\varepsilon) : n \in \mathbb{N}\}$ of $\Omega \setminus E$ by elements of Σ and a simple function $g_{n,\varepsilon} : \Omega \longrightarrow X$ defined in such a way that $g_{n,\varepsilon}(\omega) = x_i$ if $\omega \in A_i(\varepsilon)$, $1 \le i \le n$, and $g_{n,\varepsilon}(\omega) = 0$ if $\omega \notin \bigcup\{A_i : 1 \le i \le n\}$, such that for each $t \in \Omega$ there is some $m \in \mathbb{N}$ with $\|f(t) - g_{m,\varepsilon}(t)\| \le \varepsilon$. This means that f is μ-measurable. ∎

Corollary 8.1.2 *If* $f : \Omega \longrightarrow X$ *is* μ*-measurable, there is an* $E \in \Sigma$*, with* $\mu(E) = 0$*, such that, for each* $\varepsilon > 0$*, there exists a countably* X*-valued* μ*-measurable function* g_ε *with* $\sup \{\|f(\omega) - g_\varepsilon(\omega)\| : \omega \in \Omega \setminus E\} \le \varepsilon$.

Proof. Since f is strongly measurable, there is an $E \in \Sigma$ with $\mu(E) = 0$ such that $f(\Omega \setminus E)$ is separable. Denoting by (x_n) a dense sequence in $f(\Omega \setminus E)$, we have seen in the previous proof that we may construct a partition $\{A_n : n \in \mathbb{N}\}$ of $\Omega \setminus E$ by elements of Σ such that the function

$g_\varepsilon : \Omega \longrightarrow X$ defined by $g_\varepsilon(\omega) = x_n$ if $\omega \in A_n$ and $g_\varepsilon(\omega) = 0$ if $\omega \in E$ is μ-measurable and satisfies the required condition. ∎

8.2 THE BOCHNER INTEGRAL

Let $f : \Omega \longrightarrow X$ be a Σ-simple function, say $f(\omega) = \sum_{1\le j\le n} e(E_j)(\omega)x_j$. Then for each $E \in \Sigma$ we define the (Bochner) integral of f on E, which we will represent by $\int_E f \, d\mu$ or by $\int_E f(\omega) \, d\mu(\omega)$, as the vector $x_E \in X$ defined by $x_E = \sum_{1\le j\le n} \mu(E \cap E_j)x_j$. Any X-valued μ-measurable function f defined on Ω is said to be *Bochner integrable* in the completion $\hat{X}$ of X if there is a sequence (f_n) of X-valued simple functions defined on Ω such that

$$\lim_n \int_\Omega \|f_n(\omega) - f(\omega)\| \, d\mu(\omega) = 0.$$

In this case the *Bochner integral of* f on E, $\int_E f \, d\mu$, is defined by

$$\int_E f \, d\mu = \lim_n \int_E f_n \, d\mu.$$

It can be shown that this definition is independent of the particular sequence of simple functions (f_n) chosen, hence the Bochner integral is well defined. Notice that we may have $\int_E f \, d\mu \in \hat{X}$ even if f is X-valued.

The set of all (classes of) measurable functions $f : \Omega \longrightarrow X$ which are Bochner integrable is a linear space over $\mathbb{K}$ under the operations of sum of classes and product of a class of functions by a scalar, and it is denoted by $L^1(\mu,X)$. As usual we do not distinguish between a given function and the class that it represents. If $f \in L^1(\mu,X)$, there exists a simple function g such that $\int_\Omega \|f(\omega) - g(\omega)\| \, d\mu(\omega) < 1$, and therefore one has $\int_\Omega \|f(\omega)\| \, d\mu(\omega) \le 1 + \int_\Omega \|g(\omega)\| \, d\mu(\omega) < \infty$. This allows us to set $\|f\|_1 := \int_\Omega \|f(\omega)\| \, d\mu(\omega)$, and one easily verifies that this defines a norm on $L^1(\mu,X)$.

Theorem 8.2.1 *Let* $f : \Omega \longrightarrow X$ *be a (strongly) measurable function. Then* $f \in L^1(\mu,X)$ *if and only if* $\int_\Omega \|f(\omega)\| \, d\mu(\omega) < \infty$.

Proof. We have already seen that if $f \in L^1(\mu,X)$ then $\int_\Omega \|f(\omega)\|\, d\mu(\omega)$ is finite. Assume conversely that f is μ-measurable and $\omega \longrightarrow \|f(\omega)\|$ is an element of $L^1(\mu)$. According to Corollary 8.1.2 there is some $E \in \Sigma$ with $\mu(E) = 0$ such that for each *n* there exists a countably X-valued and μ-measurable function f_n on Ω such that

$$\sup \{\|f(\omega) - f_n(\omega)\| : \omega \in \Omega \setminus E\} \le \inf \{1/n, 1/(n\cdot\mu(\Omega))\}.$$

Since $\|f_n(\omega)\| \le \|f(\omega) - f_n(\omega)\| + \|f(\omega)\| \le 1/n + \|f(\omega)\|$ for each $\omega \in \Omega \setminus E$, we have that $\int_\Omega \|f_n\|\, d\mu < \infty$ for each $n \in \mathbb{N}$. Assume that we have (pointwise) $f_n(\omega) = \sum_{m=1}^{\infty} e(E_{nm})(\omega)x_{nm}$ for each $\omega \in \Omega \setminus E$, with $E_{nm} \cap E_{nk} = \emptyset$ if $m \ne k$, $E_{nm} \in \Sigma$, $\bigcup\{E_{nm} : m \in \mathbb{N}\} = \Omega \setminus E$ and $x_{nm} \in X$ for each $n, m \in \mathbb{N}$. Then, as the Lebesgue integral is countably additive,

$$\int_\Omega \|f_n(\omega)\|\, d\mu(\omega) = \int_{\Omega\setminus E} \|f_n(\omega)\|\, d\mu(\omega) = \int_{\cup_m E_{nm}} \|f_n(\omega)\|\, d\mu(\omega) =$$
$$\sum_m \int_{E_{nm}} \|f_n(\omega)\|\, d\mu(\omega) = \sum_m \int_{E_{nm}} \|x_{nm}\|\, d\mu(\omega) = \sum_m \|x_{nm}\|\mu(E_{nm})$$

Hence, because of the previous observation, $\sum_m \|x_{nm}\|\mu(E_{nm}) < \infty$ for each $n, m \in \mathbb{N}$. Let then $p(n) \in \mathbb{N}$ be such that $\sum_{m\ge p(n)} \|x_{nm}\|\mu(E_{nm}) < 1/n$ for each given *n* and set $g_n(\omega) := \sum_{1\le m\le p(n)} e(E_{nm})(\omega)x_m$. Then each g_n is simple and $\int_\Omega \|f_n(\omega) - g_n(\omega)\|\, d\mu(\omega) = \sum_{m\ge p(n)} \|x_{nm}\|\mu(E_{nm}) < 1/n$, for each $n \in \mathbb{N}$, implying that $\int_\Omega \|f(\omega) - g_n(\omega)\|\, d\mu(\omega) \le 2/n$. Hence $f \in L^1(\mu,X)$. ∎

Theorem 8.2.2 *If* $f \in L^1(\mu,X)$, *then* $\|\int_E f(\omega)\, d\mu(\omega)\| \le \int_E \|f(\omega)\|\, d\mu(\omega)$ *for each* $E \in \Sigma$. *Moreover* $\lim_{\mu(E)\to 0} \int_E f(\omega)\, d\mu(\omega) = 0$.

Proof. Supposing that $f(\omega) = \sum_{1\le j\le n} e(E_j)(\omega)x_j$ with $E_i \cap E_j = \emptyset$ if $i \ne j$ and $\bigcup_{1\le j\le n} E_j = \Omega$, where $E_j \in \Sigma$ for $1 \le j \le n$, we have $\|\int_E f(\omega)\, d\mu(\omega)\| = \|\sum_{1\le j\le n} \mu(E \cap E_j)\cdot x_j\| \le \sum_{1\le j\le n} \mu(E \cap E_j)\|x_j\| = \int_E \|f(\omega)\|\, d\mu(\omega)$ for each E belonging to Σ. In general, if $f \in L^1(\mu,X)$, there is a sequence (f_n) of simple functions with $\lim_n \int_E \|f(\omega) - f_n(\omega)\|\, d\mu(\omega) = 0$ and $\lim_n \int_E f_n\, d\mu = \int_E f\, d\mu$ for each $E \in \Sigma$. Consequently, given some $E \in \Sigma$, we have

$$\left\|\int_E f(\omega)\, d\mu(\omega)\right\| \le \left\|\int_E (f(\omega) - f_n(\omega))\, d\mu(\omega)\right\| + \int_E \|f_n(\omega)\|\, d\mu(\omega) \le$$

$$\left\|\int_E (f(\omega) - f_n(\omega))\, d\mu(\omega)\right\| + \int_E \|f(\omega) - f_n(\omega)\|\, d\mu(\omega) + \int_E \|f(\omega)\|\, d\mu(\omega)$$

Hence given $\varepsilon > 0$, choose $n \in \mathbb{N}$ such that $\left\|\int_E f\, d\mu - \int_E f_n\, d\mu\right\| < \varepsilon/2$ and $\int_\Omega \|f(\omega) - f_n(\omega)\|\, d\mu(\omega) < \varepsilon/2$, in order to get the conclusion.

The second part of the theorem is a consequence of the first. In fact, $\lim_{\mu(E)\to 0}\left\|\int_E f(\omega)\, d\mu(\omega)\right\| \le \lim_{\mu(E)\to 0}\int_E \|f(\omega)\|\, d\mu(\omega) = 0$, due to the function $\omega \longrightarrow \|f(\omega)\|$ belonging to $L^1(\mu)$ and this property holds for all Lebesgue integrable scalar functions. ■

Theorem 8.2.3 (Dominated convergence theorem) *Let* X *be a Banach space, and* (f_n) *be a sequence in* $L^1(\mu,X)$ *which converges almost everywhere to some* X*-valued function* f *defined on* Ω. *If there exists some* $g \in L^1(\mu)$ *such that* $\|f_n(\omega)\| \le g(\omega)$ *almost everywhere on* Ω *for every* $n \in \mathbb{N}$, *then* $f \in L^1(\mu,X)$, $\lim_n\int_\Omega \|f(\omega) - f_n(\omega)\|\, d\mu(\omega) = 0$ *and* $\lim_n\int_E f_n(\omega)\, d\mu(\omega) = \int_E f(\omega)\, d\mu(\omega)$ *in* X *for each* $E \in \Sigma$.

Proof. As f is the pointwise limit of a sequence (f_n) of μ-measurable functions, it follows that f is strongly measurable. Besides, given that $\|f(\omega)\| \le g(\omega)$ for almost all $\omega \in \Omega$, it follows that $\int_\Omega \|f(\omega)\|\, d\mu(\omega) < \infty$, so $f \in L^1(\mu,X)$. On the other hand, since $\|f_n(\omega) - f(\omega)\| \le 2g(\omega)$ for almost all $\omega \in \Omega$, a similar argument to that of the scalar case may be used to demonstrate that $\lim_n\int_\Omega \|f_n(\omega) - f(\omega)\|\, d\mu(\omega) = 0$. Finally, as $\left\|\int_E (f_n(\omega) - f(\omega))\, d\mu(\omega)\right\| \le \int_E \|f_n(\omega) - f(\omega)\|\, d\mu(\omega)$, as a consequence of the previous theorem, it follows that $\lim_n\int_E f_n\, d\mu = \int_E f\, d\mu$. ■

Exactly as in the scalar case one can use the preceding theorem in order to prove that, if X is a Banach space, $L^1(\mu,X)$ is a Banach space. On the other hand, defining the space $L^p(\mu,X)$ with $1 < p < \infty$ as the linear space over $\mathbb{K}$ of all those (classes of) X-valued μ-measurable functions f defined on Ω such that $\int_\Omega \|f(\omega)\|^p\, d\mu(\omega) < \infty$ endowed with the norm $\|f\|_p = \left\{\int_\Omega \|f(\omega)\|^p \, d\mu(\omega)\right\}^{1/p}$, then as in the scalar case, it can be easily shown that $L^p(\mu,X)$ is a Banach space whenever X is a Banach space. Finally, if $L^\infty(\mu,X)$ represents the linear space over the field $\mathbb{K}$ of all (classes of) X-valued μ-measurable essentially bounded functions with the essential supremum norm $\|f\|_\infty = \text{ess sup } \{\|f(\omega)\| : \omega \in \Omega\}$, then according to the Corollary 8.1.2, $L^\infty(\mu,X)$ is a dense subspace of $L^1(\mu,X)$, and $L^\infty(\mu,X)$ is a Banach space whenever X is a Banach space.

Theorem 8.2.4 (Hille) *Let* X *be a Banach space and* $f \in L^1(\mu,X)$. *If* T *is a bounded linear operator from* X *in any Banach space* Y, *the function* Tf *is Bochner integrable and* $T\int_E f(\omega)\, d\mu(\omega) = \int_E Tf(\omega)\, d\mu(\omega)$ *for each* $E \in \Sigma$.

Proof. If $f(\omega) = \sum_{1\le j\le n} e(E_j)(\omega)\cdot x_j$ is a simple function with pairwise disjoint supports in Σ, we have $T\left(\int_E f(\omega)\, d\mu(\omega)\right) = \sum_{1\le j\le n}\mu(E \cap E_j)\cdot Tx_j = \int_E Tf(\omega)\, d\mu(\omega)$. The general case can be obtained by continuity from this since, given $f \in L^1(\mu,X)$, there is a sequence (f_n) of simple functions such that $f_n \longrightarrow f$ in $L^1(\mu,X)$ and $\int_E f_n \, d\mu \longrightarrow \int_E f \, d\mu$ in X. Then from the second relation one has that $T\int_E f_n \, d\mu \longrightarrow T\int_E f \, d\mu$, and from the first one, $\lim_n \int_E \|Tf_n(\omega) - Tf(\omega)\| \, d\mu(\omega) \le \|T\|\cdot \lim_n \int_\Omega \|f_n(\mu) - f(\omega)\| \, d\mu(\omega) = 0$. Given that Tf_n is simple and Y-valued, $Tf \in L^1(\mu,Y)$ and $\lim_n \int_E Tf_n \, d\mu = \int_E Tf \, d\mu$, therefore

$$\int_E Tf \, d\mu = \lim_n \int_E Tf_n \, d\mu = \lim_n T\int_E f_n \, d\mu = T\int_E f \, d\mu. \; \blacksquare$$

8.3 BARRELLEDNESS OF THE SPACE $L^1(\mu,X)$

In this section we are going to prove that if (Ω, Σ, μ) is a finite measure space and X is a barrelled normed space, then $L^1(\mu,X)$ is barrelled. In the next sections we will see that on some occasions the space $L^p(\mu,X)$ may enjoy stronger barrelledness properties than X does, showing that $L^p(\mu,X)$, $1 \le p \le \infty$, is a barrelled space of class $\aleph_0$ whenever (Ω, Σ, μ) is an atomless finite measure space. As usual X will denote a normed space.

Lemma 8.3.1 *Assuming* X *is barrelled and* T *is a barrel in* $L^1(\mu,X)$ *which is not a neighbourhood of the origin in* $L^1(\mu,X)$, *then there exists a strictly increasing sequence* $\{n_i,\ i \in \mathbb{N}\}$ *of positive integers and a normalized sequence* $\{h_n,\ n \in \mathbb{N}\}$ *of functions in* $L^1(\mu,X)$ *whose supports are members of* Σ *and satisfy one of the following conditions:*

(i) the sequence $\{\text{supp } h_n,\ n \in \mathbb{N}\}$ *is pairwise disjoint, or*

(ii) the sequence $\{\text{supp } h_n,\ n \in \mathbb{N}\}$ *is decreasing, has void intersection and* $h_i \notin n_i T$ *for each* $i \in \mathbb{N}$.

Proof. Let us see first that there are $P_1 \in \Sigma$ and $x_1 \in X$ with $\|x_1\| = 1$ such that $e(P_1)x_1 \notin 2\mu(P_1)T$. Suppose on the contrary that for each $P \in \Sigma$ and each $x \in X$ with $\|x\| = 1$, $\frac{e(P)}{\mu(P)}x \in 2T$. Then, if $s = \sum_{i=1}^{n} y_i e(C_i)$ is a μ-simple function with $C_i \in \Sigma$, $\mu(C_i) \neq 0$, $C_i \cap C_j = \emptyset$, $1 \le i < j \le 1$, such that $\|s\| \le 1$, we have

$$\sum_{i=1}^{n} \|y_i\| \mu(C_i) \le 1$$

and hence

$$\sum_{i=1}^{n} y_i e(C_i) = \sum_{i=1}^{n} \|y_i\| \mu(C_i) \frac{e(C_i)}{\mu(C_i)} \frac{y_i}{\|y_i\|} \in 2T$$

since T is absolutely convex. As T is closed in $L^1(\mu,X)$, this implies that T absorbs the closed unit ball of $L^1(\mu,X)$, a contradiction.

Setting $f_1 := \frac{e(P_1)}{\mu(P_1)} x_1$, then $f_1 \notin 2T$.

Let μ_1 be the restriction of μ on $\Sigma_{|P_1}$, let μ_2 be the restriction of μ on $\Sigma_{|(\Omega \setminus P_1)}$ and identify the spaces $L^1(\mu_1,X)$ and $L^1(\mu_2,X)$ with their embeddings into $L^1(\mu,X)$. An analogous argument to the one given above implies that either T does not absorb

$$\left\{ \frac{e(Q)}{\mu(Q)} x,\ \|x\| = 1,\ Q \subseteq P_1,\ Q \in \Sigma \right\}$$

or T does not absorb

$$\left\{ \frac{e(Q)}{\mu(Q)} x,\ \|x\| = 1,\ Q \subseteq \Omega \setminus P_1,\ Q \in \Sigma \right\}.$$

In the first case we set $A_1 := P_1$ and $\lambda_1 := \mu_1$. In the second case we set $A_1 := \Omega \setminus P_1$ and $\lambda_1 := \mu_2$. Then, again there is some $x_2 \in X$ with

$\|x_2\| = 1$ and some $P_2 \subseteq A_1$, $P_2 \in \Sigma$, such that $e(P_2)x_2 \notin 4\mu(P_2)T$.

Set $f_2 := \dfrac{e(P_2)}{\mu(P_2)} x_2 \notin 4T$.

If μ_3 stands for the restriction of μ on $\Sigma_{|P_2}$ and μ_4 for the restriction of μ on $\Sigma_{|(A_1 \setminus P_2)}$, since $L^1(\lambda_1,X)$ is the topological direct sum of $L^1(\mu_3,X)$ and $L^1(\mu_4,X)$, it follows that either T does not absorb

$$\left\{ \frac{e(Q)}{\mu(Q)} x,\ \|x\| = 1,\ Q \subseteq P_2,\ Q \in \Sigma \right\}$$

or T does not absorb

$$\left\{ \frac{e(Q)}{\mu(Q)} x,\ \|x\| = 1,\ Q \subseteq A_1 \setminus P_2,\ Q \in \Sigma \right\}.$$

In the first case we set $A_2 := P_2$ and $\lambda_2 := \mu_3$. In the second case we set $A_2 := A_1 \setminus P_2$ and $\lambda_2 := \mu_4$.

Proceeding by recurrence we get a normalized sequence $\{f_n,\ n \in \mathbb{N}\}$ in $L^1(\mu,X)$ with $f_{n+1} \in L^1(\lambda_n,X)$, supp $f_n = P_n \in \Sigma$ and $f_n \notin 2nT$ for each $n \in \mathbb{N}$. Now if there are infinitely many pairwise disjoint elements $\{P_{n_i},\ i \in \mathbb{N}\}$ in the sequence $\{P_n,\ n \in \mathbb{N}\}$, defining $h_i := f_{n_i}$ for all $i \in \mathbb{N}$ we have case (i) of the alternative.

Otherwise there is some $m \in \mathbb{N}$ such that $P_{k+1} \subseteq P_k$ for each $k \geq m$. Hence, there is no loss of generality if we assume that $P_{i+1} \subseteq P_i$ for each $i \in \mathbb{N}$. Let $P := \bigcap\{P_i,\ i \in \mathbb{N}\}$. Then $\lim_{i\to\infty} \mu(P_i) = \mu(P)$.

Now two cases are in order

<u>Case 1</u>. $\mu(P) = 0$.

Working with the sequence $\{Q_i,\ i \in \mathbb{N}\}$, where $Q_i := P_i \setminus P$, $i \in \mathbb{N}$, instead of $\{P_i,\ i \in \mathbb{N}\}$, we have that f_i coincides with the function $\dfrac{e(P_i)}{\mu(P_i)} x_i$ in $L^1(\mu,X)$. Defining $h_i := f_i$ and $n_i := i$ for each $i \in \mathbb{N}$ we have case (ii) of the alternative.

<u>Case 2</u>. $\mu(P) \neq 0$.

In this case we have that $A_i = P_i$ for each $i \in \mathbb{N}$. As $\lim_{i\to\infty} \mu(P_i) \neq 0$ then the sequence $\{x_i/\mu(P_i),\ i \in \mathbb{N}\}$ in X is bounded. Hence we have that $e(P_n)x_n \notin 2n\mu(P)T$ for each $n \in \mathbb{N}$. As the mapping $x \longrightarrow e(A_1)x/\mu(A_1)$ is an isometry from X into $L^1(\mu,X)$, then the barrelledness of X and the boundedness of the sequence $\{x_i, i \in \mathbb{N}\}$ in X give us some $k_1 > 1$ with $e(A_1)x_{k_1} \in k_1\mu(P)T$. Since $e(P_{k_1}) = e(A_1) - e(A\setminus P_{k_1})$, this implies that $e(A_1\setminus P_{k_1})x_{k_1} \notin k_1\mu(P)T$. Then we set $g_1 := e(A_1\setminus P_{k_1})x_{k_1}$.

Again, as $x \longrightarrow e(A_{k_1})x/\mu(A_{k_1})$ is an isometry from X into the space $L^1(\mu,X)$, there is some $k_2 > k_1$ such that $e(A_{k_1})x_{k_2} \in k_2\mu(P)T$. Hence, as above we have $e(A_{k_1}\setminus P_{k_2})x_{k_2} \notin k_2\mu(P)T$. So we set $g_2 := e(A_{k_1}\setminus P_{k_2})x_{k_2}$.

Proceeding by induction we obtain a sequence $\{g_n : n \in \mathbb{N}\}$ with pairwise disjoint supports and a strictly increasing sequence (k_i) in $\mathbb{N}$ with $g_i \notin k_i\mu(P)T$ for each $i \in \mathbb{N}$.

Since for each $i \in \mathbb{N}$, $0 < \|g_i\| \le \mu(\Omega)$,

$$\frac{g_i}{\|g_i\|} \notin k_i \frac{\mu(P)}{\mu(\Omega)} T.$$

So we can select a strictly increasing sequence $\{n_i, i \in \mathbb{N}\}$ of positive integers and a normalized subsequence (h_i) of (g_i) satisfying (i). ■

Lemma 8.3.2 *If* $\{h_i, i \in \mathbb{N}\}$ *is a normalized sequence in* $L^1(\mu,X)$ *with pairwise disjoint supports in* Σ, *then its closed linear hull* $[h_i]$ *in the completion of* $L^1(\mu,X)$ *is a copy of* ℓ^1 *contained in* $L^1(\mu,X)$.

Proof. If $c_1, c_2, \ldots, c_m$ are scalars, then

$$\left\| \sum_{j=1}^{m} c_j h_j \right\| = \int_\Omega \left\| \sum_{j=1}^{m} c_j h_j(\omega) \right\|_X d\mu(\omega) = \sum_{j=1}^{m} \int_S \left\| c_j h_j(\omega) \right\|_X d\mu(\omega) = \sum_{j=1}^{m} |c_j|$$

If $\hat{X}$ denotes the completion of X, then $L^1(\mu,X)$ is a dense subspace of the Banach space $L^1(\mu,\hat{X})$ since given $g \in L^1(\mu,\hat{X})$ and $\varepsilon > 0$, choosing some $s = \sum_{i=1}^{r} y_i e(C_i)$ μ-integrable and $\hat{X}$-valued with $C_i \cap C_j = \emptyset$ if $i \ne j$ such that $\|g - s\| < \varepsilon/2$ and some $x_i \in X$ with $\|y_i - x_i\| < \varepsilon/2r\mu(C_i)$ for $1 \le i \le r$, we have that $h := \sum_{i=1}^{r} x_i e(C_i) \in L^1(\mu,X)$ and $\|g - h\| < \varepsilon$.

The above equation shows that $\{h_j, j \in \mathbb{N}\}$ is a basic sequence in $L^1(\mu,\hat{X})$ which is equivalent to the unit vector basis $\{e_j, j \in \mathbb{N}\}$ of ℓ^1. Thus $[h_j]$ is isomorphic to the Banach space ℓ^1.

Finally we are going to show that $[h_j] \subseteq L^1(\mu,X)$. In fact, given a function $g = \sum_{j=1}^{\infty} h_j^*(g)h_j$ of $[h_j]$ and $\omega \in \Omega$, if $\omega \notin \bigcup\{\text{supp } h_k,\ k \in \mathbb{N}\}$ then $g(\omega) = 0$ and if $\omega \in \text{supp } h_r$ then $g(\omega) = h_r^*(g)h_r(\omega) \in X$. Consequently, g is X-valued and therefore $g \in L^1(\mu,X)$. ■

Lemma 8.3.3 *If* $\{h_i,\ i \in \mathbb{N}\}$ *is a normalized sequence in* $L^1(\mu,X)$ *with* $\text{supp } h_{i+1} \subseteq \text{supp } h_i$ *for each* $i \in \mathbb{N}$ *and* $\bigcap\{\text{supp } h_i,\ i \in \mathbb{N}\} = \emptyset$*, then the set* $\{\sum_{i=1}^{\infty} \xi_i h_i,\ \xi \in B_{\ell^1}\}$ *is a Banach disk in* $L^1(\mu,X)$.

Proof. Since $\{h_i,\ i \in \mathbb{N}\}$ is a normalized sequence in $L^1(\mu,X)$ the series $\sum_{i=1}^{\infty} \xi_i h_i$ converges in $L^1(\mu,Y)$, where Y denotes the completion of X. Let us now see that for fixed $\xi \in \ell^1$, the function $h := \sum_{i=1}^{\infty} \xi_i h_i$ belongs to $L^1(\mu,X)$. Let $\omega \in \Omega$. If $\omega \notin S_i$ for each $i \in \mathbb{N}$, then $h(\omega) = 0$ and if ω is in some S_j it may only belong to finitely many S_i. Hence, there is some $k \in \mathbb{N}$ such that $h(\omega) = \sum_{i=1}^{k} \xi_i h_i(\omega) \in X$.

The previous argument shows that the mapping $\varphi : \ell^1 \to L^1(\mu,X)$ such that $\varphi(\xi) = \sum_{i=1}^{\infty} \xi_i h_i$ is well defined. Since φ is continuous, it follows that $D := \{\sum_{i=1}^{\infty} \xi_i h_i,\ \xi \in B_{\ell^1}\}$ coincides with the image by φ of the closed unit ball of ℓ^1. So D is a Banach disk contained in $L^1(\mu,X)$. ■

Theorem 8.3.4 *If* X *is barrelled, then* $L^1(\mu,X)$ *is barrelled.*

Proof. Suppose that $L^1(\mu,X)$ is not barrelled. Then, there is a barrel T in $L^1(\mu,X)$ which does not absorb the closed unit ball of $L^1(\mu,X)$. Hence by virtue of Lemma 8.3.1 we obtain a normalized sequence $\{h_i,\ i \in \mathbb{N}\}$ in $L^1(\mu,X)$, whose supports are all members of Σ, and a strictly increasing sequence $\{n_i,\ i \in \mathbb{N}\}$ of positive integers with $h_i \notin n_i T$ for each $i \in \mathbb{N}$.

According to Lemma 8.3.2 and Lemma 8.3.3, either $[h_i]$ is a Banach subspace of $L^1(\mu,X)$ or $\{\sum_{i=1}^{\infty} \xi_i h_i,\ \xi \in B_{\ell^1}\}$ is a Banach disk in $L^1(\mu,X)$. Since the sequence $\{n_i T,\ i \in \mathbb{N}\}$ is formed by closed absolutely convex sets and covers $L^1(\mu,X)$, the boundedness of the sequence (h_i) leads to the existence of some $q \in \mathbb{N}$ such that $\{h_i, i \in \mathbb{N}\} \subseteq n_q T$. Thus $h_q \in n_q T$, a contradiction. ∎

8.4 STRONG BARRELLEDNESS PROPERTIES OF $L^p(\mu,X)$ SPACES, $1 \le p \le \infty$

Let us recall that a set $A \in \Sigma$ is an atom if $\mu(A) > 0$ and for each $E \in \Sigma$ with $E \subseteq A$, either $\mu(E) = 0$ or $\mu(E) = \mu(A)$. The measure μ is said to be atomic if there exists at least one atom, and atomless otherwise.

We start this section with a measure theoretic result.

Lemma 8.4.1 *If μ is an atomless positive measure,* $A \in \Sigma$ *and* $\mu(A) > 0$, *then* $\alpha := \inf \{\mu(E): E \in \Sigma, E \subseteq A, \mu(E) > 0\} = 0$.

Proof. Since μ is atomless, it is clear that $\mu(A) > \alpha \ge 0$. So, assuming $\alpha > 0$ we will find a decreasing sequence $\{E_n, n \in \mathbb{N}\}$ of members of Σ such that $\lim_{n\to\infty} \mu(E_n) = \alpha$. Let $\{\varepsilon_n, n \in \mathbb{N}\}$ be a decreasing sequence of real numbers converging to 0 such that $\varepsilon_1 < \mu(A) - \alpha$. By the definition of α there is some $B \in \Sigma$, $B \subseteq A$ with $\mu(B) < \alpha + \varepsilon_1$. Now

$$\alpha_1 := \inf \{\mu(E): E \in \Sigma, E \subseteq B, \mu(E) > 0\} \ge \alpha$$

and

$$\alpha_2 := \inf \{\mu(E): E \in \Sigma, E \subseteq A \setminus B, \mu(E) > 0\} \ge \alpha$$

Let $\delta := \min \{\alpha_1, \alpha_2\}$. If $\delta \ne \alpha$, by the definition of α there is some $C \in \Sigma$, $C \subseteq A$ such that $\mu(C) < \delta \le \alpha_i$, $i = 1, 2$. By the definition of α_i, $i = 1, 2$, $\mu(C \cap B) = 0$ and $\mu(C \cap (A\setminus B)) = 0$.

Thus, $\mu(C) = \mu(C \cap B) + \mu(C \cap (A\setminus B)) = 0$, a contradiction.

Therefore $\delta = \alpha$. If $\alpha_1 = \alpha$ set $E_1 := B$, if not set $E_1 := A \setminus B$.

Repeating the above argument with E_1 instead of A, we can find some $E_2 \in \Sigma$, $E_2 \subseteq E_1$ with $\mu(E_2) < \alpha + \varepsilon_2$ and

$$\inf \{\mu(E): E \in \Sigma, E \subseteq E_2, \mu(E) > 0\} \ge \alpha.$$

By recurrence we can find a decreasing sequence (E_n) of members of Σ such that $\lim_{n\to\infty} \mu(E_n) = \alpha$. Considering $E := \bigcap\{E_n, n \in \mathbb{N}\}$ then $E \in \Sigma$, $E \subseteq A$ and $E \subseteq E_n$ $\forall n \in \mathbb{N}$, so $\mu(E) = \alpha$. Now, E is an atom since if $F \subseteq E$, $F \in \Sigma$ and $\mu(F) > 0$, then by the definition of α, $\alpha \le \mu(F) \le \mu(E) = \alpha$, hence $\mu(F) = \mu(E)$. But this contradicts the atomless character of μ. ∎

Proposition 8.4.2 *If μ is an atomless positive finite measure and $\alpha \in [0, \mu(\Omega)]$, then there is some $A \in \Sigma$ such that $\mu(A) = \alpha$.*

Proof. The cases $\alpha = 0$ and $\alpha = \mu(\Omega)$ are trivial, so we will assume that $0 < \alpha < \mu(\Omega)$. We define in Σ the equivalence relation $A \simeq B$ if

$$\mu((A \setminus B) \cup (B \setminus A)) = 0.$$

In the quotient set $\hat{\Sigma} = \Sigma/\simeq$, we define the order relation $\hat{A} \le \hat{B}$ if $\mu(A\setminus B) = 0$. So, $\hat{\Sigma}$ becomes a non-empty partially ordered set. Let us consider the subset of $\hat{\Sigma}$, $\xi = \{\hat{A}, \mu(A) \le \alpha\}$, and let us see that every totally ordered subset of ξ has an upper bound.

Let $\{\hat{A}_i, i \in I\}$ be a chain of $\{\xi, \le\}$ and set

$$\rho := \sup \{\mu(A_i): \mu(A_i) \le \alpha, i \in I\}.$$

Now given $n = 1$ there is some $i_1 \in I$ such that $\mu(A_{i_1}) > \rho - 1$, and for each positive integer $n \ge 2$ there is some $i_n \in I$ such that $\hat{A}_{i_{n-1}} \le \hat{A}_{i_n}$ and $\mu(A_{i_n}) > \rho - \frac{1}{n}$. Set $B := \bigcup\{A_{i_n}, n \in \mathbb{N}\} \in \Sigma$; then it follows that $\mu(\bigcup\{A_{i_j}, 1 \le j \le n+1\}) = \mu(A_{i_{n+1}})$ and hence

$$\mu(B) = \lim_n \mu(A_{i_n}) = \rho \le \alpha.$$

Consequently $B \in \xi$.

Besides B is an upper bound of ξ. In fact given $i \in I$, if there is some $n \in \mathbb{N}$ such that $\hat{A}_i \le \hat{A}_{i_n}$, then $\hat{A}_i \le \hat{B}$, and if $\hat{A}_i$ is not smaller than any $\hat{A}_{i_n}$ then $\hat{A}_{i_n} \le \hat{A}_i$ $\forall n \in \mathbb{N}$. So,

$$\mu(B \setminus A_i) = \mu(\bigcup\{A_{i_n} \setminus A_i, n \in \mathbb{N}\}) \le \sum_{n=1}^{\infty} \mu(A_{i_n} \setminus A_i) = 0.$$

So $\hat{B} \le \hat{A}_i$ and $\rho = \mu(B) = \mu(B \cap A_i) \le \mu(A_i) \le \rho$. This implies that $\rho = \mu(B) = \mu(B \cap A_i) = \mu(A_i)$, hence $\mu(A_i\setminus B) = 0$ and therefore $\hat{B} = \hat{A}_i$.

Hence, by Zorn's lemma there exists a maximal element of $\{\xi, \le\}$, say $\hat{A}$. Let us see that $\mu(A) = \alpha$. Clearly $\mu(A) \le \alpha$. Assume $\mu(A) < \alpha$.

Then, $\mu(\Omega\setminus A) = \mu(\Omega) - \mu(A) \ge \alpha - \mu(A) > 0$, and by the previous lemma there must be some $E \in \Sigma$, $E \subseteq \Omega \setminus A$ with $0 < \mu(E) < \alpha - \mu(A)$.

Hence $\mu(A \cup E) = \mu(A) + \mu(E) < \alpha$ and $A \hat{\cup} E \in \xi$. From $\hat{A} \le A \hat{\cup} E$, it follows that $\hat{A} = A \hat{\cup} E$, since $\hat{A}$ is maximal. Consequently $\mu(E) = 0$. This is a contradiction. ∎

Theorem 8.4.3 *If μ is atomless, then $L^p(\mu,X)$, $1 \le p \le \infty$, is barrelled.*

Proof. Let us assume $L^p(\mu,X)$ is not barrelled. Then, there is a barrel T in $L^p(\mu,X)$ which is not a neighbourhood of the origin. We are going to define inductively a sequence $\{(\Omega_k,\Sigma_k,\mu_k),\ k \in \mathbb{N}\}$ of atomless finite measure spaces satisfying the following properties

(a) $\Omega_k \in \Sigma_{k-1}$, $\mu(\Omega_k) = \mu(\Omega_{k-1})/2$, $\Sigma_k = \Sigma_{k-1}|_{\Omega_k}$, $\mu_k = \mu_{k-1}|_{\Sigma_k}$

(b) T does not absorb the closed unit ball V_k of $L^p(\mu_k,X)$.

First of all, by Proposition 8.4.2 we may assume the existence of some $A \in \Sigma$ such that $\mu(A) = \mu(\Omega)/2$. By restricting Σ and μ to A and its complement $\Omega \setminus A$ we obtain the atomless measure spaces $(A,\mathcal{A},\lambda)$ and $(\Omega \setminus A,\mathcal{A}',\lambda')$ so that $L^p(\mu,X)$ is the topological direct sum of $L^p(\lambda,X)$ and $L^p(\lambda',X)$. If W and W' denote the closed unit balls of $L^p(\lambda,X)$ and $L^p(\lambda',X)$, respectively, then T either does not absorb W or it does not absorb W'. Consequently, if we set $(\Omega_1,\Sigma_1,\mu_1)$ to be either $(A,\mathcal{A},\lambda)$ or $(\Omega \setminus A,\mathcal{A}',\lambda')$, depending on whether T does not absorb W or does not absorb W', we have completed the first step of our induction process.

Now assuming that $(\Omega_i,\Sigma_i,\mu_i)$, $1 \le i \le k$, have already been defined we proceed to build $(\Omega_{k+1},\Sigma_{k+1},\mu_{k+1})$ with the previous requirements.

The atomless character of μ_k provides some $A_k \in \Sigma_k$ such that $\mu_k(A_k) = \mu_k(\Omega_k)/2$. Exactly as before, $L^p(\mu_k,X)$ may be decomposed as the topological direct sum of $L^p(\lambda_k,X)$ and $L^p(\lambda'_k,X)$, where $(A_k,\mathcal{A}_k,\lambda_k)$ and $(\Omega_k \setminus A_k,\mathcal{A}'_k,\lambda'_k)$ are the respective restrictions of Σ_k and μ_k to A_k and $\Omega_k \setminus A_k$. Since T does not absorb the closed unit ball V_k of $L^p(\mu_k,X)$ it cannot absorb the closed unit balls W_k and W'_k of $L^p(\lambda_k,X)$ and $L^p(\lambda'_k,X)$, respectively. So, choosing again the trio $(\Omega_{k+1},\Sigma_{k+1},\mu_{k+1})$ to be either $(A_k,\mathcal{A}_k,\lambda_k)$ or $(\Omega_k \setminus A_k,\mathcal{A}'_k,\lambda'_k)$ according to whether T does not absorb W_k or does not absorb W'_k, we are through with our induction process.

For each $k \in \mathbb{N}$ we choose some $g_k \in V_k \setminus kT$. Then the sequence $\{g_k,\ k \in \mathbb{N}\}$ is contained in the unit ball of $L^p(\mu,X)$, the support of each g_k being contained in Ω_k.

Now, if $\xi \in \ell^1$, $\sum_{k=1}^{\infty} \xi_k g_k$ converges in the completion $L^p(\mu,\hat{X})$ of

$L^p(\mu,X)$. But the function $\sum_{k=1}^{\infty} \xi_k g_k$ takes its values in X outside the zero measure set $\bigcap\{\Omega_k,\ k \in \mathbb{N}\}$. Consequently, $\sum_{k=1}^{\infty} \xi_k g_k \in L^p(\mu,X)$, and the continuity of the mapping from ℓ^1 into $L^p(\mu,X)$ defined by $\xi \longrightarrow \sum_{k=1}^{\infty} \xi_k g_k$ turns the set $D = \{\sum_{k=1}^{\infty} \xi_k g_k,\ \xi \in B_{\ell^1}\}$, where B_{ℓ^1} is the closed unit ball of ℓ^1, into a Banach disk. Since D is contained in $L^p(\mu,X)$ we have that T must contain a neighbourhood of the origin $(L^p(\mu,X))_D$. Hence there is some $n_0 \in \mathbb{N}$ such that $D \subseteq n_0 T$. But $g_{n_0} \in D \setminus n_0 T$, a contradiction.

This completes the proof of the theorem. ∎

Theorem 8.4.4 *If μ is atomless,* $L^p(\mu,X)$, $1 \le p \le \infty$, *is suprabarrelled.*

Proof. Suppose $L^p(\mu,X)$ is not a suprabarrelled space. Then there is an increasing sequence of non-barrelled subspaces $\{E_n,\ n \in \mathbb{N}\}$ which covers $L^p(\mu,X)$. For each $n \in \mathbb{N}$, assume T_n is a barrel in E_n which is not a neighbourhood of the origin in E_n. Let B_n denote the closure of T_n in $L^p(\mu,X)$ and let $H_n = \bigcap\{\langle B_m\rangle,\ m \ge n\}$.

Mimicking what was done in Theorem 8.4.3, a sequence $(\Omega_k,\Sigma_k,\mu_k)$ of atomless measure spaces can be constructed satisfying the same properties except the last one, which must be replaced by

(b') V_k is not contained in H_n for all k and n.

The only non-trivial step of the inductive process is the first one, which we reproduce now.

Since μ has no atoms, we split $L^p(\mu,X)$ in two halves, namely $L^p(\lambda,X)$ and $L^p(\lambda',X)$, corresponding to the atomless measure spaces $(A,\mathcal{A},\lambda)$ and $(\Omega \setminus A,\mathcal{A}',\lambda')$, where $A \in \Sigma$ and $\mu(A) = \mu(\Omega)/2$. If W and W' denote the closed unit balls of $L^p(\lambda,X)$ and $L^p(\lambda',X)$ respectively, we have that either W is not contained in H_n for each $n \in \mathbb{N}$ or W' is not contained in H_n for each $n \in \mathbb{N}$. Otherwise $L^p(\mu,X) \subseteq H_{n_0}$ for some n_0 and thus $L^p(\mu,X) \subseteq \langle B_{n_0}\rangle$. Hence, making use of Theorem 8.4.3, it is clear that B_{n_0} is a neighbourhood of zero in $L^p(\mu,X)$ and $T_{n_0} = B_{n_0} \cap E_{n_0}$ is a neighbourhood of the origin in E_{n_0}, a contradiction. So we now know how $(\Omega_1,\Sigma_1,\mu_1)$ has to be chosen.

For each $k \in \mathbb{N}$, select an element $g_k \in V_k$, $g_k \notin H_k$. The sequence (g_k) is contained in the closed unit ball of $L^p(\mu,X)$, with supp $g_k \subseteq \Omega_k$ for each $k \in \mathbb{N}$. Thus, as in the previous theorem, $D := \{\sum_{k=1}^{\infty} \xi_k g_k, \ \xi \in B_{\ell^1}\}$ is a Banach disk contained in $L^p(\mu,X)$. Due to the fact that the subspaces H_n cover $L^p(\mu,X)$ there exists some $n_0 \in \mathbb{N}$ such that, for each $n \geq n_0$, $H_n \cap E_D$ is barrelled and dense in $E_D := (L^p(\mu,X))_D$. Hence, since $H_n \subseteq \langle B_n \rangle$, we have that $\langle B_n \rangle \cap E_D$ is barrelled and dense in E_D for each $n \geq n_0$. Thus, $D \subseteq \langle B_n \rangle$ for each $n \geq n_0$ and $D \subseteq H_{n_0}$, contradicting the assumption that $g_{n_0} \in D \setminus H_{n_0}$. ■

Theorem 8.4.5 *If μ is atomless, then $L^p(\mu,X)$, $1 \leq p \leq \infty$, is barrelled of class $\aleph_0$.*

Proof. We already know that $L^p(\mu,X)$ is barrelled of class 1. Next we will assume that $L^p(\mu,X)$ is barrelled of class s-1 for some integer $s \geq 2$, and we are going to show that then it is barrelled of class s, too.

If this is not the case, we have an s-net W in $L^p(\mu,X)$, namely $W = \{E_{n_1\ldots n_p}, \ n_r \in \mathbb{N}, \ 1 \leq r \leq p \leq s\}$, of dense subspaces such that no $F \in W_s$ is barrelled. Suppose that, for each $(n_1,n_2,\ldots,n_s) \in \mathbb{N}^s$, $T_{n_1\ldots n_s}$ is a barrel in $E_{n_1\ldots n_s}$ which is not a neighbourhood of the origin in $E_{n_1\ldots n_s}$. Let $B_{n_1\ldots n_s}$ denote the closure of $T_{n_1\ldots n_s}$ in $L^p(\mu,X)$ and let $Z_{n_1\ldots n_s} := \langle B_{n_1\ldots n_s} \rangle$.

Going backwards, we define the following subspaces

$$H_{n_1\ldots n_s} = \bigcap\{Z_{n_1\ldots n_{s-1}m}, \ m \geq n_s\},$$

and for $j = s-1, s-2, \ldots, 1$,

$$Z_{n_1\ldots n_j} = \bigcup\{H_{n_1\ldots n_j m}, \ m \in \mathbb{N}\}$$

and

$$H_{n_1\ldots n_j} = \bigcap\{Z_{n_1\ldots n_{j-1}m},\ m \in \mathbb{N}\}.$$

It is plain that for $j = 1, 2, \ldots, s$, $E_{n_1\ldots n_j} \subseteq H_{n_1\ldots n_j}$ and the increasing sequence $\{H_{n_1\ldots n_j},\ n_j \in \mathbb{N}\}$ covers the subspace $E_{n_1\ldots n_{j-1}}$, taking $L^p(\mu,X)$ as E_{n_0}.

As before, our main point is to achieve a sequence $(\Omega_k,\Sigma_k,\mu_k)$ of atomless measure spaces with the same requirements as in the previous theorem, the last one still being that V_k is not contained in H_n, for all $k, n \in \mathbb{N}$. We will take some time to describe the first step of the induction process. With identical notation to that used in the previous theorem, we want to show that either W is not contained in H_n for each $n \in \mathbb{N}$ or W' is not contained in H_n for each $n \in \mathbb{N}$. Otherwise, there is an $n_1 \in \mathbb{N}$ such that $L^p(\mu,X) \subseteq H_{n_1} \subseteq Z_{n_1} = \bigcup\{H_{n_1m},\ m \in \mathbb{N}\}$. The induction hypothesis then plays its role to guarantee the existence of $n_2 \in \mathbb{N}$ for which $H_{n_1n_2}$ is dense and barrelled of class s-2, and the same can be said for $Z_{n_1n_2} = \bigcup\{H_{n_1n_2m},\ m \in \mathbb{N}\}$. Now, continuing this selecting argument, we can assert that there is some $(n_1,n_2,\ldots,n_s) \in \mathbb{N}^s$ such that $H_{n_1\ldots n_s}$ is dense and barrelled, hence $Z_{n_1\ldots n_s}$ is also dense and barrelled and $B_{n_1\ldots n_s}$ is a neighbourhood of zero in $Z_{n_1\ldots n_s}$. This implies that $T_{n_1\ldots n_s} = B_{n_1\ldots n_s} \cap E_{n_1\ldots n_s}$ is a neighbourhood of the origin in $E_{n_1\ldots n_s}$, a contradiction.

For each $k \in \mathbb{N}$ we choose an element $g_k \in V_k \setminus H_k$. Using again the previous notation, the set $D := \{\sum_{k=1}^{\infty} \xi_k g_k,\ \xi \in B_{\ell^1}\}$ is a Banach disk contained in $L^p(\mu,X)$.

Now, there is some $j_0 \in \mathbb{N}$ such that for each $n_1 \geq j_0$, $H_{n_1} \cap E_D$ is barrelled of class s-1 and dense in E_D. Since $\{H_{n_1n_2},\ n_2 \in \mathbb{N}\}$ is an increasing covering of H_{n_1}, there is some $j_{n_1} \in \mathbb{N}$ such that, for each $n_2 \geq j_{n_1}$, $H_{n_1n_2} \cap E_D$ is barrelled of class s-2 and dense in E_D. An appropriate number of repetitions tells us that, whenever $n_1 \geq j_0$, $n_2 \geq j_{n_1}$, ..., $n_{s-1} \geq j_{n_1 \ldots n_{s-2}}$, there is some $j_{n_1 \ldots n_{s-2} n_{s-1}} \in \mathbb{N}$ such that for $n_s \geq j_{n_1 \ldots n_{s-2} n_{s-1}}$, $H_{n_1 \ldots n_{s-2} n_{s-1} n_s} \cap E_D$ is barrelled and dense in E_D. Then $Z_{n_1 \ldots n_{s-2} n_{s-1} n_s} \cap E_D$ is also barrelled and dense in E_D for $n_1 \geq j_0$, $n_2 \geq j_{n_1}$, ..., $n_s \geq j_{n_1 \ldots n_{s-2} n_{s-1}}$, and $D \subseteq \langle B_{n_1 \ldots n_{s-1} n_s} \rangle = Z_{n_1 \ldots n_s}$. So, $D \subseteq \bigcap \{Z_{n_1 \ldots n_s}, n_1 \geq j_0, n_2 \geq j_{n_1}, \ldots, n_s \geq j_{n_1 \ldots n_{s-1}}\}$, which is a subset of H_{j_0}. This contradicts the choice of g_{j_0}. ■

Corollary 8.4.6 *Let* f *be a mapping with closed graph from* $L^p(\mu,X)$ *into the space* F. *Let* W *be an s-net in* F *and suppose that every* $L \in W_s$ *has a finer topology* $\mathcal{T}_L$ *finer than that induced by* F *such that* $L(\mathcal{T}_L)$ *is a* Γ_r*-space. Then there is some* $L \in W_s$ *such that* f *is a continuous mapping from* $L^p(\mu,X)$ *in* $L(\mathcal{T}_L)$.

Proof. This is an easy consequence of the last theorem and Exercise 3 of section 5.4. ■

8.5 EXERCISES

1. *An alternative proof of the barrelledness of the space* $L^p(\mu,X)$ *with* $1 \leq p < \infty$. Let (Ω,Σ,μ) be a finite measure space and let X be a normed space. The measure μ is called purely atomic if there is a countable family $\{E_n, n \in J\}$ of pairwise atoms of μ with $\mu\left(\Omega \setminus \bigcup\{E_n, n \in J\}\right) = 0$.

(i) If (Ω,Σ,μ) is a finite measure space and μ has some atoms there are two finite measures ν and λ on Σ such that ν is atomless, λ is purely atomic and $\nu + \lambda = \mu$.

(ii) If (Ω,Σ,μ) is a finite measure space and X is a normed space there are two subsets Ω_1 and Ω_2 in Σ and two finite measures μ_1 and μ_2 on $\Sigma_1 = \Sigma_{|A_1}$ and $\Sigma_2 = \Sigma_{|A_2}$ respectively, μ_1 atomless and μ_2 purely atomic, such that $L^p(\mu,X)$ is isometric to the product space $L^p(\mu_1,X) \times L^p(\mu_2,X)$ for each $p \in [1, \infty]$.

(iii) Let (Ω,Σ,μ) be a finite measure space, X be a normed space and assume $1 \leq p < \infty$. If μ is purely atomic, $L^p(\mu,X)$ is isometric either to

X^n for some $n \in \mathbb{N}$ or to the space $\ell^p\{X\}$ of all the sequences (x_n) in X such that $\sum \|x_n\|^p < \infty$ provided with the norm

$$\|(x_n)\| = \left(\sum_{n=1}^{\infty} \|x_n\|^p\right)^{1/p}.$$

(iv) Let X be a normed space. If $1 \le p < \infty$, then $\ell^p\{X\}$ is barrelled if and only if X is barrelled.

(v) If $1 \le p < \infty$ and X is a normed space the space $L^p(\mu,X)$ is barrelled whenever X is barrelled.

2. *On the structure of the barrels of* $L^1(\mu,X)$. Assume that (Ω,Σ,μ) is a finite measure space and X is a normed space. Assume that T is a barrel in $L^1(\mu,X)$.

(i) T is not a neighbourhood of the origin in $L^1(\mu,X)$ if and only if there exists a strictly increasing sequence (n_i) of positive integers and a normalized sequence (h_n) of functions of $L^1(\mu,X)$ such that $h_i \notin n_i T$ for each $i \in \mathbb{N}$, whose supports (S_n), all of them members of Σ, satisfy one of the following conditions: (1) the sequence (S_n) is pairwise disjoint or, (2) the sequence (S_n) is decreasing.

(ii) If X is barrelled, then T is not a neighbourhood of the origin in $L^1(\mu,X)$ if and only if there exists a strictly increasing sequence (n_i) of positive integers and a normalized sequence (h_n) of functions of $L^1(\mu,X)$ such that $h_i \notin n_i T$ for each $i \in \mathbb{N}$ whose supports (S_n), all them members of Σ, satisfy one of the following conditions: (1) the sequence (S_n) is pairwise disjoint, or (2) the sequence (S_n) is decreasing and verifies that $\bigcap\{S_n,\ n \in \mathbb{N}\} = \emptyset$.

3. *Barrelledness of the space* $L^\infty(\mu,X)$. Assume that (Ω,Σ,μ) is a finite measure space and let X be a normed space.

(i) If μ is purely atomic, then the space $L^\infty(\mu,X)$ is isometric either to X^n for some $n \in \mathbb{N}$ or to the space $\ell^\infty(X)$ of all bounded sequences in X endowed with the supremum norm.

(ii) The space $\ell^\infty(X)$ is barrelled if and only if X is barrelled.

(iii) Suppose that μ is arbitrary. Then the space $L^\infty(\mu,X)$ is barrelled whenever X is barrelled.

4. *Strong barrelledness properties in* $L^p(\mu,X)$*, with* $1 \le p \le \infty$. Suppose (Ω,Σ,μ) is a finite measure space and let X be a normed space. Given a positive integer n, if $1 \le p < \infty$ we shall denote by $\ell^p\left\{\{n+1,n+2,\dots\},X\right\}$ the subspace of $\ell^p\{X\}$ of all sequences whose support is contained in $\{n+1,n+2,\dots\}$ of $\mathbb{N}$. If $p = \infty$ we write $\ell^\infty\left(\{n+1,n+2,\dots\},X\right)$ and $\ell^\infty(X)$.

(i) If $\{B_n,\ n \in \mathbb{N}\}$ is a sequence of closed absolutely convex sets in $\ell^p\{X\}$ or in $\ell^\infty(X)$ covering X, there is a $q \in \mathbb{N}$ such that $\langle B_q\rangle$ contains $\ell^p\left\{\{q+1,q+2,\dots\},X\right\}$ or $\ell^\infty\left(\{q+1,q+2,\dots\},X\right)$, respectively.

(ii) If X is barrelled of class $s \in \mathbb{N}$, then both $\ell^p\{X\}$ and $\ell^\infty(X)$ are barrelled of class s.

(iii) If X is a barrelled space of class s, then the space $L^p(\mu,X)$ is barrelled of class s for each $p \in [1, \infty]$.

(iv) If X is ultrabornological, then $L^p(\mu,X)$ is ultrabornological.

5. *When is* $L^p(\mu,X)$ *an unordered Baire-like space?* Assume (Ω,Σ,μ) is a finite measure space, and let X be a normed space.

(i) If X is unordered Baire-like, then $L^1(\mu,X)$ is unordered Baire-like.

(ii) If $L^p(\mu,X)$, with $1 < p < \infty$, is totally barrelled, then $L^p(\mu,X)$ is unordered Baire-like whenever X is unordered Baire-like.

6. *Barrelledness of the space* $bvca(\Sigma,X)$. Assume X is a normed space and Σ is a σ-algebra of subsets of a set Ω. Denote by $bvca(\Sigma,X)$ the linear space over $\mathbb{K}$ of all X-valued countably additive measures of bounded variation provided with the variation norm $\|F\| = |F|(\Omega)$, where for each $E \in \Sigma$ we have

$$|F|(E) = \sup\{\textstyle\sum_{A\in\pi}\|F(A)\|,\ \pi \in \Pi(E)\}.$$

If μ is a positive countably additive measure, i.e., if $\mu \in ca^+(\Sigma)$, we denote by $bvca(\Sigma,\mu,X)$ the linear subspace of $bvca(\Sigma,X)$ of all μ-continuous measures. When $\mu \in ca^+(\Sigma)$ and Z is a Banach space, define $T : L^1(\mu,Z) \longrightarrow bvca(\Sigma,\mu,Z)$ by $Tf(E) = (B)\int_E f\, d\mu$ for $E \in \Sigma$.

(i) T is a (linear) isometry into a subspace of $bvca(\Sigma,\mu,Z)$.

(ii) A Banach space Z is said to have the Radon-Nikodym property with respect to a positive measure μ if $T : L^1(\mu,Z) \longrightarrow bvca(\Sigma,\mu,Z)$ is an onto isometry. Assuming the completion $\hat{X}$ of X has the Radon-Nikodym property with respect to each $\mu \in ca^+(\Sigma)$, then $bvca(\Sigma,X)$ is barrelled if and only if X is barrelled.

8.6 NOTES AND REMARKS

Sections 8.1 and 8.2 are based on Diestel [D] and Diestel and Uhl [DU]. Our proof of the barrelledness of the space $L^p(\mu,X)$ with p = 1, μ being an arbitrary positive finite scalar measure and X a barrelled normed space, has been given since it seems to provide certain information on the topological structure of this space. This has been emphasized in Exercise 2. The case $p = \infty$ has been considered in Exercise 3. An analogous treatment to the one followed in section 8.3 for p = 1 seems not be possible for p > 1. However, this can by done by using the methods developed by Ferrando [Fe3] for the case $p = \infty$.

The use of our Theorem 8.4.5 along with the techniques described in Exercise 4 allows us to conclude that the space $L^p(\mu,X)$ with $1 \le p \le \infty$ and (Ω,Σ,μ) an arbitrary finite measure space is barrelled of class $\aleph_0$ whenever X is a barrelled space of class $\aleph_0$. A proof of this fact for $p = \infty$ may be found in [FS5], while part (iv) of Exercise 4 is from [DFFP2]. Other results concerning ultrabornological properties of Bochner integrable function spaces can be found in [DFFP2] and [Fe4].

Our Theorem 8.4.3 provides a unitary proof of the following two results.

Theorem 8.6.1 *Let* (Ω,Σ,μ) *be a finite measure space and suppose that* X *is a normed space. If the measure* μ *is atomless, then the space* $L^p(\mu,X)$ *with* $1 \le p < \infty$ *is barrelled.*

Theorem 8.6.2 *Let* (Ω,Σ,μ) *be a finite measure space and let* X *be a normed space. If* μ *is atomless, then* $L^\infty(\mu,X)$ *is barrelled.*

Theorem 8.6.1 has been proved in Drewnowski, Florencio and Paúl [DrFP1] and Theorem 8.6.2 can be found in Díaz, Florencio and Paúl [DFP]. Other results concerning the space $L^\infty(\mu,E)$, E being a locally convex space and (Ω,Σ,μ) a Radon measure space, where μ is a positive regular Borel measure defined on a complete σ-algebra Σ containing the Borel subsets of a Hausdorff locally compact and σ-compact topological space Ω, may be found in [DFFP1]. Assuming (Ω,Σ,μ) is a finite measure space and X a normed space, a study of the barrelledness properties of the space $\Lambda(X)$, $\Lambda(X)$ being a complete solid lattice of scalar functions defined on Ω, of strongly measurable X-valued functions f such that $\|f(\cdot)\| \in \Lambda$, is performed in [DrFP2].

The use of a Banach disk that is a continuous image of the closed unit ball of ℓ^1 has been taken from Díaz et al. [DFP]. Our Lemma 8.4.1 has been chosen from De Guzmán and Rubio's text [DGR], as well as Proposition 8.4.2, although these are two well-known facts. The exposition of the chapter from Theorem 8.4.3 until the end has been taken from Ferrando, Ferrer and López-Pellicer [FFL].

Barrelledness properties in certain vector-valued sequence spaces have been investigated by M. Florencio and P.J. Paúl in [FP] in the context of locally convex spaces. One of their main results is going to be enunciated after some preliminaries: let λ a perfect sequence space and suppose that X is a normed space. A sequence (x_n) in X is said to be absolutely λ-summable if $(\|x_n\|) \in \lambda$, where $\|\cdot\|$ stands for the norm of X. We denote by $\lambda\{X\}$ the linear space of all absolutely λ-summable sequences from X provided with the topology defined by the family of seminorms $\{p_M, \ M \in \mathcal{M}\}$, where $\mathcal{M}$ is the family of all normal $\sigma(\lambda^\times,\lambda)$-bounded subsets of $\lambda^\times$ and

$$p_M((x_n)) := \sup \{\textstyle\sum_{n\geq 1} |\alpha_n| \|x_n\| : (\alpha_n) \in M\}$$

By $\varphi(X)$ we shall denote the linear subspace of $X^{\mathbb{N}}$ of all those sequences with finite support. As λ is perfect, we have that $\varphi(X) \subseteq \lambda\{X\}$. Finally, we set λ_r to denote the closure of φ in λ under the strong topology $\beta(\lambda,\lambda^\times)$. Under all these circumstances, Florencio and Paúl's result can be enunciated as follows.

Theorem 8.6.3 *The space $\lambda_r(X)$ is barrelled if and only if X is a barrelled space.*

If $\lambda = \ell^p$ with $1 \leq p < \infty$, then λ_r coincides with ℓ^p and hence the previous result implies that each space $\ell^p\{X\}$ is barrelled if and only if X is barrelled. This is just part (iv) of our Exercise 1. In our Exercise 4 we propose a technique to prove that $\ell^p\{X\}$ is in fact a barrelled space of class n if and only if X is barrelled of class n. Exercise 5 comes from [Fe7].

9. The spaces $\ell_0^\infty(\Sigma, E)$ and $B(\Sigma, X)$

9.1 INTRODUCTION

Throughout this chapter, Ω will be a set and Σ an algebra of subsets of Ω. Up to now we have studied $\ell_0^\infty(\Sigma)$ spaces of scalar-valued functions. In this chapter we will investigate the barrelledness properties of some vector valued function spaces. In section 9.2 we will study $\ell_0^\infty(\Sigma, E)$ spaces of Σ-simple functions with values in a Hausdorff locally convex space E, Σ being an infinite algebra, while in 9.3 we will study the space of X-valued functions defined on a set Ω which are the uniform limit of a sequence of elements of $\ell_0^\infty(\Sigma, X)$, Σ being a σ-algebra of subsets of Ω and X being a normed space.

9.2 BARRELLEDNESS CONDITIONS OF THE SPACE $\ell_0^\infty(\Sigma, X)$

Given a (Hausdorff locally convex) space E, a sequence $\{x_n, n \in \mathbb{N}\}$ in E is said to be absolutely summable if the series $\sum_{n=1}^{\infty} p(x_n)$ converges for each continuous seminorm p on E. We will denote by $\ell^1(E)$ the set of all the absolutely summable sequences of E. In what follows Φ_0 will stand for the subspace of c_0 of all those elements with only a finite number of non-null coordinates, provided with the topology induced by c_0. We identify ℓ^1 with the dual space of Φ_0 and, as usual, $\{e_n, n \in \mathbb{N}\}$ will denote the canonical basis of c_0 and ℓ^1. On the other hand, given two spaces E and F and two neighbourhoods of the origin U and V in E and F respectively, for each $z = \sum_{i=1}^{n} x_i \otimes y_i \in E \otimes F$ we write

$$\varepsilon_{UV}(z) = \sup \left\{ \left| \sum_{i=1}^{n} \langle x_i, u\rangle \langle y_i, v\rangle \right|, \ u \in U^\circ, \ v \in V^\circ \right\}.$$

Lemma 9.2.1 *Let* E *and* F *be two spaces and assume* F *contains a copy of* Φ_0*. Let* D *be a subset of* $\ell^1(E'(\sigma(E', E)))$ *such that*

$$\sup \left\{\sum_{n=1}^{\infty} |v_n(x)|,\ (v_n) \in D\right\} = M_x < \infty$$

for every $x \in E$. *If* $E \otimes_{\varepsilon} F$ *is barrelled then there exists a neighbourhood of the origin* U *in* E *such that* $\sum_{n=1}^{\infty} p_{U^{\circ}}(v_n) \leq 1$ *for every* $(v_n) \in D$, *the polar of* U *being taken in* E'.

Proof. Let F_0 be a subspace of F such that there is a topological isomorphism T of F_0 into Φ_0 and let T' be the transpose of T from ℓ^1 into F_0'. Let W be a neighbourhood of the origin in F such that $W \cap F_0 = T^{-1}(B)$, where B denotes the closed unit ball of Φ_0. Now, for each $n \in \mathbb{N}$, set $f_n^0 := T^{-1}(e_n)$ and $f_n^{0\prime} := T'(e_n)$. Then for every $f_0 \in F_0$,

$$|f_n^{0\prime}(f_0)| = |e_n(Tf_0)| \leq \|Tf_0\|_{\infty} = p_W(f_0).$$

Hence, by the Hahn-Banach theorem, there exists some $f_n' \in F'$ such that its restriction to F_0 coincides with $f_n^{0\prime}$ and

$$|f_n'(f)| \leq p_W(f)$$

for each $f \in F$.

Now we consider the subset of $E' \otimes F' \subseteq (E \otimes_{\varepsilon} F)'$

$$Q := \left\{\sum_{n=1}^{m} v_n \otimes f_n',\ (v_n) \in D,\ m \in \mathbb{N}\right\},$$

which is $\sigma((E \otimes_{\varepsilon} F)', E \otimes_{\varepsilon} F)$-bounded since for each $x \in E$, $f \in F$ and $q = \sum_{n=1}^{m} v_n \otimes f_n' \in Q$ we have

$$|q(x \otimes f)| \leq \sum_{n=1}^{m} |v_n(x) f_n'(f)| \leq p_W(f) \sum_{n=1}^{\infty} |v_n(x)| \leq p_W(f) \cdot M_x.$$

By hypothesis $E \otimes_{\varepsilon} F$ is barrelled, so Q is equicontinuous and there must be some neighbourhood of the origin U in E and V in F such that $|q(z)| \leq 1$ for every $q \in Q$ and $z \in E \otimes F$ if $\varepsilon_{UV}(z) \leq 1$.

Assume λ is a positive real number such that $W \cap F_0 \subseteq \lambda(V \cap F_0)$, and given $(v_n) \in D$, for each $n \in \mathbb{N}$ pick some $u_n \in U$ such that $p_{U^{\circ}}(v_n) < v_n(u_n) + 2^{-n}$. This is possible because

$$p_{U^{\circ}}(v_n) = \sup \left\{ |\langle u, v_n \rangle|,\ u \in U \right\}.$$

Then for every $m \in \mathbb{N}$,

$$\varepsilon_{UV}\left(\sum_{n=1}^{m} u_n \otimes f_n^0\right) = \sup \left\{ \left|\sum_{n=1}^{m} \langle u_n, u \rangle \langle f_n^0, v \rangle\right|,\ u \in U^{\circ},\ v \in V^{\circ} \right\} =$$

$$\sup \{|v(\sum_{n=1}^{m} u(u_n)f_n^0)|,\ u \in U^o,\ v \in V^o\} \leq \lambda \sup \{p_W(\sum_{n=1}^{m} u(u_n)f_n^0),\ u \in U^o\} =$$

$$\lambda \cdot \sup \{\|\sum_{=1}^{m} u(u_n)e_n\|_\infty,\ u \in U^o\} = \lambda \cdot \sup \{|u(u_n)|,\ u \in U^o,\ 1 \leq n \leq m\} \leq \lambda.$$

On the other hand, if $q = \sum_{n=1}^{m} v_n \otimes f_n{}' \in Q$, $q(\sum_{n=1}^{m} u_n \otimes f_n^0) = \sum_{n=1}^{m} v_n(u_n)$.

Therefore,

$$\sum_{n=1}^{m} p_{U^\circ}(v_n) < \sum_{n=1}^{m} (v_n(u_n) + 2^{-n}) = q(\sum_{n=1}^{m} u_n \otimes f_n^0) + \sum_{n=1}^{m} 2^{-n} < \lambda + 1.$$

Thus, $\sum_{n=1}^{\infty} p_{U^\circ}(v_n) \leq \lambda + 1$ for every $(v_n) \in D$. Hence, if we consider the neighbourhood of the origin $U' = (1 + \lambda)^{-1}U$ instead of U, the proof is complete. ∎

Let P be the family of all the continuous seminorms on E. We shall represent by $\ell_\pi^1(E)$ the space $\ell^1(E)$ provided with the locally convex topology generated by the family of seminorms $\{q_p,\ p \in P\}$ such that for each $p \in P$ and $x = (x_n) \in \ell^1(E)$,

$$q_p(x) := \sum_{n=1}^{\infty} p(x_n).$$

The space E is said to have property (B) if for each bounded subset $D \subseteq \ell_\pi^1(E)$ there exists some closed absolutely convex bounded subset B in E such that, for every $(v_n) \in D$,

$$\sum_{n=1}^{\infty} p_B(v_n) \leq 1.$$

Lemma 9.2.2 *Let* E *and* F *be two spaces and suppose that* F *contains a copy of* Φ_0. *If* $E \otimes_\varepsilon F$ *is barrelled, then* $E'(\beta(E',E))$ *has property (B).*

Proof. Let D be a bounded subset of $\ell_\pi^1(E'(\beta(E',E)))$. According to the previous proposition there is a neighbourhood of the origin U in E such that $\sum_{n=1}^{\infty} p_{U^\circ}(v_n) \leq 1$ for every $(v_n) \in D$, where U^o is an equicontinuous $\sigma(E',E)$-closed absolutely convex subset of E'. As E is barrelled since it is isomorphic to a complemented subspace of $E \otimes_\varepsilon F$, it follows that U^o is strongly bounded. Hence U^o is a closed absolutely convex subset of $E'(\beta(E',E))$ and consequently $E'(\beta(E',E))$ has property (B). ∎

Proposition 9.2.3 *Let* E *and* F *be two spaces and suppose* F *is a normed barrelled space which contains a copy of* Φ_0. *Then,* $E \otimes_\varepsilon F$ *is barrelled if and only if* E *is barrelled and nuclear.*

Proof. If $E \otimes_\varepsilon F$ is barrelled then E is barrelled since it is a copy of a complemented subspace of $E \otimes_\varepsilon F$. Therefore, in order to prove that E is nuclear, by [Me2, 4.2.11, 4.3.3] and Lemma 9.2.2, it suffices to show that $\ell^1(E'(\beta(E',E))) = \ell^1(E'(\sigma(E',E)))$. Now, since it is obvious that $\ell^1(E'(\beta(E',E))) \subseteq \ell^1(E'(\sigma(E',E)))$, we only have to prove the other inclusion. Let $v = (v_n) \in \ell^1(E'(\sigma(E',E)))$, then

$$\sum_{n=1}^{\infty} |v_n(x)| = M_x < \infty$$

for each $x \in E$ and, by Lemma 9.2.1 considering $D = \{v\}$, there is some neighbourhood of the origin U in E such that

$$\sum_{n=1}^{\infty} p_{U^\circ}(v_n) \leq 1.$$

On the other hand, given a bounded subset A of E, there is some $\lambda > 0$ such that $A \subseteq \lambda U$, thus $\lambda A^\circ \supseteq U^\circ$ and

$$\sum_{n=1}^{\infty} p_{A^\circ}(v_n) \leq \lambda \sum_{n=1}^{\infty} p_{U^\circ}(v_n) \leq \lambda.$$

Hence $v \in \ell^1(E'(\beta(E',E)))$.

Conversely, if E is nuclear, $E \otimes_\varepsilon F = E \otimes_\pi F$ and the conclusion is immediate since the projective tensor product of a normed barrelled space and a barrelled space is barrelled. ■

Given an algebra Σ of subsets of a set Ω, denote by $\mathcal{U}$ a base of absolutely convex neighbourhoods of the origin of a space E, and let us denote by $\{p_U,\ U \in \mathcal{U}\}$ the family of gauges of all the elements of $\mathcal{U}$. We shall suppose $\ell_0^\infty(\Sigma,E)$ is equipped with the uniform convergence topology defined by the family $\{q_U,\ U \in \mathcal{U}\}$ such that, for each $f \in \ell_0^\infty(\Sigma,E)$,

$$q_U(f) := \sup\,\{p_U(f(t)),\ t \in X\}.$$

Lemma 9.2.4 *Given an algebra* Σ, $\ell_0^\infty(\Sigma) \otimes_\varepsilon E$ *is topologically isomorphic to* $\ell_0^\infty(\Sigma,E)$.

Proof. The linear mapping $\phi : \ell_0^\infty(\Sigma)\otimes_\varepsilon E \to \ell_0^\infty(\Sigma,E)$ such that

$$\phi(f \otimes x)(t) = f(t)\cdot x$$

for every $f \in \ell_0^\infty(\Sigma)$ and every $x \in E$ is surjective. We are going to show that ϕ is also injective. Let $z = \sum_{i=1}^{n} f_i \otimes x_i \in \ell_0^\infty(\Sigma)\otimes E$ such that $\phi(z) = 0$. Assuming the x_i's are linearly independent, for every $t \in X$ we have that $\sum_{i=1}^{n} f_i(t)x_i = 0$ implies that $f_i(t) = 0$ for $i = 1, 2, \ldots, n$. Hence $f_i \equiv 0$ for $1 \le i \le n$ and therefore $z = 0$.

In order to show that ϕ is a topological isomorphism, let B be the closed unit ball of $\ell_0^\infty(\Sigma)$, assume $\mathcal{U}$ is a base of absolutely convex neighbourhoods of the origin in E and let $\{\varepsilon_{BU}, U \in \mathcal{U}\}$ denote a fundamental system of seminorms in $\ell_0^\infty(\Sigma)\otimes_\varepsilon E$, with

$$\varepsilon_{BU}\left(\sum_{i=1}^{n} f_i \otimes x_i\right) = \sup\left\{\left|\sum_{i=1}^{n} \langle f_i,\mu\rangle\langle x_i,u\rangle\right|, \mu \in B^o, u \in U^o\right\}$$

As B^o is the weak* closure of the absolutely convex hull of the set $H = \{\delta_t, t \in X\}$ of the Dirac deltas and given that, for every fixed $u \in U$, $\sum_{i=1}^{n} \langle f_i,\mu\rangle\langle x_i,u\rangle$ is a convex function of μ, we have that

$$\varepsilon_{BU}\left(\sum_{i=1}^{n} f_i \otimes x_i\right) = \sup\left\{\left|\sum_{i=1}^{n} \langle f_i,\mu\rangle\langle x_i,u\rangle\right|, \mu \in H, u \in U^o\right\} =$$

$$\sup\left\{\left|\sum_{i=1}^{n} \langle f_i,\delta_t\rangle\langle x_i,u\rangle\right|, t \in X, u \in U^o\right\} =$$

$$\sup\left\{\left|\left\langle\sum_{i=1}^{n} f_i(t)x_i,u\right\rangle\right|, t \in X, u \in U^o\right\} = \sup\left\{p_U\left(\sum_{i=1}^{n} f_i(t)x_i\right), t \in X\right\} =$$

$$q_U\left(\phi\left(\sum_{i=1}^{n} f_i \otimes x_i\right)\right). \blacksquare$$

Theorem 9.2.5 *Let* E *be a space and let* Σ *be an infinite algebra of subsets of* Ω. *Then* $\ell_0^\infty(\Sigma,E)$ *is barrelled if and only if* $\ell_0^\infty(\Sigma)$ *and* E *are barrelled and* E *is nuclear.*

Proof. The result is an immediate consequence of Lemma 9.2.4, Proposition 9.2.3 and the fact that $\ell_0^\infty(\Sigma)$ is normed and contains a copy of Φ_0 (consider $\langle\{e(A_n), n \in \mathbb{N}\}\rangle$, where $\{A_n, n \in \mathbb{N}\}$ is a pairwise disjoint sequence in Σ). ■

Corollary 9.2.6 *Let* E *be a space,* s *a positive integer and* Σ *an infinite algebra of subsets of* Ω*. Then* $\ell_0^\infty(\Sigma,E)$ *is barrelled of class* s *if and only if* $\ell_0^\infty(\Sigma)$ *and* E *are barrelled of class* s *and* E *is nuclear.*

Proof. If $\ell_0^\infty(\Sigma)$ and E are barrelled of class s and E is nuclear, the conclusion follows from Remark 4.3.2. Conversely, if $\ell_0^\infty(\Sigma,E)$ is barrelled of class s then $\ell_0^\infty(\Sigma)$ is also barrelled of class s, since it is a quotient of $\ell_0^\infty(\Sigma,E)$. On the other hand E is nuclear because of Theorem 9.2.5. ∎

Corollary 9.2.7 *Let* E *be a space and* Σ *an infinite algebra of subsets of* Ω*. Then* $\ell_0^\infty(\Sigma,E)$ *is barrelled of class* $\aleph_0$ *if and only if* $\ell_0^\infty(\Sigma)$ *and* E *are barrelled of class* $\aleph_0$ *and* E *is nuclear.*

Proposition 9.2.8 *Given an infinite algebra* Σ*,* $\ell_0^\infty(\Sigma,E)$ *is covered by a sequence* $\{F_n,\ n \in \mathbb{N}\}$ *of closed linear subspaces such that the codimension of* F_n *in* $\ell_0^\infty(\Sigma,E)$ *is greater than or equal to* dim E.

Proof. Let $\{A_n,\ n \in \mathbb{N}\}$ be an infinite sequence of non-empty pairwise disjoint subsets of X such that $A_n \in \Sigma$ for every $n \in \mathbb{N}$ and let $\{t_n,\ n \in \mathbb{N}\}$ be a sequence of points of X with $t_n \in A_n$ for each $n \in \mathbb{N}$. Let $\{S_n,\ n \in \mathbb{N}\}$ be the sequence of all pairs (p,q) of positive integers with $p > q$ with the diagonal ordering. If $S_n := \{t_p, t_q\}$ we denote by F_n the closed linear subspace of $\ell_0^\infty(\Sigma,E)$ formed by all $f \in \ell_0^\infty(\Sigma,E)$ such that $f(t_p) = f(t_q)$.

If $S_n = \{t_p,\ t_q\}$ and $\{x_i,\ i \in I\}$ is an algebraic basis of E, the functions $\{e(A_p)x_i,\ i \in I\}$ are linearly independent with respect to F_n and thus codim $F_n \geq$ dim E. The sequence $\{F_n,\ n \in \mathbb{N}\}$ covers $\ell_0^\infty(\Sigma,E)$ since if $f \in \ell_0^\infty(\Sigma,E)$ there are two different positive integers r and s, with $r > s$, and some vector $x \in E$ such that $f(t_r) = f(t_s) = x$. Now if $S_m = \{t_r, t_s\}$, then $f \in F_m$. ∎

9.3 BARRELLEDNESS CONDITIONS OF THE SPACE B(Σ,X)

Next we will obtain some results concerning strong barrelledness properties of the space $B(\Sigma,X)$ of the X-valued functions defined on a set Ω that are the uniform limit of a sequence of elements of $\ell_0^\infty(\Sigma,X)$, Σ being a σ-algebra of subsets of Ω and X being a normed space, these two spaces being endowed with the topology defined by the usual supremum-norm $\|f\| = \sup\{\|f(\omega)\|, \omega \in \Omega\}$.

In the following we will denote by $K(\Sigma,X)$ the dense linear subspace of $B(\Sigma,X)$ formed by those functions that take at most a countable number of different values. Given $A \in \Sigma$, the spaces $\ell_0^\infty(\Sigma_{|A},X)$ and $K(\Sigma_{|A},X)$ will be denoted by $\ell_0^\infty(A,X)$ and $K(A,X)$, respectively.

If $f \in \ell_0^\infty(\Sigma,X)$, obviously there is a partition $\{A_1, A_2, \ldots, A_n\}$ of Ω of non-empty elements of Σ and different vectors $\{x_1, x_2, \ldots, x_n\}$ of X such that $f(\omega) = x_i$ if $\omega \in A_i$, $1 \le i \le n$. A similar result will be deduced for $f \in K(\Sigma,X) \setminus \ell_0^\infty(\Sigma,X)$ after the following lemma.

Lemma 9.3.1 *Let* $f \in K(\Sigma,X) \setminus \ell_0^\infty(\Sigma,X)$. *If* $x \in \operatorname{Im} f$, *then* $f^{-1}(x) \in \Sigma$.

Proof. Let $\{f_n, n \in \mathbb{N}\}$ be a sequence in $\ell_0^\infty(\Sigma,X)$ which converges uniformly to f and consider $Y := \langle\{f(\omega), \omega \in \Omega\} \cup \{f_n(\omega), n \in \mathbb{N}, \omega \in \Omega\}\rangle = \langle\{\operatorname{Im} f, \operatorname{Im} f_n, n \in \mathbb{N}\}\rangle$. Then Y' is $\sigma(Y',Y)$-separable and if $\{y_n^*, n \in \mathbb{N}\}$ is a weak*-total sequence in Y', each function y_n^*f is Σ-measurable since it is the pointwise limit of the sequence of Σ-simple functions $\{y_n^*f_m, m \in \mathbb{N}\}$. Thus $A_n := \{\omega \in \Omega, y_n^*f(\omega) = y_n^*x\} = (y_n^*f)^{-1}(y_n^*x) \in \Sigma$.

If we show that $f^{-1}(x)$ coincides with $A := \bigcap\{A_n, n \in \mathbb{N}\}$ the proof will be finished. But given $\omega \in f^{-1}(x)$, then $f(\omega) = x$ and hence $y_n^*f(\omega) = y_n^*x$ for each $n \in \mathbb{N}$. Thus $\omega \in A_n$ for each $n \in \mathbb{N}$ and so $\omega \in A$.

Conversely, if $\omega \in A$, then $y_n^*f(\omega) = y_n^*x$ for each $n \in \mathbb{N}$, so we have $y_n^*(f(\omega) - x) = 0$ for $n \in \mathbb{N}$. Hence $\{y_n^*, n \in \mathbb{N}\}$ is total and consequently $x = f(\omega)$, that is, $\omega \in f^{-1}(x)$. ∎

Hence if $f \in K(\Sigma,X) \setminus \ell_0^\infty(\Sigma,X)$ and $\{x_n, n \in \mathbb{N}\}$ denotes the bounded sequence of the different values taken by f, setting $A_n := f^{-1}(x_n)$ for each $n \in \mathbb{N}$, then $\{A_n, n \in \mathbb{N}\}$ becomes an infinite partition of Ω formed by non-empty elements of Σ such that $f(\omega) = x_n$ if $\omega \in A_n$ for $n \in \mathbb{N}$.

Theorem 9.3.2 *Let* Σ *be a* σ*-algebra. Then* $K(\Sigma,X)$ *is barrelled if and only if* X *is barrelled.*

Proof. We need only to show that if X is barrelled then $K(\Sigma,X)$ is barrelled, since X is topologically isomorphic to the subspace F of $K(\Sigma,X)$ formed by the constant functions, which is complemented in $K(\Sigma,X)$ because the map δ_ω of $K(\Sigma,X)$ into F such that $\delta_\omega(f) = f(\omega)\cdot e(\Omega)$ is a projection for each $\omega \in \Omega$.

Assume that X is barrelled but there is a barrel T in $K(\Sigma,X)$ which is not a neighbourhood of the origin in $K(\Sigma,X)$. Then T cannot absorb the unit sphere S_1 of $\ell_0^\infty(\Sigma,X)$ since if it did it would also absorb the closed unit ball of $\ell_0^\infty(\Sigma,X)$ and, consequently, of $K(\Sigma,X)$. Therefore, there must be some $f_1 \in S_1$ such that $f_1 \notin 2T$. Let $\{Q_{11}, Q_{12}, \ldots, Q_{1k_1}\}$ be a partition of Ω formed by non-empty elements of Σ such that $f_1(\omega) = x_{1i} \in X$ if $\omega \in Q_{1i}$, $1 \le i \le k_1$, and $x_{1i} \ne x_{1j}$ for $1 \le i < j \le k_1$.

Now given that $\ell_0^\infty(\Sigma,X)$ is the topological direct sum of the subspaces $\{\ell_0^\infty(Q_{1i},X),\ 1 \le i \le k_1\}$, T cannot absorb the unit spheres of all of them, and there must be some $m_1 \in \{1, \ldots, k_1\}$ and $f_2 \in S_2$, the unit sphere of $\ell_0^\infty(Q_{1m_1},X)$, such that $f_2 \notin 4T$. Let $\{Q_{21}, Q_{22}, \ldots, Q_{2k_2}\}$ be a partition of Q_{1m_1} by non-empty elements of Σ so that $f_2(\omega) = x_{2i} \in X$ if $\omega \in Q_{2i}$, $1 \le i \le k_2$, and $x_{2i} \ne x_{2j}$ for $1 \le i < j \le k_2$.

Now given that $\ell_0^\infty(Q_{1m_1},X)$ is the topological direct sum of the subspaces $\{\ell_0^\infty(Q_{2i},X),\ 1 \le i \le k_2\}$, T cannot absorb the unit spheres of all of them, so there must be some $m_2 \in \{1, \ldots, k_2\}$ and $f_3 \in S_3$, the unit sphere of $\ell_0^\infty(Q_{2m_1},X)$, such that $f_3 \notin 6T$.

Continuing by recurrence we obtain a normalized sequence $\{f_n,\ n \in \mathbb{N}\}$ of Σ-simple functions, a sequence $\{k_n,\ n \in \mathbb{N}\}$ of positive integers and a countable family $\{Q_{ni},\ n \in \mathbb{N},\ 1 \le i \le k_n\}$ of non-empty elements of Σ such that, for each $n \in \mathbb{N}$, $f_n(\omega) = x_{ni} \in X$ if $\omega \in Q_{ni}$, $1 \le i \le k_n$, and $x_{ni} \ne x_{nj}$ for $1 \le i < j \le k_n$, in such a way that, for each $n \in \mathbb{N}$,

(i) $\operatorname{supp} f_{n+1} \subseteq Q_{nm_n}$ for some $m_n \in \{1, \ldots, k_n\}$.

(ii) f_n is constant in $\operatorname{supp} f_m$ for every $m > n$.

(iii) $Q_{n+1,m_{n+1}} \subseteq Q_{nm_n}$.

(iv) $f_n \notin 2nT$.

Set $P := \bigcap\{Q_{nm_n}, n \in \mathbb{N}\}$. Assuming $P \neq \emptyset$, for each $n \in \mathbb{N}$ we write f_n as $f_n = \sum\{x_{ni} e(Q_{ni}), i \in \{1, \ldots, k_n\} \setminus \{m_n\}\} + x_{nm_n} e(Q_{nm_n} \setminus P) + x_{nm_n} e(P)$, then since $\{f_n, n \in \mathbb{N}\}$ is normalized we deduce that $\|x_{nm_n}\| \leq 1$ for each $n \in \mathbb{N}$. Now the mapping $x \longrightarrow e(P)x$, being an isometry from X onto its image, the barrelledness of X implies that there must be some $n_0 \in \mathbb{N}$ such that $x_{nm_n} e(P) \in n_0 T$ for each $n \in \mathbb{N}$, and consequently $x_{nm_n} e(P) \in nT$ for each $n \geq n_0$.

For each $n \in \mathbb{N}$ define $g_n := f_n - x_{nm_n} e(P)$ which does not belong to nT for $n \geq n_0$, since $f_n = g_n + x_{nm_n} e(P)$ and $x_{nm_n} e(P) \in nT$ for each $n \geq n_0$. Then the g_n verify that $\bigcap\{\text{supp } g_n, n \geq n_0\} \subseteq \bigcap\{Q_{nm_n} \setminus P, n \geq n_0\} = \emptyset$. If $P = \emptyset$, we take $n_0 = 1$ and $g_n = f_n$ for each $n \in \mathbb{N}$. Clearly we also have that $\bigcap\{\text{supp } g_n, n \geq n_0\} = \emptyset$.

The sequence $\{g_n, n \geq n_0\}$ is bounded in $K(\Sigma,X)$ since $\|g_n\| \leq \|f_n\| = 1$ for each $n \geq n_0$. Thus if $\xi \in \ell^1$, $\sum_{n \geq n_0} \xi_n g_n$ converges in the completion of $K(\Sigma,X)$ and takes at most a countable number of values in X. Indeed if $\omega \in P$, $\sum_{j \geq n_0} \xi_j g_j(\omega) = 0$, and if $\omega \notin P$, there exists some positive integer $m_0 \geq n_0$ with $\omega \notin Q_{nm_n}$ for all $n > m_0$, and so

$$\sum_{n=n_0}^{\infty} \xi_n g_n(\omega) = \sum_{n=n_0}^{\infty} \xi_n f_n(\omega) = \sum_{n=n_0}^{m_0} \xi_n f_n(\omega) \in X.$$

Moreover, the sequence $\{\sum_{n_0 \leq j \leq m} \xi_j g_j, m \in \mathbb{N}\} \subseteq \ell_0^{\infty}(\Sigma,X)$ converges to $h_\xi = \sum_{j \geq n_0} \xi_j g_j$ in the completion $K(\Sigma,\hat{X})$ of $K(\Sigma,X)$, i.e, $h_\xi \in K(\Sigma,X)$. Hence, if B_{ℓ^1} denotes the unit ball of ℓ^1, the Banach disk $D := \{h_\xi, \xi \in B_{\ell^1}\}$ of $K(\Sigma,\hat{X})$ is contained in $K(\Sigma,X)$. Thus there is a positive integer $q \geq n_0$ such that $D \subseteq qT$ and consequently $g_q \in qT$, a contradiction. ∎

In the following three results suppose s is any positive integer and let $\{E_{m_1 \ldots m_p}, m_r \in \mathbb{N}, 1 \leq r \leq p \leq s\}$ be an s-net in $K(\Sigma,X)$ formed by dense subspaces, with Σ a σ-algebra. For each $m_1, \ldots, m_s \in \mathbb{N}$ assume

$T_{m_1 \ldots m_s}$ is a barrel of $E_{m_1 \ldots m_s}$, $B_{m_1 \ldots m_s}$ its closure in $K(\Sigma,X)$ and $L_{m_1 \ldots m_s} := \langle B_{m_1 \ldots m_s} \rangle$. By decreasing recurrence, for $p = s-1, \ldots, 1$, define the subspaces $F_{m_1 \ldots m_{p+1}} := \bigcap\{L_{m_1 \ldots m_p m}, m \geq m_{p+1}\}$, $L_{m_1 \ldots m_p} := \bigcup\{F_{m_1 \ldots m_p m}, m \in \mathbb{N}\}$, and $F_{m_1} := \bigcap\{L_m, m \geq m_1\}$. Notice that $\{F_m, m \in \mathbb{N}\}$ and $\{F_{m_1 \ldots m_p m}, m \in \mathbb{N}\}$ are 1-nets in $K(\Sigma,X)$ and $L_{m_1 \ldots m_p}$ respectively, for $m_r \in \mathbb{N}$, $1 \leq r \leq p \leq s-1$, and $E_{m_1 \ldots m_p} \subseteq F_{m_1 \ldots m_p}$, for each $m_r \in \mathbb{N}$, with $1 \leq r \leq p \leq s$.

Lemma 9.3.3 *If $\{A_n, n \in \mathbb{N}\}$ is a sequence of non-void pairwise disjoint elements of Σ, there is some $n_0 \in \mathbb{N}$ such that $K(\bigcup\{A_n, n > n_0\},X) \subseteq F_{n_0}$.*

Proof. Assume the lemma is false and that for each $p \in \mathbb{N}$ there is some $f_p \in K(\bigcup\{A_n, n > p\},X) \setminus F_p$ with $\|f_p\|_\infty = 1$. Then $\{f_n, n \in \mathbb{N}\}$ is bounded in $K(\Sigma,X)$ and if $\xi \in \ell^1$, $\sum_{n=1} \xi_n f_n$ converges in the completion of $K(\Sigma,X)$.

Now $\sum_n \xi_n f_n$ takes at most countably many values in X, since if $\omega \in \Omega \setminus \bigcup\{A_n, n \in \mathbb{N}\}$ then $\sum_n \xi_n f_n(\omega) = 0$ and if $\omega \in \bigcup\{A_n, n \in \mathbb{N}\}$ there is some $q \in \mathbb{N}$ so that $\omega \in A_q$, i.e. $\omega \notin \bigcup\{A_n, n > q\}$. As supp $f_n \subseteq \bigcup\{A_i, i > n\}$, then $\sum_n \xi_n f_n(\omega) = \sum_{1 \leq n \leq q} \xi_n f_n(\omega)$. Moreover, $\{\sum_{1 \leq n \leq m} \xi_n f_n, m \in \mathbb{N}\} \subseteq K(\Sigma,X)$ converges to $\sum_n \xi_n f_n$ in the completion of $K(\Sigma,X)$ and, since each $\sum_{1 \leq n \leq m} \xi_n f_n$ is the uniform limit of Σ-simple functions, $\sum_n \xi_n f_n$ is the uniform limit of Σ-simple functions.

Therefore, $D := \{\sum_n \xi_n f_n, \xi \in B_{\ell^1}\}$ is a Banach disk and, denoting by E_D the normed space $\langle D \rangle$ whose norm is the gauge of D, there is an $m'_1 \in \mathbb{N}$ such that $F_{m_1} \cap E_D$ is a dense Baire subspace of E_D for each $m_1 \geq m'_1$. By finite induction, suppose that we have found m'_1 and the functions $m'_i(m_1, \ldots, m_{i-1})$, $2 \leq i \leq p \leq s-1$, such that for any positive integer $m_1 \geq m'_1$, $m_i \geq m'_i(m_1, \ldots, m_{i-1})$, $2 \leq i \leq p$, $F_{m_1 \ldots m_i} \cap E_D$ is a dense Baire subspace of E_D. Then, for any $m_1 \geq m'_1$, $\ldots$, $m_p \geq m'_p(m_1, \ldots, m_{p-1})$, given that $\{F_{m_1 \ldots m_p m}, m \in \mathbb{N}\}$ covers $F_{m_1 \ldots m_p}$ there is some $m'_{p+1}(m_1, \ldots, m_p) \in \mathbb{N}$ such that $F_{m_1 \ldots m_{p+1}} \cap E_D$ is a dense Baire subspace of E_D for each

$m_{p+1} \geq m'_{p+1}(m_1,\dots,m_p)$. Hence $D \subseteq L_{m_1\dots m_s}$ if $m_1 \geq m'_1$, ..., $m_s \geq m'_s(m_1,\dots,m_{s-1})$, since $B_{m_1\dots m_s} \cap L_{m_1\dots m_s} \cap E_D$ is a barrel and hence a neighbourhood of the origin in the Baire space $L_{m_1\dots m_s} \cap E_D$ for $m_1 \geq m'_1$, ..., $m_s \geq m'_s(m_1,\dots,m_{s-1})$. It follows from this that $D \subseteq F_{m_1\dots m_s}$ for $m_1 \geq m'_1$, ..., $m_s \geq m'_s(m_1,\dots,m_{s-1})$ and therefore $D \subseteq L_{m_1\dots m_{s-1}}$ if $m_1 \geq m'_1,\dots, m_{s-1} \geq m'_{s-1}(m_1,\dots,m_{s-2})$. This implies that $D \subseteq F_{m_1\dots m_{s-1}}$ for $m_1 \geq m'_1$, ..., $m_{s-1} \geq m'_{s-1}(m_1,\dots,m_{s-2})$. Going on in this way, we obtain that $D \subseteq F_{m_1}$ for $m_1 \geq m'_1$, and consequently $f_{m_1} \in F_{m_1}$, which is a contradiction. ■

Lemma 9.3.4 *If* $X \in \mathscr{C}_s$, *there is some* $n_0 \in \mathbb{N}$ *such that* $\ell_0^\infty(\Sigma,X) \subseteq F_{n_0}$.

Proof. Suppose that the lemma is not true and there is some $f_1 \in \ell_0^\infty(\Sigma,X)\setminus F_1$ such that $\|f_1\| = 1$. Let $\{Q_{11}, Q_{12}, \dots, Q_{1k_1}\}$ be a partition of Ω formed by non-empty elements of Σ such that $f_1(\omega) = x_{1i} \in X$ if $\omega \in Q_{1i}$, $1 \leq i \leq k_1$, and $x_{1i} \neq x_{1j}$ for $1 \leq i < j \leq k_1$.

Given that $\ell_0^\infty(\Sigma,X)$ is the topological direct sum of the subspaces $\{\ell_0^\infty(Q_{1i},X),\ 1 \leq i \leq k_1\}$, there must be some $m_1 \in \{1, \dots, k_1\}$ such that $\ell_0^\infty(Q_{1m_1},X)$ is not contained in F_n for each $n \in \mathbb{N}$ and consequently there is some $f_2 \in \ell_0^\infty(Q_{1m_1},X)\setminus F_2$ so that $\|f_2\| = 1$. Let $\{Q_{21}, Q_{22}, \dots, Q_{2k_2}\}$ be a partition of Q_{1m_1} formed by non-empty elements of Σ such that $f_2(\omega) = x_{2i} \in X$ if $\omega \in Q_{2i}$, $1 \leq i \leq k_2$, and $x_{2i} \neq x_{2j}$ for $1 \leq i < j \leq k_2$. There exists some $m_2 \in \{1, \dots, k_2\}$ such that $\ell_0^\infty(Q_{2m_2},X)$ is not contained in F_n for each $n \in \mathbb{N}$.

Assume that we have obtained by induction a sequence $\{f_n,\ n \in \mathbb{N}\}$ of Σ-simple functions, a sequence of positive integers $\{k_n,\ n \in \mathbb{N}\}$, and a countable family $\{Q_{ni},\ n \in \mathbb{N},\ 1 \leq i \leq k_n\}$ of non-empty elements of Σ such that, for each $n \in \mathbb{N}$, $f_n(\omega) = x_{ni} \in X$ if $\omega \in Q_{ni}$, $1 \leq i \leq k_n$, and $x_{ni} \neq x_{nj}$ for $1 \leq i < j \leq k_n$, with the following conditions holding for each $n \in \mathbb{N}$

(i) $\|f_n\| = 1$.

(ii) supp $f_{n+1} \subseteq Q_{nm_n}$ for some $m_n \in \{1, \ldots, k_n\}$.

(iii) $Q_{n+1,m_{n+1}} \subseteq Q_{nm_n}$.

(iv) $f_n \notin F_n$.

Set $P := \bigcap\{Q_{nm_n}, n \in \mathbb{N}\}$ and for each $n \in \mathbb{N}$ define

$$g_n := \begin{cases} f_n \text{ for } P = \emptyset \\ f_n - x_{nm_n} e(P) \text{ for } P \neq \emptyset \end{cases}$$

In the second case the mapping $x \longrightarrow e(P)x$ of X into $K(\Sigma,X)$ is an isometry, and as $X \in \mathscr{C}_s$ there must be an $m'_1 \in \mathbb{N}$ with $F_{m_1} \cap e(P)X \in \mathscr{C}_{s-1}$ and is dense in $e(P)X$ for $m_1 \geq m'_1$ and a fortiori $L_{m_1} \cap e(P)X \in \mathscr{C}_{s-1}$ and is dense in $e(P)X$ for each $m_1 \geq m'_1$. By finite induction suppose we have found m'_1 and the functions $m'_i(m_1,\ldots,m_{i-1})$, $2 \leq i \leq p \leq s-1$, such that for $m_1 \geq m'_1$, $m_i \geq m'_i(m_1,\ldots,m_{i-1})$, $2 \leq i \leq p$, $L_{m_1 \ldots m_i} \cap e(P)X \in \mathscr{C}_{s-i}$ and is dense in $e(P)X$. Then, for any $m_1 \geq m'_1, \ldots, m_p \geq m'_p(m_1,\ldots,m_{p-1})$ given that $\{F_{m_1 \ldots m_p m}, m \in \mathbb{N}\}$ covers $L_{m_1 \ldots m_p}$, then there exists some $m'_{p+1}(m_1,\ldots,m_p) \in \mathbb{N}$ such that $F_{m_1 \ldots m_{p+1}} \cap e(P)X \in \mathscr{C}_{s-p-1}$ and is dense in $e(P)X$ for $m_{p+1} \geq m'_{p+1}(m_1,\ldots,m_p)$. Consequently, $L_{m_1 \ldots m_{p+1}} \cap e(P)X \in \mathscr{C}_{s-p-1}$ and is dense in $e(P)X$ for each $m_{p+1} \geq m'_{p+1}(m_1,\ldots,m_p)$.

Therefore, $e(P)X \subseteq L_{m_1 \ldots m_s}$ if $m_1 \geq m'_1, \ldots, m_s \geq m'_s(m_1,\ldots,m_{s-1})$ since $L_{m_1 \ldots m_s}$ is the linear hull $B_{m_1 \ldots m_s}$, and $L_{m_1 \ldots m_s} \cap e(P)X$, being barrelled, it is a closed subspace of $e(P)X$ for $m_1 \geq m'_1, \ldots, m_s \geq m'_s(m_1,\ldots,m_{s-1})$. Thus, $e(P)X \subseteq F_{m_1 \ldots m_s}$ for all $m_1 \geq m'_1, \ldots, m_s \geq m'_s(m_1,\ldots,m_{s-1})$, implying $e(P)X \subseteq L_{m_1 \ldots m_{s-1}}$ if $m_1 \geq m'_1, \ldots, m_{s-1} \geq m'_{s-1}(m_1,\ldots,m_{s-2})$. This shows that $e(P)X \subseteq F_{m_1 \ldots m_{s-1}}$ for $m_1 \geq m'_1, \ldots, m_{s-1} \geq m'_{s-1}(m_1,\ldots,m_{s-2})$. Continuing in this way, we obtain that $e(P)X \subseteq F_{m_1}$ for all $m_1 \geq m'_1$, and consequently $e(P)x_{jm_j} \in F_{m_1}$ for each $j \in \mathbb{N}$ and for all $m_1 \geq m'_1$. Thus, $g_{m_1} \notin F_{m_1}$ for each $m_1 \geq m'_1$.

As in Theorem 9.3.2, from $\bigcap\{\text{supp } g_n,\ n \in \mathbb{N}\} = \emptyset$ it follows that $D := \{\sum_n \xi_n g_n,\ \xi \in B_{\ell^1}\}$ is a Banach disk of the completion of $K(\Sigma,X)$ which is contained in $K(\Sigma,X)$ and there must be some positive integer $p \geq m_1'$ such that $D \subseteq F_p$. Hence $g_p \in F_p$, a contradiction. ∎

Theorem 9.3.5 *If* $X \in \mathscr{C}_s$, *then there exists some* $n_0 \in \mathbb{N}$ *such that* $K(\Sigma,X)$ *coincides with* F_{n_0}.

Proof. By Lemma 9.3.4 without loss of generality we may assume that $\ell_0^\infty(\Sigma,X) \subseteq F_n$ for every $n \in \mathbb{N}$. Suppose that $K(\Sigma,X)$ is not contained in any F_n, $n \in \mathbb{N}$. Let $f_1 \in K(\Sigma,X)\setminus F_1$ such that $\|f_1\| = 1$ and let $\{Q_{1i},\ i \in \mathbb{N}\}$ be the partition of Ω formed by non-empty elements of Σ defined by the function f_1 in such a way that $f_1(\omega) = x_{1i} \in X$ if $\omega \in Q_{1i}$, $i \in \mathbb{N}$, and $x_{1i} \neq x_{1j}$ for $i \neq j$.

Then by Lemma 9.3.3 there exists some positive integer $n_2 > n_1 = 1$ such that $K(\bigcup\{Q_{1i},\ i > n_2\},X) \subseteq F_{n_2}$. Therefore, setting $\Omega_1 := \bigcup\{Q_{1i},\ 1 \leq i \leq n_2\}$, $K(\Omega_1,X)$ cannot be contained in any F_n, $n \geq n_2$, and there must be some $f_2 \in K(\Omega_1,X)\setminus F_{n_2}$ such that $\|f_2\| = 1$. Let $\{Q_{2i},\ i \in \mathbb{N}\}$ be the partition of Ω_1 formed by non-empty elements of Σ defined by the Σ-measurable function f_2.

Again, since $\{F_n,\ n > n_2\}$ covers $K(\Omega_1,X)$, by Lemma 9.3.3 there is some integer $n_3 > n_2$ such that $K(\bigcup\{Q_{2i},\ i > n_3\},X) \subseteq F_{n_3}$. Setting $\Omega_2 := \bigcup\{Q_{2i},\ 1 \leq i \leq n_3\}$, $K(\Omega_2,X)$ cannot be contained in any F_n, $n \geq n_3$, and we take $f_3 \in K(\Omega_2,X) \setminus F_{n_3}$ such that $\|f_3\| = 1$.

Assume that, by this means, we have obtained a sequence of positive integers $\{n_i,\ i \in \mathbb{N}\}$ and a sequence $\{f_n,\ n \in \mathbb{N}\}$ of functions of $K(\Sigma,X)$ which determine a countable family $\{Q_{ni},\ n,\ i \in \mathbb{N}\}$ formed by non-empty elements of Σ such that, for each $n \in \mathbb{N}$, $f_n(\omega) = x_{ni} \in X$ if $\omega \in Q_{ni}$, $i \in \mathbb{N}$, and $x_{ni} \neq x_{nj}$ for $i \neq j$, in such a way that, setting $\Omega_n := \bigcup\{Q_{ni},\ 1 \leq i \leq n_{i+1}\}$ $\forall n \in \mathbb{N}$, for each $i \in \mathbb{N}$ we have that,

(i) $\text{supp } f_{i+1} \subseteq \Omega_i$.

(ii) $e(\Omega_i)f_i \in \ell_0^\infty(\Omega_i,X) \subseteq \ell_0^\infty(\Sigma,X)$.

(iii) $\Omega_{i+1} \subseteq \Omega_i$.

(iv) $f_i \notin F_{n_i}$.

Let $g_i := f_i - e(\Omega_i)f_i$ for each $i \in \mathbb{N}$. Then $g_i \notin F_{n_i}$ for each $i \in \mathbb{N}$ and $\operatorname{supp} g_i \cap \operatorname{supp} g_j = \emptyset$ for $i \neq j$. Denoting by G the completion of $K(\Sigma,X)$, $\overline{<\{g_n, n \in \mathbb{N}\}>}^G$ is a copy of c_0 in G, since it can be easily seen that the sequence $\{g_n/\|g_n\| : n \in \mathbb{N}\}$ is equivalent to the unit vector basis of c_0. Moreover, it is also easy to see that

$$\overline{<\{g_n, n \in \mathbb{N}\}>}^G \subseteq K(\Sigma,X)$$

In fact,

$$\sum_{n=1}^{\infty} c_n g_n(\omega) = \begin{cases} 0 \text{ for } \omega \notin \bigcup\{\operatorname{supp} g_n, n \in \mathbb{N}\} \\ c_m g_m(\omega) \in X \text{ for } \omega \in \operatorname{supp} g_m, m \in \mathbb{N} \end{cases}$$

So $\sum_n c_n g_n \in K(\Sigma,X)$ since it takes at most countably many values in X and is the uniform limit of Σ-simple functions. Consequently, there must exist some $q \in \mathbb{N}$ such that $\{g_n, n \in \mathbb{N}\} \subseteq F_k$ for each $k \geq n_q$. Hence $g_q \in F_{n_q}$, a contradiction. ∎

Theorem 9.3.6 *Let* Σ *be a* σ*-algebra. Given* $s \in \mathbb{N}$*, then* $K(\Sigma,X)$ *is barrelled of class* s *if and only if* X *is barrelled of class* s.

Proof. We have only to show that if $X \in \mathscr{C}_s$ then $K(\Sigma,X) \in \mathscr{C}_s$. By Theorem 9.3.2, we already know that $K(\Sigma,X)$ is a metrizable barrelled space and, consequently, Baire-like.

By recurrence, let $p \in \{1, \ldots, s\}$ and assume $K(\Sigma,X) \in \mathscr{C}_{p-1}$. Let $W := \{E_{m_1\ldots m_i}, m_r \in \mathbb{N}, 1 \leq r \leq i \leq p\}$ be a p-net in $K(\Sigma,X)$ of dense subspaces. We will show that some $E_{m_1\ldots m_p} \in W_p$ is barrelled. Assuming that every subspace of W is dense there is no loss of generality since $K(\Sigma,X) \in \mathscr{C}_{p-1}$.

For each $m_1, \ldots, m_p \in \mathbb{N}$, suppose $T_{m_1\ldots m_p}$ is a barrel of $E_{m_1\ldots m_p}$ which is not a neighbourhood of the origin in $E_{m_1\ldots m_p}$, $B_{m_1\ldots m_p}$ is its closure in $K(\Sigma,X)$ and $L_{m_1\ldots m_p} := <B_{m_1\ldots m_p}>$. By decreasing recurrence, for $i = p-1, \ldots, 1$ define the subspaces $F_{m_1\ldots m_{i+1}} := \bigcap\{L_{m_1\ldots m_i m}, m \geq$

$m_{i+1}\}$, $L_{m_1 \ldots m_i} := \bigcup\{F_{m_1 \ldots m_i m}, \ m \in \mathbb{N}\}$, and $F_{m_1} := \bigcap\{L_m, \ m \geq m_1\}$. Then, $\{F_m, \ m \in \mathbb{N}\}$ and $\{F_{m_1 \ldots m_i m}, \ m \in \mathbb{N}\}$ are 1-nets in $K(\Sigma,X)$ and $L_{m_1 \ldots m_i}$, respectively, for each $m_r \in \mathbb{N}$, $1 \leq r \leq i \leq p-1$ and $E_{m_1 \ldots m_i} \subseteq F_{m_1 \ldots m_i}$, for each $m_r \in \mathbb{N}$, $1 \leq r \leq i \leq p$.

By Theorem 9.3.5, there must exist some $m_1 \in \mathbb{N}$ such that $K(\Sigma,X)$ coincides with F_{m_1}. Then clearly $L_{m_1} \in \mathscr{C}_{p-1}$ and there is some $m_2 \in \mathbb{N}$ such that $F_{m_1 m_2} \in \mathscr{C}_{p-2}$ and is dense in $K(\Sigma,X)$. A fortiori, $L_{m_1 m_2} \in \mathscr{C}_{p-2}$ and is dense in $K(\Sigma,X)$. Continuing in this way we are able to find some $F_{m_1 \ldots m_p} \in \mathscr{C}_0$. So $B_{m_1 \ldots m_p} \cap F_{m_1 \ldots m_p}$ is a neighbourhood of the origin in $F_{m_1 \ldots m_p}$ and hence $T_{m_1 \ldots m_p}$ is a neighbourhood of the origin in $E_{m_1 \ldots m_p}$, which is a contradiction. Thus $K(\Sigma,X) \in \mathscr{C}_p$.

Hence $K(\Sigma,X) \in \mathscr{C}_s$ and the proof is complete. ∎

Theorem 9.3.7 *Let* Σ *be a* σ*-algebra (and let* s *be any positive integer). Then* $B(\Sigma,X)$ *is barrelled (of class* s*) if and only if* X *is barrelled (of class* s*).*

Proof. It is an obvious consequence of Theorems 9.3.2 and 9.3.6, since $K(\Sigma,X)$ is a dense subspace of $B(\Sigma,X)$. ∎

Corollary 9.3.8 *Let* Σ *be a* σ*-algebra. Then* $K(\Sigma,X)$ *(resp.* $B(\Sigma,X)$*) is barrelled of class* $\aleph_0$ *if and only if* X *is barrelled of class* $\aleph_0$*.*

9.4 EXERCISES

1. *Non-barrelledness of the space $\ell_0^\infty(\Sigma,F)$, F being a normed space.* Let Σ be a σ-algebra of subsets of a set Ω and suppose that F is a normed space. A Banach space E has the approximation property if given any $\varepsilon > 0$ and any compact set K in E, there exists a continuous linear operator T into E such that $\|Tx - x\| \le \varepsilon$ for all $x \in K$. A Banach space G is said to be an S_p-space if it contains a copy of the space $\ell_p^{(n)}$ such that there are continuous linear mappings P_n and J_n from G into $\ell_p^{(n)}$ and from $\ell_p^{(n)}$ into G respectively, such that P_nJ_n is the identity operator on $\ell_p^{(n)^p}$, and this for all $n \in \mathbb{N}$.

(i) If $\ell_0^\infty(\Sigma,F)$ is barrelled, then $\ell^\infty(\Sigma)\hat{\otimes}_\varepsilon F$ is barrelled.

(ii) $\ell^\infty(\Sigma)$ has the approximation property.

(iii) $\ell^\infty(\Sigma)\hat{\otimes}_\varepsilon F = \ell^\infty(\Sigma)\hat{\otimes}_\pi F$

(iv) If Σ is infinite, then $c_0\otimes_\varepsilon F = c_0\otimes_\pi F$

(v) As c_0 is an S_p-space, then F is nuclear.

2. *The dual of $\ell_0^\infty(\Sigma,X)$ and $B(\Sigma,X)$.* Let Σ be a σ-algebra of subsets of a set Ω and let X be a normed space. If X' denotes the strong dual of X, denote by $bva(\Sigma,X')$ the linear space of all X'-valued vector measures F of bounded variation defined on Σ. If $E \in \Sigma$ and $F : \Sigma \longrightarrow X^*$, define the weak* semivariation of F by $\|F\|(E) := \sup\left\{\sum_{A\in\pi}|xF(A)| : \|x\| \le 1,\ \pi \in \mathcal{P}\right\}$, where $\mathcal{P}$ is the family of all finite Σ-partitions of E.

(i) The linear space $M(\Sigma,X')$ of all those X'-valued measures F on Σ such that $\|F\|(\Omega) < \infty$ is a normed space under $\|\cdot\|$.

(ii) The mapping $\phi : M(\Sigma,X') \longrightarrow B(\Sigma,X)'$ defined by $\left(\phi(F)\right)(f) = \int_\Omega f\, dF$ for each $F \in M(\Sigma,X')$ and each $f \in B(\Sigma,X)$ is an isomorphism onto from the algebraic point of view.

(iii) If $f \in \ell^\infty(\Omega,\Sigma)$ and $x \in X$, then $\left(\phi(F)\right)(xf) = \int_\Omega f\, d(xF)$.

(iv) If, given $F \in M(\Sigma,X')$, we denote by $|F|$ the variation of F, then

$$|F|(\Omega) = \sup\left\{\left|\left(\phi(F)\right)(f)\right| : f \in B(\Sigma,X),\ \|f\| \le 1\right\}.$$

(v) $M(\Sigma,X')$ is a linear subspace of $bva(\Sigma,X')$.

(vi) The strong dual $B(\Sigma,X)'$ of $B(\Sigma,X)$ is topologically isomorphic to the space $M(\Sigma,X')$ of all X'-valued measures of finite weak* semi-variation provided with the variation norm of $bva(\Sigma,X')$.

3. *The space* $BCV(\Sigma,X)$. If we represent by $\ell^\infty(\Sigma,X)$ the linear space of all bounded X-valued Σ-measurable functions defined on Ω provided with the supremum-norm, we shall denote by $BCV(\Sigma,X)$ the linear subspace of $\ell^\infty(\Sigma,X)$ of all functions of countable range. Clearly, $\ell_0^\infty(\Sigma,X) \subseteq K(\Sigma,X) \subseteq BCV(\Sigma,X) \subseteq \ell^\infty(\Sigma,X)$ and $B(\Sigma,X)$ is the closure of $\ell_0^\infty(\Sigma,X)$ in $\ell^\infty(\Sigma,X)$.

(i) $BCV(\Sigma,X)$ is a dense subspace of $\ell^\infty(\Sigma,X)$.

(ii) If X is an infinite-dimensional Banach space and Σ is infinite, there is a normalized basic sequence (x_n) in X such that $f := \sum_n x_n e(E_n)$ is, for each infinite partition (E_n) of Ω formed by non-empty sets of Σ, an element of $BCV(\Sigma,X)$ which does not belong to $B(\Sigma,X)$. *[Hint: By Mazur's theorem there is some normalized basic sequence* (z_n) *in X. If* (z_n) *is equivalent to the unit vector basis of* ℓ^1 *we are done; if this is not the case, use Rosenthal's* ℓ^1*-theorem.]*

4. *The space* $\ell_0^\infty(\Sigma,X)_Q$. Let X be a normed space, Σ a σ-algebra of subsets of some non-empty set Ω and denote by Q the absolutely convex hull of the set $\{x\chi_A : \|x\| \le 1,\ A \in \Sigma\}$. Denote by $\ell_0^\infty(\Sigma,X)_Q$ the linear hull of Q with the norm of Minkowski's functional of Q.

(i) $\ell_0^\infty(\Sigma,X)_Q$ is a barrelled space.

(ii) $\ell_0^\infty(\Sigma,X)_Q$ is isomorphic to $\ell_0^\infty(\Sigma) \otimes_\pi X$.

5. *Weak barrelledness of* $\ell_0^\infty(\Sigma,E)$. Let Σ be a σ-algebra on a set Ω and E a metrizable barrelled space. As usual denote by $\ell_0^\infty(\Sigma,E)$ the space of all E-valued Σ-simple functions provided with the uniform convergence topology.

(i) Let F be a space. If there is a locally convex topology τ on F finer than the original topology such that $F(\tau)$ is quasi-Baire (or *d* space) then F is weakly barrelled.

(ii) $\ell_0^\infty(\Sigma,E)$ is weakly barrelled.

9.5 NOTES AND REMARKS

Our presentation of the first part of this chapter has been strongly motivated by Freniche's paper [Fr1]. His Theorem 1 gives necessary and sufficient conditions for $\ell_0^\infty(\Sigma,E)$ to be barrelled. In particular, it follows that if Σ is a σ-algebra of subsets of Ω and X is a normed space, $\ell_0^\infty(\Sigma,X)$ is barrelled if and only if X is finite-dimensional. We give an independent proof of this result in our Exercise 1. Related to the above is the following. Let Σ be a Boolean ring of subsets of a set Ω and $\mu : \Sigma \longrightarrow [0, +\infty]$ denote a finitely additive non-trivial measure on Σ, and if E is a (Hausdorff) locally convex space over $\mathbb{K}$, denote by $Z(\mu)$ the (closed) subspace of $\ell_0^\infty(\Sigma,E)$ of all those functions which are equivalent to the zero function. Then if $S(\mu,E)$ represents the quotient $\ell_0^\infty(\Sigma,E)/Z(\mu)$ provided with the corresponding quotient topology, [DDFFP] shows that $S(\mu,E)$ is barrelled if and only if $\ell_0^\infty(\Sigma)$ is barrelled and E is barrelled and nuclear.

Let us remark that property (B) used in section 9.2 can be found in Pietsch [Pi].

Proposition 9.2.8 is taken from Ferrando and López Pellicer [FL7].

Assuming that Σ is an algebra of subsets of a set Ω, E is a Hausdorff locally convex space and $B(\Sigma,E)$ is endowed with the uniform convergence topology, J. Mendoza [Me1] proves the following theorem.

Theorem 9.5.1 *The space* $B(\Sigma,E)$ *is barrelled if and only if* E *is barrelled and the strong dual* $E'(\beta(E',E))$ *of* E *has property (B).*

If X is a normed space, its strong dual satisfies property (B) and then 9.5.1 states that, if Σ is an algebra of subsets of Ω, then $B(\Sigma,X)$ is barrelled if and only if X is barrelled. When Σ is a σ-algebra, we show in Theorem 9.3.2 that the dense subspace $K(\Sigma,X)$ of $B(\Sigma,X)$ is barrelled if and only if X is a barrelled space. In our main step, we show in Theorem 9.3.6, under the same hypotheses, that $K(\Sigma,X)$ is barrelled of class n if and only if X is barrelled of class n, $n \in \mathbb{N}$. This allows us to conclude in Theorem 9.3.7 that $K(\Sigma,X)$ is barrelled of class n (of class $\aleph_0$) if and only if X does share this same property. This entails an extension of the quoted result of Mendoza in case X is normed and Σ is a σ-algebra.

The techniques used in our proof of Theorem 9.3.7 have been taken from Ferrando and Sánchez Ruiz [FS7] and are different from those used by J. Mendoza in [Me1], based on duality theory. Our Exercise 2 gathers precisely some results of [Me1].

We do not wish to finish these notes without mentioning some barrelledness properties concerning ℓ^p-direct sums of locally convex spaces. If $\{X_i : i \in I\}$ is a family of normed spaces, denote by $(\oplus_{i\in I} X_i)_p$, with $1 \le p < \infty$, the ℓ^p-direct sum of the spaces X_i, i.e.

$$(\oplus_{i\in I}X_i)_p = \left\{\mathbf{x} = (x_i) \in \prod\{X_j,\ j \in I\} : (\|x_i\|) \in \ell^p(I)\right\}$$

equipped with the norm $\|\mathbf{x}\| = \|(\|x_i\|)\|_p$. If $p = \infty$, then

$$(\oplus_{i\in I}X_i)_\infty = \left\{\mathbf{x} = (x_i) \in \prod\{X_j,\ j \in I\} : (\|x_i\|) \in \ell^\infty(I),\ |\mathrm{supp}\ \mathbf{x}| \le \aleph_0\right\}$$

provided with the norm $\|\mathbf{x}\| = \sup\{\|x_i\|,\ i \in I\}$.

According to a well-known result of Lurje, (cf. [PB2, 4.9.17]), $(\oplus_{n\in\mathbb{N}}X_n)_p$ is barrelled if and only if each X_n is barrelled. This result has been extended by Drewnowski et al. [DrFP3] and by Kakol and Roelcke [KR2] by showing that whenever each X_i is seminormed, then $(\oplus_{i\in I}X_i)_p$ is barrelled (ultrabarrelled) if and only if each space X_i is barrelled (ultrabarrelled). Ferrando [Fe7] has shown that assuming each X_i is a normed space, $(\oplus_{i\in I}X_i)_p$, $1 \le p \le \infty$, is unordered Baire-like if and only if each X_i is unordered Baire-like.

If Σ is a σ-algebra of subsets of a set Ω and X is a normed space, denote by $\ell^\infty(\Sigma,X)$ the linear space over $\mathbb{K}$ of all those bounded X-valued Σ-measurable functions defined on Ω provided with the supremum norm. Notice that $B(\Sigma,X)$ is just the closure of $\ell_0^\infty(\Sigma,X)$ in $\ell^\infty(\Sigma,X)$. Using some methods related to those of section 9.3, it is proved by Ferrando [Fe5] that $\ell^\infty(\Sigma,X)$ is barrelled if and only if X is barrelled. This result has been extended in [FS6] and [Fe8]. On the other hand if $\ell^\infty(\Omega,X)$ denotes the linear space of all bounded X-valued functions, X being a normed space, defined on Ω equipped with the supremum norm, it can be shown [Fe5] that assuming X separable, $\ell^\infty(\Omega,\Sigma)$ is barrelled whenever X is barrelled. This last result can also be obtained as a consequence of a theorem of Drewnowski, Florencio and Paúl [DrFP5] which states that if either $|\Omega|$ or $|X|$ are non-measurable cardinals, then $\ell^\infty(\Omega,X)$ is barrelled whenever X is a barrelled normed space.

10. The Pettis integrable function space

10.1 DEFINITION OF THE PETTIS INTEGRAL

In what follows (Ω, Σ, μ) will stand for a finite measure space, X will denote a Banach space, and f some X-valued weakly μ-measurable function defined on Ω. Throughout this chapter we will follow the notation of Banach space analysts and will denote by X^* the topological dual of X.

Lemma 10.1.1 (Dunford) *If $x^*f \in L^1(\mu)$ for each $x^* \in X^*$ then given some $E \in \Sigma$, the mapping $x^* \longrightarrow \int_E x^*f\, d\mu$ is a continuous linear form defined on X^*, i.e., there is an $x_E^{**} \in X^{**}$ such that $x_E^{**}x^* = \int_E x^*f\, d\mu$ $\forall x^* \in X^*$.*

Proof. Given some fixed $E \in \Sigma$ define the mapping $T : X^* \longrightarrow L^1(\mu)$ by $Tx^* = x^*f\cdot e(E)$. Clearly T is a well-defined linear operator on X^* since $\int_\Omega x^*f\cdot e(E)\, d\mu = \int_E x^*f\, d\mu$, and we are assuming that $x^*f \in L^1(\mu)$.

Let us see that T has a closed graph. In fact, if $\lim_n x_n^* = x^*$ and $\lim_n Tx_n^* = g$, then $\lim_n x_n^*f\cdot e(E) = g$ in $L^1(\mu)$; i.e,

$$\lim_n \int_\Omega |x_n^*f(\omega)\cdot e(E)(\omega) - g(\omega)|\, d\mu(\omega) = 0$$

As is well known this implies there is a subsequence $(e(E)x_{n_j}^* f)$ of $(e(E)x_n^*f)$ which converges to g in almost all points $\omega \in \Omega$. Thus there is an $F \in \Sigma$ with $\mu(F) = 0$ such that $\lim_j x_{n_j}^* f(\omega)\cdot e(E)(\omega) = g(\omega)$ for each ω in $\Omega \setminus F$. But since $\lim_n x_n^* = x^*$ in X^*, then $x_n^* \longrightarrow x^*$ under the weak* topology of X^*; consequently $\lim_n x_n^*f(\omega)e(E)(\omega) = x^*f(\omega)$ for each $\omega \in \Omega$, and in particular for each $\omega \in \Omega \setminus F$. So, $e(E)x^*f$ is equivalent to g in $L^1(\mu)$ and $\lim_n e(E)x_n^*f = e(E)x^*f$ in $L^1(\mu)$, showing that T has a closed graph in $X^* \times L^1(\mu)$. Then by the closed graph theorem T is a continuous operator and consequently $\|T\| < \infty$ and $\|Tx^*\|_1 \le \|T\|\|x^*\|$. This means that $\int_E |x^*f|\, d\mu \le \|T\|\|x^*\|$ for $x^* \in X^*$, implying that $|\int_E x^*f\, d\mu| \le \|T\|\|x^*\|$. ■

If f is a weakly μ-measurable function defined on Ω with values in a Banach space X and such that $x^*f \in L^1(\mu)$ for each $x^* \in X^*$, the above lemma allows us to define the so-called *Dunford integral* of f on $E \in \Sigma$ as the linear form $x_E^{**} \in X^{**}$ provided in it, which in the following will be denoted by $(D)\int_E f\, d\mu$. Thus, $(D)\int_E f\, d\mu \in X^{**}$ is such that

$$\langle x^*, (D)\int_E f\, d\mu\rangle = \int_E x^*f\, d\mu$$

for each $x^* \in X^*$. Now, it is obvious that if $f : \Omega \longrightarrow X$ is weakly μ-measurable, f is Dunford integrable on each $E \in \Sigma$ if and only if x^*f belongs to $L^1(\mu)$ for each $x^* \in X^*$. If, besides, we have $(D)\int_E f\, d\mu \in X$ for each $E \in \Sigma$, then f is said to be *Pettis integrable*. In this case, the vector $(D)\int_E f\, d\mu$ of X is called the *Pettis integral* of f on E, and it is denoted by $(P)\int_E f\, d\mu$.

10.2 PETTIS INTEGRABLE FUNCTIONS

If X is a reflexive Banach space, it is obvious that each $f : \Omega \longrightarrow X$ which is Dunford integrable is Pettis integrable although this need not be true in general.

Theorem 10.2.1 (Pettis) *If* $f : \Omega \longrightarrow X$ *is Pettis integrable, then the mapping* $F : \Sigma \longrightarrow X$ *defined by* $F(E) = (P)\int_E F\, d\mu$ *for each* $E \in \Sigma$ *is a* μ*-continuous countable additive vector measure.*

Proof. Let $\{E_n, n \in \mathbb{N}\}$ be a sequence of pairwise disjoint elements of Σ. For each $x^* \in X^*$ one has that

$$x^*F(\cup_n E_n) = x^*\left((P)\int_{\cup E_n} f\, d\mu\right) = \int_{\cup E_n} x^*f\, d\mu = \sum_n \int_{E_n} x^*f\, d\mu =$$

$$\sum_n x^*\left((P)\int_{E_n} f\, d\mu\right) = x^*\left(\sum_{n=1}^{\infty} (P)\int_{E_n} f\, d\mu\right) = x^*\left(\sum_n F(E_n)\right).$$

This shows that F is weak countably additive. Hence, according to the Orlicz-Pettis theorem F is countably additive in norm. On the other hand, if $E \in \Sigma$ is such that $\mu(E) = 0$, then $\int_E x^*f\, d\mu = 0$ for each $x^* \in X^*$. Consequently, $F(E) = (P)\int_E f\, d\mu = 0$ and therefore $F \ll \mu$. ∎

Corollary 10.2.2 *Let* f : Ω ⟶ X *be Dunford integrable. If* f *is strongly measurable,* f *is Pettis integrable if and only if the vector measure* F *given by* $F(E) = (D)\int_E f(\omega)\, d\mu(\omega)$ *for each* $E \in \Sigma$, *is countably additive.*

Proof. Assuming f is Pettis integrable, regardless of whether f is μ-measurable or not, we have already shown that F is countably additive.

Assume that f is μ-measurable and F is countably additive in X**.

Since f is μ-measurable, according to Corollary 8.1.2 there is a countably X-valued μ-measurable function g such that

$$\text{ess sup } \{\|f(\omega) - g(\omega)\|, \ \omega \in \Omega\} < 1.$$

Hence, there exists a partition $\{A_n, \ n \in \mathbb{N}\}$ of Ω by non-empty elements of Σ such that $g(\omega) = \sum_n x_n e(A_n)(\omega)$ for each $\omega \in \Omega$. Setting $E_n := \bigcup\{A_j, \ 1 \le j \le n\}$ for each $n \in \mathbb{N}$, then (E_n) is an increasing sequence of members of Σ with $\bigcup\{E_n, \ n \in \mathbb{N}\} = \Omega$ such that

$$\text{ess sup } \{\|f(\omega)\|, \ \omega \in E_n\} \le$$
$$\text{ess sup } \{\|f(\omega) - g(\omega)\|, \ \omega \in E_n\} + \text{ess sup } \{\|g(\omega)\|, \ \omega \in E_n\} \le 1 + \delta,$$

where $\delta := \text{máx } \{\|x_j\|, \ 1 \le j \le n\}$. Thus, f is essentially bounded in each E_n. Hence, for each $E \in \Sigma$ and $n \in \mathbb{N}$, there exists the Bochner integral on $E \cap E_n$. But as x* is bounded on X, by Hille's theorem one has that

$$x^*\left((B)\int_{E\cap E_n} f \, d\mu\right) = \int_{E\cap E_n} x^*f \, d\mu = x^*\left((D)\int_{E\cap E_n} f \, d\mu\right)$$

for each $x^* \in X^*$. So, $(D)\int_{E\cap E_n} f \, d\mu = (B)\int_{E\cap E_n} f \, d\mu \in X$, and consequently $F(E \cap E_n) \in X$ for $n \in \mathbb{N}$. Since, by hypothesis, F is countably additive, $F(E) = F(\bigcup\{E \cap E_n, \ n \in \mathbb{N}\}) = \sum_n F(E \cap E_n) = \lim_n F(E \cap E_n) \in X$ since X is a Banach space. Therefore $(D)\int_E f \, d\mu \in X$ for each $E \in \Sigma$, and f is Pettis integrable. ∎

Remark. It is obvious that if f : Ω ⟶ X is Bochner integrable, then f is Pettis integrable and both integrals coincide on each $E \in \Sigma$. In fact if f is Bochner integrable f is μ-measurable, and as $\int_E |x^*f(\omega)| \, d\mu(\omega) \le \|x^*\|\int_E \|f(\omega)\| \, d\mu(\omega)$ for each $x^* \in X^*$ and $E \in \Sigma$, f is Dunford integrable.

Then, from the proof above, $(D)\int_E f\,d\mu = (B)\int_E f\,d\mu \in X$ for each $E \in \Sigma$, hence f is Pettis integrable and $(P)\int_E f\,d\mu = (B)\int_E f\,d\mu$ for each $E \in \Sigma$.

Theorem 10.2.3 (Huff) *A Dunford integrable function* $f: \Omega \longrightarrow X$ *is Pettis integrable if and only if the linear operator* $T : X^* \longrightarrow L^1(\mu)$ *defined by* $Tx^* = x^*f$ *is weak*-weak continuous. Hence if* f *is Pettis integrable,* T *is weakly compact.*

Proof. If f is Pettis integrable, for each $E \in \Sigma$, the linear form u_E on X^* defined by $u_E(x^*) = \int_E x^*f\,d\mu$, belongs to X. Consequently, if (x_d^*) is a net in X^* which converges to x^* under the weak* topology of X^*, then $u_E(x_d^*) \longrightarrow u_E(x^*)$. So $\int_E x_d^*f\,d\mu \to \int_E x^*f\,d\mu$, or even better

$$\int_\Omega e(E)x_d^*f\,d\mu \longrightarrow \int_\Omega e(E)x^*f\,d\mu.$$

Now, as is well known, this implies that (x_d^*f) converges weakly to x^*f in $L^1(\mu)$. Hence, $\lim_d Tx_d^* = Tx^*$ weakly in $L^1(\mu)$.

Assume conversely that T is weak*-weak continuous. Then, if (x_d^*) converges to x^* under the weak* topology of X^*, one has that $Tx_d^* \to Tx^*$ weakly in $L^1(\mu)$, i.e. $\int_\Omega e(E)x_d^*f\,d\mu \to \int_\Omega e(E)x^*f\,d\mu$ for each $E \in \Sigma$. This means that $u_E(x_d^*) \longrightarrow u_E(x^*)$ in $\mathbb{K}$ for each $E \in \Sigma$, which shows that $u_E \in X$ for each $E \in \Sigma$. Hence f is Pettis integrable.

For the second part of the theorem, note that the closed unit ball of X^* is weak* compact and T is weak*-weak continuous. ∎

10.3 PROPERTIES OF THE PETTIS INTEGRAL

Some of the most interesting properties concerning the Pettis integral are stated in the following result.

Theorem 10.3.1 *Assume* f *is Dunford integrable and define* $F : \Sigma \longrightarrow X^{**}$ *by* $F(E) = (D)\int_E f\,d\mu$ *for each* $E \in \Sigma$. *Then we have that:*

(i) F *is a bounded vector measure.*

(ii) If f *is Pettis integrable, then* $F(\Sigma)$ *is a weakly relatively*

compact set. The converse affirmation is also true if we assume that f *is μ-measurable.*

(iii) If f *is Bochner integrable,* $F(\Sigma)$ *is relatively compact.*

Proof. Define $T : X^* \longrightarrow L^1(\mu)$ by $Tx^* = x^*f$. We have already seen in Lemma 10.1.1 that T is a bounded linear operator, hence $\|T\| < \infty$, i.e.

$$\sup \left\{ \int_\Omega |x^*f(\omega)| \; d\mu(\omega), \; x^* \in X^*, \; \|x^*\| \le 1 \right\} < \infty.$$

Hence, for each $E \in \Sigma$,

$$\|F(E)\| = \sup \left\{ |x^*F(E)|, \; x^* \in X^*, \; \|x^*\| \le 1 \right\} = \sup \left\{ \left| \int_E x^*f \; d\mu \right|, \; \|x^*\| \le 1 \right\}$$
$$\le \sup \left\{ \int_\Omega |x^*f(\omega)| \; d\mu(\omega), \; x^* \in X^*, \; \|x^*\| \le 1 \right\} < \infty.$$

This shows (i).

The first part of (ii) is a consequence of Theorem 10.2.1 and measure theory. Indeed, since F is countably additive and μ-continuous, we can apply a theorem from Bartle-Dunford-Schwartz (see Exercise 7 of this chapter) which shows that F has a weakly relatively compact range.

For the second part of (ii), assume f is μ-measurable and F has a weakly relatively compact range. We are going to demonstrate that f is Pettis integrable. In order to do that, it suffices to prove that F is countably additive and use Corollary 10.2.2.

Let us see first that F is strongly additive. Let S be the linear operator from $\ell_0^\infty(\Sigma)$ to X^{**} defined by $Se(E) = (D)\int_E f(\omega) \; d\mu(\omega)$ for each $E \in \Sigma$. Since the closed unit ball B of $\ell_0^\infty(\Sigma)$ is contained in the subset $4\Gamma\{e(E), \; E \in \Sigma\}$ of $\ell_0^\infty(\Sigma)$ and $\|Se(E)\| = \sup \left\{ |\langle x^*, (D)\int_E f \; d\mu\rangle|, \; \|x^*\| \le 1 \right\}$ $\le \sup \left\{ \int_\Omega |x^*f| \; d\mu, \; \|x^*\| \le 1 \right\} = \|T\|$, we have that $\|Sh\| \le 4\|T\|$ for each h of B, hence S is bounded. Let K be the extension of S to $\ell^\infty(\Sigma)$ with values in X^{**}. As $F(\Sigma)$ is weakly relatively compact in X^{**}, according to the theorem of Krein-Smulian $\overline{\Gamma(F(\Sigma))}$ is weakly relatively compact in X^{**}, and given that $Kh = Sh \in 4\overline{\Gamma(F(\Sigma))}$ for each h belonging to the unit ball of $\ell_0^\infty(\Sigma)$, it follows that $Kg \in 4\overline{\Gamma(F(\Sigma))}$ for each g of the closed unit ball of $\ell^\infty(\Sigma)$. Therefore, K is a weakly compact linear operator. Let $\{E_n, \; n \in \mathbb{N}\}$ be a sequence of pairwise disjoint elements of Σ, and let L be the subspace of $\ell^\infty(\Sigma)$ (isometric to ℓ^∞) of all those functions which take a constant value on each E_n. Denote by P the restriction of K into L. Then, if Q denotes the isometry from ℓ^∞ into L, the mapping $J := P \circ Q$ is a weakly compact linear operator from ℓ^∞ into X^{**}. Hence the formal series $\sum_n Je_n$ is unconditionally convergent in X^{**}. Consequently, $\sum_n P(e(E_n)) = \sum_n F(E_n)$ is unconditionally convergent in X^{**} and, hence, F is strongly additive.

Let us prove now that F is countably additive in norm; this will complete the proof of (ii). Note that assuming $x^* \in X^*$, then

$$x^*F(\cup E_n) = \langle x^*, (D)\int_{\cup E_n} f \, d\mu \rangle = \int_{\cup E_n} x^*f \, d\mu = \sum_n \int_{E_n} x^*f \, d\mu = \sum_n x^*F(E_n)$$

and hence F is weak* countably additive. If $x^{**} = \sum_n F(E_n)$, then we have $x^*x^{**} = \langle x^*, \sum_n F(E_n) \rangle = \sum_n x^*F(E_n) = x^*F(\cup E_n)$ for each $x^* \in X^*$. Therefore $x^{**} = F(\cup E_n)$ and consequently F is countably additive in norm.

For the proof of (iii) assume that f is Bochner integrable and let (f_n) be a sequence of simple functions from $L^1(\mu,X)$ which converges to f in $L^1(\mu,X)$, that is, such that $\lim_n \int_\Omega \|f(\omega) - f_n(\omega)\| \, d\mu(\omega) = 0$.

Define $K_n : \ell^\infty(\Sigma) \longrightarrow X$ by $K_n e(E) = (B)\int_E f_n \, d\mu$. Since each K_n is a bounded operator of finite-dimensional range, each K_n is compact. Set

$$Ke(E) = (B)\int_E f \, d\mu.$$

Then K is bounded, since its restriction to $\ell_0^\infty(\Sigma)$ is bounded (note that $\|Ke(E)\| = \|(B)\int_E f \, d\mu\| \le \int_E \|f\| \, d\mu \le \|f\|_1 \ \forall E \in \Sigma$).

Given $\varepsilon > 0$, there is some $n_0 \in \mathbb{N}$ such that $\|f - f_n\|_1 \le \varepsilon$ for each $n \ge n_0$. Consequently, for each $n \ge n_0$ one has that

$$\|(K - K_n)e(E)\| \le \int_\Omega \|f(\omega) - f_n(\omega)\| \, d\mu(\omega) \le \varepsilon$$

for $E \in \Sigma$. This implies that for each h belonging to the closed unit ball of $\ell_0^\infty(\Sigma)$, then $\|(K - K_n)h\| \le 4\varepsilon$ for each $n \ge n_0$. But as the closed unit ball of $\ell^\infty(\Sigma)$ is the closure of the ball of $\ell_0^\infty(\Sigma)$ and $K - K_n$ is a continuous mapping from $\ell^\infty(\Sigma)$ into X for each $n \in \mathbb{N}$, it follows that $\|(K - K_n)g\| \le 4\varepsilon$ for each g of the closed unit ball of $\ell^\infty(\Sigma)$ provided $n \ge n_0$. This establishes that $\lim_n \|K - K_n\| = 0$; i.e, $K_n \longrightarrow K$ under the uniform convergence topology of operators of $\mathcal{L}(\ell^\infty(\Sigma),X)$. Hence K is a compact operator, because it is the uniform limit of a sequence of compact operators. ■

10.4 DUNFORD INTEGRABLE FUNCTION SPACES

In the following we shall denote by $\mathcal{D}_1(\mu,X)$ the linear space over $\mathbb{K}$ of all those functions $f : \Omega \longrightarrow X$ which are Dunford integrable. We define in $\mathcal{D}_1(\mu,X)$ the norm

$$\|f\| = \sup \left\{ \int_\Omega |x^*f(\omega)| \, d\mu(\omega), \ x^* \in X^*, \ \|x^*\| \le 1 \right\}$$

for each $f \in \mathcal{D}_1(\mu,X)$. Let us see that, in fact, this defines a norm on $\mathcal{D}_1(\mu,X)$. In the first place, notice that by the proof of Lemma 10.1.1, the linear operator $T_f : X^* \longrightarrow L^1(\mu)$, defined by $T_f x^* = x^*f$ for each x^* of X^*, is bounded. Hence, $\|T_f\|_{\mathcal{L}} < \infty$, and consequently

$$\sup \left\{ \int_\Omega |x^*f(\omega)| \, d\mu(\omega), \ x^* \in X^*, \ \|x^*\| \le 1 \right\} < \infty.$$

Moreover, since $\|f\| = \|T_f\|_{\mathcal{L}}$ for each $f \in \mathcal{D}_1(\mu,X)$, and $\|\cdot\|_{\mathcal{L}}$ is a norm on $\mathcal{L}(X^*,L^1(\mu))$, then $\|\cdot\|$ is a norm on $\mathcal{D}_1(\mu,X)$.

The linear subspace of $\mathcal{D}_1(\mu,X)$ of all Pettis integrable functions will be denoted by $\mathcal{P}_1(\mu,X)$, and the subspaces of $\mathcal{D}_1(\mu,X)$ and $\mathcal{P}_1(\mu,X)$ of all μ-measurable functions will be represented by $D_1(\mu,X)$ and $P_1(\mu,X)$, respectively.

Lemma 10.4.1 *Assuming that* $f \in \mathcal{D}_1(\mu,X)$ *and* $F : \Sigma \longrightarrow X^{**}$ *is defined by* $F(E) := (D)\int_E f \, d\mu$, *then* $\|e(E)f\| = \|F\|(E)$, *where the first is the norm of* $\mathcal{D}_1(\mu,X)$ *and the second the semivariation norm in* $ba(\Sigma,X)$. *In particular* $\|f\| = \|F\|$.

Proof. Let $E \in \Sigma$ and denote by $\Pi(E)$ the family of all finite partitions of E by elements of Σ. Since $x^*F(E) = \int_E x^*f \, d\mu$ for each $x^* \in X^*$ one has that $|x^*F|(E) = \int_E |x^*f(\omega)| \, d\mu(\omega)$, where $|\cdot|$ stands for the variation of the scalar measure x^*F. Consequently,

$$\sup \left\{ |x^*F|(E), \ \|x^*\| \le 1 \right\} = \sup \left\{ \int_\Omega |e(E)x^*f| \, d\mu, \ \|x^*\| \le 1 \right\}.$$

That is,

$\sup\{\sum_{A\in\pi}|x^*F(A)|,\ \pi \in \Pi(E),\ \|x^*\| \le 1\} = \sup\{\int_\Omega |x^*fe(E)|\ d\mu,\ \|x^*\| \le 1\}$

hence, $\|F\|(E) = \|e(E)f\|$. ■

Lemma 10.4.2 *If* $f \in \mathcal{P}_1(\mu,X)$, *then* $\lim_{\mu(E)\to 0} \|e(E)f\| = 0$.

Proof. Since $f \in \mathcal{P}_1(\mu,X)$, according to the Theorem 10.2.1, $F : \Sigma \longrightarrow X$, defined by $F(E) = (P)\int_E f\ d\mu$ for each $E \in \Sigma$, is a countably additive and μ-continuous measure. Hence, $\lim_{\mu(E)\to 0}\|F(E)\| = 0$ and then, given $\varepsilon > 0$, there is a $\delta > 0$ such that if $\mu(E) < \delta$, then $\|F(E)\| < \varepsilon/4$. So, assuming $\mu(E) < \delta$, then $\mu(E \cap A) < \delta$ for each $A \in \Sigma$, and hence $\|F(E \cap A)\| < \varepsilon/4$. Therefore, $\|F\|(E) \le 4\sup\{\|F(E \cap A)\|,\ A \in \Sigma\} < \varepsilon$. Since $\|F\|(E) = \|fe(E)\|$ as a consequence of the previous lemma, then $\|fe(E)\| < \varepsilon$. ■

Lemma 10.4.3 *If* $S(\mu,X)$ *denotes the linear subspace of* $P_1(\mu,X)$ *formed by the simple functions, then* $S(\mu,X)$ *is dense in* $P_1(\mu,X)$.

Proof. Since $L^1(\mu,X) \subseteq P_1(\mu,X)$, $S(\mu,X)$ is a linear subspace of $P_1(\mu,X)$. If $f \in P_1(\mu,X)$, then f is μ-measurable and due to Corollary 8.1.2 given $\varepsilon > 0$ there is a countably X-valued μ-measurable function g_ε such that $\|f - g_\varepsilon\|_\infty < \varepsilon/\mu(\Omega)$. Since for finite measure spaces $L^\infty(\mu,X) \subseteq L^1(\mu,X) \subseteq P_1(\mu,X)$, then $f - g_\varepsilon \in P_1(\mu,X)$. Moreover, $\|f - g_\varepsilon\| \le \|f - g_\varepsilon\|_\infty\mu(\Omega) < \varepsilon$. Since $g_\varepsilon = (g_\varepsilon - f) + f$ and $g_\varepsilon - f$, $f \in P_1(\mu,X)$, then $g_\varepsilon \in P_1(\mu,X)$.

Assume that $g_\varepsilon = \sum_n x_n e(E_n)$, with $E_n \cap E_m = \emptyset$ for $n \ne m$, and define $A_n := \bigcup\{E_j,\ 1 \le j \le n\}$ for each $n \in \mathbb{N}$. By the previous lemma there is a $n_0 \in \mathbb{N}$ such that $\|e(\Omega \setminus A_n)g_\varepsilon\| < \varepsilon$ for each $n \ge n_0$. Therefore,

$$\|g_\varepsilon - \sum_{j=1}^{n} x_j e(E_j)\| < \varepsilon$$

for each $n \ge n_0$. This implies that

$$\|f - \sum_{j=1}^{n} x_j e(E_j)\| < 2\varepsilon$$

for each $n \ge n_0$, which completes the proof. ■

Theorem 10.4.4 *Assume* $f : \Omega \longrightarrow X$ *is Dunford integrable. If* f *is also* μ*-measurable,* f *is Pettis integrable if and only if* $F(\Sigma)$ *is relatively compact.*

Proof. If $F(\Sigma)$ is relatively compact, it is weakly relatively compact, so f is Pettis integrable by virtue of part (ii) of Theorem 10.3.1. For the converse assume f is μ-measurable and Pettis integrable, i.e, let $f \in P_1(\mu,X)$. We are going to see that $F(\Sigma)$ is relatively compact in norm.

Using the previous result, choose $(f_n) \subseteq S(\mu,X)$ such that $f_n \longrightarrow f$ in $P_1(\mu,X)$ and define $F_n(E) = (P)\int_E f_n \, d\mu$ for each $n \in \mathbb{N}$ and each $E \in \Sigma$. If $f_n := \sum_{1\le m\le r(n)} x_{nm} e(E_{nm})$, with $\bigcup\{E_{nm}, 1 \le m \le r(n)\} = \Omega$ and $E_{nm} \cap E_{nk} = \emptyset$ whenever $m \ne k$ for each $n \in \mathbb{N}$, then $F_n(E) = \sum_{1\le m\le r(n)} x_{nm}\mu(E \cap E_{nm})$ for $n \in \mathbb{N}$. Since $F_n(\Sigma) \subseteq \langle\{x_{nm}, 1 \le m \le r(n)\}\rangle$ each F_n has a relatively compact range in norm. Hence, given $\varepsilon > 0$, let $\{z_{n1}, z_{n2}, \ldots, z_{nk(n)}\}$ be an $\varepsilon/2$-net in $F_n(\Sigma)$ and choose some $n \in \mathbb{N}$ such that

$$\sup\{\|F(E) - F_n(E)\|, E \in \Sigma\} \le \|F - F_n\| = \|f - f_n\| < \varepsilon/2.$$

Given $E \in \Sigma$, let z_{nj} be such that $\|F_n(E) - z_{nj}\| < \varepsilon/2$. Then

$$\|F(E) - z_{nj}\| \le \|F(E) - F_n(E)\| + \|F_n(E) - z_{nj}\| < \varepsilon.$$

This shows that $\{z_{n1}, z_{n2}, \ldots, z_{nk(n)}\}$ is an ε-net in $F(\Sigma)$, thus $F(\Sigma)$ is precompact in X, i.e, relatively compact in norm. ■

Theorem 10.4.5 *Let* $f \in \mathcal{D}_1(\mu,X)$. *If* $f \in L^1(\mu,X)$, *there is* $z^* \in X^*$ *with* $\|z^*\| \le 1$ *such that* $\|f\| = \int_\Omega |z^*f(\omega)| \, d\mu(\omega)$. *That is,* $\|f\| = \|z^*f\|_1$.

Proof. If $f \in L^1(\mu,X)$ and B denotes the closed unit ball of X^* the map $\psi : B(weak^*) \longrightarrow \mathbb{R}$ defined by $\psi(x^*) = \int_\Omega |x^*f| \, d\mu$ is continuous. In fact, as f is μ-measurable, by Theorem 8.1.1 f has an essentially separable range. Given that we are going to work in the range of f, there is not loss of generality by assuming that X is separable. In this case, B is weakly* metrizable. If $(x_n^*) \subseteq B$ converges to $x^* \in B$ under the weak* topology of B, then $x_n^*f(\omega) \longrightarrow x^*f(\omega)$, and, since $|x_n^*f(\omega)| \le \|f(\omega)\|$ for each $n \in \mathbb{N}$ and $\omega \in \Omega$, and since $\omega \longrightarrow \|f(\omega)\|$ belongs to $L^1(\mu)$, then by the dominated convergence theorem, $\lim_n \int_\Omega |x_n^*f| \, d\mu = \int_\Omega |x^*f| \, d\mu$, which establishes the continuity of ψ. As B is a compact metric space under the weak* topology, there is some x^* where $\psi(z^*)$ reaches its maximum. ■

10.5 BARRELLEDNESS OF THE PETTIS INTEGRABLE FUNCTION SPACE

Let (Ω,Σ,μ) be a finite measure space. A normed space E is said to admit a (Ω,Σ,μ)-Boolean algebra of projections if there is a family $\{P_A,\ A \in \Sigma\}$ of projections on E which satisfy the following conditions:

(1) P_Ω is the identity on E

(2) $P_A = 0$ for each $A \in \Sigma$ such that $\mu(A) = 0$

(3) $P_{A\cap B} = P_A \circ P_B$ for each $A, B \in \Sigma$

(4) $P_{A\cup B} = P_A + P_B$ for each $A, B \in \Sigma$ such that $A \cap B = \emptyset$

(5) P_A is continuous for each $A \in \Sigma$

(6) $\lim_{\mu(A)\to 0} P_A(x) = 0$ for each $x \in E$.

Note that the family $\{P_A,\ A \in \Sigma\}$ is equicontinuous. In fact, given $x \in E$, define $F_x : \Sigma \longrightarrow E$ by $F_x(A) = P_A(x)$. Then according to property (2), $F_x(\emptyset) = 0$, and by property (4), $F_x(A \cup B) = F_x(A) + F_x(B)$ whenever $A \cap B = \emptyset$. This proves that F_x is an E-valued vector measure. Property (6) says that F_x is μ-continuous. As μ is countably additive, if (A_n) is a sequence of pairwise disjoint elements of Σ whose union is A, then

$$\lim_n \mu(A \setminus \bigcup_{j=1}^{n} A_j) = \lim_n |\mu(A) - \textstyle\sum_{1\le j\le n}\mu(A_j)| = 0,$$

which implies that $\lim_n F_x(A \setminus \bigcup_{j=1}^{n} A_j) = 0$. Consequently, $F_x(A) = \sum_n F_x(A_n)$ and F_x is countably additive. Thus $\{F_x,\ x \in E\}$ is a family of E-valued countably additive measures defined in Σ. Now, if $\mathcal{B}$ denotes the closed unit ball of E, $\sup\{\|F_x(Q)\|,\ x \in \mathcal{B}\} < \infty$ for each $Q \in \Sigma$ since otherwise there would exist an $A \in \Sigma$ such that $\sup\{\|F_x(A)\|,\ x \in \mathcal{B}\} = \infty$, that is, there would exist a sequence (x_n) in $\mathcal{B}$ with $\|F_{x_n}(A)\| > n$, $n \in \mathbb{N}$, i.e., such that $\|P_A(x_n)\| > n$ for each $n \in \mathbb{N}$. Hence, setting $y_n := x_n/n$, $n \in \mathbb{N}$, then $y_n \longrightarrow 0$ while $\|P_A(y_n)\| > 1$ for $n \in \mathbb{N}$, contradicting the continuity of P_A. Thus, from Nikodym's boundedness theorem for vector measures, we have that $\sup\{\|F_x(Q)\|,\ x \in \mathcal{B},\ Q \in \Sigma\} < +\infty$. So, $\sup\{\|P_Q\|,\ Q \in \Sigma\} < \infty$, and consequently $\{P_Q,\ Q \in \Sigma\}$ is uniformly bounded, or equicontinuous.

Remark 10.5.1 We have seen that (6) implies that each F_x is a countable additive vector measure. If instead of (6) we request that each F_x be countably additive, a well-known result of Pettis concerning vector measures shows that (2) and (6) are equivalent. Thus we can replace (6) by the requirement that each F_x be countably additive.

Theorem 10.5.2 *Assume* E *admits a* (Ω,Σ,μ)*-Boolean algebra of projections* $\{P_A,\ A \in \Sigma\}$ *satisfying the following condition:*

(7) *If* (A_n) *is a sequence of pairwise disjoint elements of* Σ *and* (x_n) *is a null sequence in* E *such that* $P_{A_n}(x_n) = x_n$ *for each* $n \in \mathbb{N}$, *then there is a subsequence* $(x_{n(k)})$ *such that* $\sum_k x_{n(k)}$ *converges in* E.

If (Ω,Σ,μ) *is atomless, then* E *is barrelled.*

Proof. Assuming that E is not barrelled there is a weak* bounded subset M of E*, which is not bounded under the norm of E*. Hence, we have that $\sup\{\|v\|,\ v \in M\} = \infty$. Therefore we can choose $v_n \in M$ with $\|v_n\| > n^2$ for each $n \in \mathbb{N}$. Now, if $\mathcal{B}$ denotes the closed unit ball E let $z_n \in \mathcal{B}$ be such that $|\langle v_n,z_n\rangle| > n^2$, and define $x_n := z_n/n$ for each $n \in \mathbb{N}$. Then $x_n \longrightarrow 0$ in E and $|\langle v_n,x_n\rangle| > n$, $n \in \mathbb{N}$. This shows that

$$\sup\{|\langle u,x_n\rangle|,\ n \in \mathbb{N},\ u \in M\} = +\infty.$$

Setting $\Omega_0 := \Omega$ and using (1) we may write

$$\sup\{|\langle u,P_{\Omega_0}(x_n)\rangle|,\ n \in \mathbb{N},\ u \in M\} = +\infty \qquad (*)$$

and select $u_1 \in M$ and $n(1) \in \mathbb{N}$ such that

$$\gamma_1 = |\langle u_1,\ P_{\Omega_0}(x_{n(1)})\rangle| > 1.$$

According to (6), $\lim_{\mu(A)\to 0}\|P_A(x_{n(1)})\| = 0$. Therefore, there is a $\delta > 0$ such that $\|P_A(x_{n(1)})\| < \gamma_1 - 1$ for each $A \in \Sigma$ with $\mu(A) \le \delta$.

Since μ is atomless, there are C_1, $C_2 \in \Sigma$, with $\mu(C_1) = \mu(C_2) = \delta$ and $C_1 \cup C_2 = \Omega_0$. Hence, $|\langle u_1,P_{C_1}(x_{n(1)})\rangle|$, $|\langle u_1,P_{C_2}(x_{n(1)})\rangle| < \gamma_1 - 1$. But, on the other hand, from (*), and using (4), it follows that either for C_1 or for C_2, say, for Ω_1, it holds that

$$\sup\{|\langle u,P_{\Omega_1}(x_n)\rangle|,\ n > n(1),\ u \in M\} = +\infty. \qquad (**)$$

Set $A_1 := \Omega_0 \setminus \Omega_1$. Then, as $\mu(\Omega_1) \le \delta$, we have

$$|\langle u_1,P_{A_1}(x_{n(1)})\rangle| = |\langle u_1,P_{\Omega_0}(x_{n(1)})\rangle - \langle u_1,P_{\Omega_1}(x_{n(1)})\rangle| \ge$$

$$|\langle u_1, P_{\Omega_0}(x_{n(1)})\rangle| - |\langle u_1, P_{\Omega_1}(x_{n(1)})\rangle| \geq \gamma_1 - (\gamma_1 - 1) = 1.$$

Now, using (**), choose $u_2 \in M$ and $n(2) > n(1)$ such that

$$|\langle u_2, P_{\Omega_1}(x_{n(2)})\rangle| > 2$$

and, reasoning as before, find some $\Omega_2 \in \Sigma$ with $\Omega_2 \subseteq \Omega_1$, such that

$$\sup \{ |\langle u, P_{\Omega_2}(x_n)\rangle|,\ n > n(2),\ u \in M\} = +\infty$$

and such that, if $A_2 := \Omega_1 \setminus \Omega_2$, then $|\langle u_2, P_{A_2}(x_{n(2)})| > 2$.

Proceeding by recurrence we obtain a sequence $(u_n) \subseteq M$, a sequence (A_n) of pairwise disjoint elements of Σ, and a subsequence $(x_{n(j)})$ of (x_n) such that

$$|\langle u_j, P_{A_j}(x_{n(j)})\rangle| > j$$

for each $j \in \mathbb{N}$.

Since $\{P_A,\ A \in \Sigma\}$ is equicontinuous, there is an $\eta > 0$ with $\|P_A\| < \eta$ for each $A \in \Sigma$. On the other hand, given $\varepsilon > 0$, there is some $n_0 \in \mathbb{N}$ such that, if $j \geq n_0$, then $\|x_{n(j)}\| < \varepsilon/\eta$ since $x_{n(j)} \longrightarrow 0$. Hence, if $j \geq n_0$, one has that $\|P_{A_j}(x_{n(j)})\| = \|P_{A_j}\| \|x_{n(j)}\| \leq \eta \|x_{n(j)}\| < \varepsilon$.

This shows that $\lim_j P_{A_j}(x_{n(j)}) = 0$. Hence setting $z_j := P_{A_j}(x_{n(j)})$ for each $j \in \mathbb{N}$, (z_j) is a null sequence of E such that $P_{A_j}(z_j) = z_j$ for each $j \in \mathbb{N}$. Therefore by property (7) there is a subsequence $(x_{n(j(k))})$ of $(x_{n(j)})$ such that the series $\sum_k P_{A_k}(x_{n(j(k))})$ converges in E to some vector $y \in E$. Now define

$$w_k := u_{j(k)},\quad y_k := x_{n(j(k))},\quad B_k := A_{j(k)}$$

for each $k \in \mathbb{N}$. Then,

$$y = \sum_k P_{B_k}(y_k)$$

and, on the other hand,

$$|\langle w_k, P_{B_k}(y_k)\rangle| > k \tag{***}$$

for $k \in \mathbb{N}$.

From the previous equation, if follows from (3) and (5) that

$$P_{B_k}(y) = P_{B_k}(y_k)$$

for each $k \in \mathbb{N}$. As M is weak* bounded, the scalar measures $m_k : \Sigma \longrightarrow \mathbb{K}$

defined by $m_k(A) = \langle w_k, P_A(y)\rangle$ for each $A \in \Sigma$ verify that

$$\sup \{ |m_k(A)|,\ k \in \mathbb{N} \} < +\infty$$

for $A \in \Sigma$. Therefore, according to Nikodym's boundedness theorem,

$$\sup \{ |m_k(A)|,\ k \in \mathbb{N},\ A \in \Sigma \} < +\infty.$$

This implies that there is some $r \in \mathbb{N}$ such that $|\langle w_k, P_{B_k}(y)\rangle| < r$ for each $k \in \mathbb{N}$. Consequently $|\langle w_r, P_{B_r}(y_r)\rangle| < r$, contradicting (***). ∎

Corollary 10.5.3 *Assume that* E *is a normed space that admits a* (Ω,Σ,μ)-*Boolean algebra of projections* $\{P_A,\ A \in \Sigma\}$ *which satisfies condition* (7) *and suppose that* F *is a closed linear subspace of* E. *If* (Ω,Σ,μ) *is atomless and* $P_A(F) \subseteq F$ *for each* $A \in \Sigma$, *then* F *is barrelled.*

Proof. The argument of the proof of the previous theorem is also true if we consider the subspace F instead of E. The requirement that F must be closed ensures that the series of condition (7) does converge to an element of F. ∎

Theorem 10.5.4 *The normed space* $\mathcal{P}_1(\mu,X)$ *is barrelled.*

Proof. Let us assume in the first place that the measure space (Ω,Σ,μ) is atomless. For each $A \in \Sigma$ y each $f \in \mathcal{P}_1(\mu,X)$, define $P_A(f) = e(A)\cdot f$.

Clearly the family $\{P_A,\ A \in \Sigma\}$ satisfies conditions (1)-(5), and, on the other hand, according to Lemma 10.4.2, $\lim_{\mu(E)\to 0}\|P_E(f)\| = 0$ for each $f \in \mathcal{P}_1(\mu,X)$, which shows that condition (6) also holds. Therefore, we have only to check condition (7) in order to finish the proof, since the conclusion follows from the previous theorem.

Let (A_n) be a sequence of pairwise disjoint elements of Σ and let (f_n) denote a null sequence in $\mathcal{P}_1(\mu,X)$ such that $P_{A_n}(f_n) = f_n$ for each $n \in \mathbb{N}$, i.e., such that f_n has its support contained in A_n, and this for each $n \in \mathbb{N}$. We have to show that there is a subsequence $(f_{n(j)})$ of (f_n) such that $\sum_j f_{n(j)}$ converges in $\mathcal{P}_1(\mu,X)$ to some vector of $\mathcal{P}_1(\mu,X)$.

Now, since $\|f_n\| \to 0$ we may choose $(n(j)) \subseteq \mathbb{N}$ with $\|f_{n(j)}\| \le 1/2^j$. Thus, $\sum_j f_{n(j)}$ converges absolutely to some vector of the completion of $\mathcal{P}_1(\mu,X)$. Set $h_j := f_{n(j)}$ for each $j \in \mathbb{N}$, and notice that the series

$\sum_j h_j(\omega)$ converges in X for each $\omega \in \Omega$ since the A_n are pairwise disjoint and so the sum is finite in each point. Defining $h : \Omega \longrightarrow X$ by $h(\omega) = \sum_j h_j(\omega)$ for each $\omega \in \Omega$, we need demonstrate that $x^*h \in L^1(\mu)$ for each $x^* \in X^*$, and $(D)\int_E h \, d\mu \in X$ for each $E \in \Sigma$.

Since $\sum_n \int_\Omega |x^*h_n(\omega)| \, d\mu(\omega) \le \|x^*\| \cdot \sum_n \|h_n\| \le \|x^*\|$, the series $\sum_n x^*h_n$ converges to x^*h almost everywhere in Ω and $x^*h \in L^1(\mu)$. Moreover, one has that $\int_E x^*f(\omega) \, d\mu(\omega) = \sum_n \int_E x^*h_n(\omega) \, d\mu(\omega)$ for each $E \in \Sigma$ and $x^* \in X^*$. That is, $(D)\int_E h \, d\mu = \sum_n (P)\int_E h_n \, d\mu$ for the weak* topology of X^{**}. Since

$$\|(P)\int_E h_n \, d\mu\| = \sup \{ |\langle x^*, (P)\int_E h_n \, d\mu\rangle|, \ \|x^*\| \le 1 \} =$$
$$\sup \{ |\int_E x^*h_n \, d\mu|, \ \|x^*\| \le 1 \} \le \sup \{ \int_E |x^*h_n| \, d\mu, \ \|x^*\| \le 1 \} \le$$
$$\sup \{ \int_\Omega |x^*h_n| \, d\mu, \ \|x^*\| \le 1 \} = \|h_n\|$$

for all $n \in \mathbb{N}$ it follows that $\sum_n \|(P)\int_E h_n \, d\mu\| \le \sum_n \|h_n\| \le 1$. Consequently $\sum_n (P)\int_E h_n \, d\mu$ is absolutely convergent in X to some vector $x_E \in X$. But, because of what we have seen above, necessarily $x_E = (D)\int_E h \, d\mu$, which shows that $h \in \mathcal{P}_1(\mu,X)$.

If the measure space (Ω,Σ,μ) contains atoms, we know there are two subsets Ω_a and Ω_{pa} such that the restriction of μ to $\Sigma_{|\Omega_a}$ is atomless while the restriction of μ to $\Sigma_{|\Omega_{pa}}$ is purely atomic. Moreover, $\mathcal{P}_1(\mu,X)$ is the topological direct sum of $\mathcal{P}_1(\Omega_a,\Sigma_{|\Omega_a},\mu,X)$ and $\mathcal{P}_1(\Omega_{pa},\Sigma_{|\Omega_{pa}},\mu,X)$.

According to what we have proved before, $\mathcal{P}_1(\Omega_a,\Sigma_{|\Omega_a},\mu,X)$ is barrelled. On the other hand, as the number of atoms of (Ω,Σ,μ) is countable, the space $\mathcal{P}_1(\Omega_{pa},\Sigma_{|\Omega_{pa}},\mu,X)$ is isometric either to X^m (if μ has m atoms) or to the Banach space $\ell^1\langle X\rangle$ of all unconditionally convergent sequences of X provided with the norm

$$\|(x_n)\| = \sup \{ \sum_n |x^*x_n|, \ \|x^*\| \le 1 \}.$$

In any case $\mathcal{P}_1(\Omega_{pa},\Sigma_{|\Omega_{pa}},\mu,X)$ is a Banach space, hence $\mathcal{P}_1(\mu,X)$ is barrelled, since it is the product of two barrelled spaces. ■

Remark 10.5.5 Although in general $P_1(\mu,X)$ is not a closed subspace of $\mathcal{P}_1(\mu,X)$, we may reason as in the previous theorem to show that $P_1(\mu,X)$ is barrelled, since $P_A(P_1(\mu,X))$ is contained in $P_1(\mu,X)$ for each $A \in \Sigma$ and h is μ-measurable.

10.6 ULTRABORNOLOGICAL PROPERTIES OF DUNFORD INTEGRABLE FUNCTION SPACES

In this final section we will consider a normed space E which admits a (Ω,Σ,μ)-Boolean family of projections (which we will continue calling a (Ω,Σ,μ)-Boolean algebra of projections) satisfying exclusively the following conditions:

(1) P_Ω is the identity on E

(2) $P_A = 0$ for each $A \in \Sigma$ such that $\mu(A) = 0$

(3) $P_{A\cap B} = P_A \circ P_B$ for each $A, B \in \Sigma$

(4) $P_{A\cup B} = P_A + P_B$ for each $A, B \in \Sigma$ such that $A \cap B = \emptyset$

(5) P_A is continuous for each $A \in \Sigma$.

Theorem 10.6.1 *Assume* E *has a* (Ω,Σ,μ)*-Boolean algebra of projections* $\{P_A,\ A \in \Sigma\}$ *and the family* $\{P_A,\ A \in \Sigma\}$ *is equicontinuous. We will also assume that this family satisfies the following condition:*

(7) If (Ω_n) *is a decreasing sequence of elements of* Σ *such that* $\mu(\bigcap_{n=1}^{\infty} \Omega_n) = 0$ *and* (x_n) *is a bounded sequence of vectors of* E *such that* $P_{\Omega_n}(x_n) = x_n$ *for each* $n \in \mathbb{N}$*, then the series* $\sum_n \xi_n x_n$ *(which converges in the completion of* E*) converges in* E *for each* $\xi \in \ell^1$.

If (Ω,Σ,μ) *is atomless, then* E *is ultrabornological.*

Proof. Suppose T is an absolutely convex subset of E which absorbs each Banach disk in E. We are going to show that T is a neighbourhood of the origin in E.

Let $\mathcal{B}$ be the closed unit ball of E. Assume by contradiction that T does not absorb $\mathcal{B}$. Let $x_1 \in \mathcal{B} \setminus T$ be such that $\|x_1\| = 1$ and set $\Omega_1 := \Omega$. Since μ is atomless we can split Ω_1 into two disjoint subsets belonging to Σ, both of measure $\mu(\Omega)/2$. For one of these two, say Ω_2, it must happen that T does not absorbs $P_{\Omega_2}(\mathcal{B})$. Then take some $x_2 \in P_{\Omega_2}(\mathcal{B}) \setminus 2T$.

Since $\{P_A, A \in \Sigma\}$ is equicontinuous there is a $\delta > 1$ with $\|P_A\| < \delta$ for each $A \in \Sigma$, which implies that $\|x_2\| < \delta$. Split Ω_2 into two disjoint sets of Σ, both of measure $\mu(\Omega)/4$, and choose one, say Ω_3, such that T does not absorb $P_{\Omega_3}(P_{\Omega_2}(\mathcal{B}))$ — which, using (3), coincides with $P_{\Omega_3}(\mathcal{B})$. Then we choose $x_3 \in P_{\Omega_3}(\mathcal{B}) \setminus 3T$, noticing that $\|x_3\| < \delta$.

By recurrence we obtain a decreasing sequence (Ω_n) of elements of Σ verifying that $\mu(\bigcap_n \Omega_n) = 0$, and a normalized sequence (x_n) in E such that $x_n \in P_{\Omega_n}(\mathcal{B})$, $\|x_n\| < \delta$ and $x_n \notin nT$ for every $n \in \mathbb{N}$. If $y_n \in \mathcal{B}$ is such that $x_n = P_{\Omega_n}(y_n)$ for each $n \in \mathbb{N}$, as $x_n = P_{\Omega_n}(y_n) = P_{\Omega_n}(P_{\Omega_n}(y_n)) = P_{\Omega_n}(x_n)$ for all $n \in \mathbb{N}$, the condition (7) allows us to conclude that $\sum_n \xi_n x_n$ converges in E for each $\xi \in \ell^1$. Since the mapping $\psi : \ell^1 \longrightarrow E$ defined by $\psi(\xi) = \sum_n \xi_n x_n$ is continuous, $D := \{\sum_n \xi_n x_n, \ \xi \in \ell^1 \ \|\xi\|_1 \le 1\}$ is a Banach disk contained in E. As T absorbs each Banach disk of E, T absorbs D and hence there is some $m \in \mathbb{N}$ such that $D \subseteq mT$. This implies that $x_m \in mT$, a contradiction. ■

Corollary 10.6.2 *If* (Ω,Σ,μ) *is atomless, then* $\mathcal{D}_1(\mu,X)$ *and* $\mathcal{P}_1(\mu,X)$ *are ultrabornological spaces.*

Proof. Let us prove that $\mathcal{D}_1(\mu,X)$ satisfies condition (7) with $P_A(f) = e(A)\cdot f$ for each $A \in \Sigma$. Notice that, as $\|P_A\| = \sup\{\|e(A)f\|, \ \|f\| \le 1\}$ then one has $\|P_A\| = \sup\left\{\int_A |x^*f(\omega)| \ d\mu(\omega), \ \|x^*\| \le 1, \ \|f\| \le 1\right\} \le \sup\left\{\int_\Omega |x^*f(\omega)| \ d\mu(\omega), \ \|x^*\| \le 1, \ \|f\| \le 1\right\} \le \sup\{\|f\|, \ \|f\| \le 1\} \le 1$, thus the family $\{P_A, A \in \Sigma\}$ is equicontinuous. Now assume that $\{\Omega_n, n \in \mathbb{N}\}$ is a decreasing sequence of members of Σ such that $\mu\left(\bigcap_{n=1}^{\infty} \Omega_n\right) = 0$ and suppose (f_n) is a bounded sequence of elements of $\mathcal{D}_1(\mu,X)$ — assume for instance that $\|f_n\| \le 1$ for each $n \in \mathbb{N}$ — and let $\xi \in \ell^1$ (without loss of generality we assume that $\|\xi\|_1 \le 1$). If $\omega \in \Omega$, either $\omega \in \bigcap\{\Omega_n, n \in \mathbb{N}\}$, or $\omega \notin \bigcap\{\Omega_n, n \in \mathbb{N}\}$. In the first case, since $\mu\left(\bigcap_{n=1}^{\infty} \Omega_n\right) = 0$, we may agree that $f_n(\omega) = 0$ for all $n \in \mathbb{N}$, which implies that $\sum_n \xi_n f_n(\omega) = 0$. In the second case there is a $m \in \mathbb{N}$ such that $\omega \notin \Omega_m$, so $\sum_n \xi_n f_n(\omega) =$

$\sum_{n\le m}\xi_n f_n(\omega) \in X$. This shows that $\sum_n \xi_n f_n(\omega)$ converges in X for almost all $\omega \in \Omega$.

If f denotes the pointwise limit of the series considered above, f is weakly μ-measurable. On the other hand, for every $x^* \in X^*$, $\|x^*\| \le 1$, one has $\|x^*f_n\|_1 = \int_\Omega |x^*f_n(\omega)|\, d\mu(\omega) \le \|f_n\| \le 1$ for all $n \in \mathbb{N}$, therefore $\sum_n \int_\Omega |\xi_n x^*f_n|\, d\mu \le \sum_n |\xi_n|\, \|x^*f_n\|_1 \le 1$. Since $\sum_n \xi_n x^*f_n(\omega) = x^*f(\omega)$ almost everywhere in Ω, the dominated convergence theorem yields $x^*f \in L^1(\mu)$ and $\sum_n \xi_n x^*f_n$ converges to x^*f in $L^1(\mu)$. But, given $\varepsilon > 0$ there is a $k \in \mathbb{N}$ such that $\sum_{n\le j\le m} |\xi_j| < \varepsilon$ for $m \ge n \ge k$, hence

$$\|\textstyle\sum_{1\le j\le m}\xi_j x^*f_j - \sum_{1\le j\le n}\xi_j x^*f_j\|_1 = \|\sum_{n<j\le m}\xi_j x^*f_j\|_1 =$$

$$\int_\Omega |\textstyle\sum_{n\le j\le m}\xi_j x^*f_j(\omega)|\, d\mu(\omega) \le \sum_{n\le j\le m}|\xi_j|\, \|x^*f_j\|_1 \le \sum_{n\le j\le m}|\xi_j| < \varepsilon$$

for each $x^* \in x^*$ with $\|x^*\| \le 1$. Letting $m \longrightarrow +\infty$ and keeping $n \ge k$ it follows that, for each vector x^* of the closed unit ball of X^*

$$\|x^*f - \textstyle\sum_{1\le j\le n}\xi_j x^*f_j\|_1 \le \varepsilon$$

which shows that $\sum_n \xi_n f_n$ converges to f in $\mathcal{D}_1(\mu,X)$.

In order to show that $\mathcal{P}_1(\mu,X)$ satisfies property (7) it suffices to note that $\mathcal{P}_1(\mu,X)$ is a closed subspace of $\mathcal{D}_1(\mu,X)$, hence, reasoning as before with $f_n \in \mathcal{P}_1(\mu,X)$, we have that $f = \sum_n \xi_n f_n \in \mathcal{P}_1(\mu,X)$. ■

Theorem 10.6.3 *Let* X *be a Banach space. Then* $P_1(\mu,X)$ *is a udb space if and only if it is TB.*

Proof. If $P_1(\mu,X)$ is a *udb* space, it is obviously TB. Assume $P_1(\mu,X)$ is TB and let $\{W_n,\ n \in \mathbb{N}\}$ be a sequence of closed absolutely convex sets of $P_1(\mu,X)$ covering $P_1(\mu,X)$. We are going to show in the first place that there is some $i \in \mathbb{N}$ such that W_i absorbs to the family

$$\left\{e(A)x/\{\mu(A)\},\ \|x\| = 1,\ A \in \Sigma,\ \mu(A) \ne 0\right\}.$$

Define the following closed absolutely convex subsets of X

$$V_{nm} = \left\{x \in X : e(A)x/\{\mu(A)\} \in mW_n \text{ for each } A \in \Sigma \text{ with } \mu(A) \ne 0\right\}$$

for all $n,\ m \in \mathbb{N}$.

Given $z \in X$, $z \neq 0$, then $L(z) := \{f(\cdot)z : f \in L^1(\mu)\}$ is a subspace of $P_1(\mu,X)$ (linearly) isometric to $L^1(\mu)$, and hence there are $r, s \in \mathbb{N}$ such that $e(A)z/\{\mu(A)\} \in sW_r$ for each $A \in \Sigma$ with $\mu(A) \neq 0$. This implies that $z \in V_{rs}$ and consequently $\bigcup\{V_{nm} : n, m \in \mathbb{N}\} = X$. Since X is a Baire space there are i, j, $k \in \mathbb{N}$ such that kV_{ij} contains the unit sphere of X. Hence, $e(A)x/\{\mu(A)\} \in jkW_i$ $\forall x \in X$ with $\|x\| = 1$ and $\forall A \in \Sigma$ such that $\mu(A) \neq 0$. Consequently $e(A)x \in \langle W_i \rangle$ for each $x \in X$ and each $A \in \Sigma$, i.e. $S(\mu,X) \subseteq \langle W_i \rangle$. Given that $S(\mu,X)$ is a dense linear subspace of $P_1(\mu,X)$ according to Lemma 10.4.3, it follows that $\langle W_i \rangle$ is dense in $P_1(\mu,X)$.

Since we are supposing that $P_1(\mu,X)$ is totally barrelled, we may assume that each subspace $\langle W_j \rangle$ is barrelled, which implies that $\langle W_i \rangle$ is a closed subspace of $P_1(\mu,X)$. Therefore $\langle W_i \rangle$, being closed and dense in $P_1(\mu,X)$, must coincide with $P_1(\mu,X)$, and hence W_i is a neighbourhood of the origin in $P_1(\mu,X)$. So, $P_1(\mu,X)$ is a *udb* space. ■

10.7 EXERCISES

1. *A Dunford integrable function which is not Pettis integrable.* Let us consider the mapping $f : [0, 1] \longrightarrow c_0$ defined by

$$f(t) = \Big(e((0, 1])(t), 2e((0,1/2])(t), \ldots, ne((0,1/n]), \ldots\Big)$$

for each $t \in [0, 1]$.

(i) $x^*f = \sum_n \xi_n ne((0,1/n]) \in L^1(\mu)$ for all $x^* = (\xi_1, \xi_2, \ldots) \in c_0^* = \ell^1$.

(ii) $\int_0^1 x^*f \, d\mu = \langle x^*, \eta \rangle$ where $\eta = (1, 1, 1, \ldots)$.

(iii) Since $\eta \notin c_0$, f is not Pettis integrable.

2. *The space* $K(\mu,X)$. Assume Ω is a set, Σ is a σ-algebra of subsets of Ω and X is a Banach space. We shall denote by $ca(\Sigma,X)$ the Banach space of all X-valued countably additive measures defined in Σ, equipped with the semivariation norm $\|F\|(E) = \sup\{\sum_{A\in\pi}|x^*F(A)| : x^* \in X^*, \|x^*\| \leq 1, \pi \in \Pi(E)\}$, for each $E \in \Sigma$, where $\Pi(E)$ represents the family of finite partitions of E by elements of Σ. If μ is a positive countably additive measure on Σ, $K(\mu,X)$ stands for the linear subspace of $ca(\Sigma,X)$ of all μ-continuous measures and $S(\mu,X)$ denotes the subspace of $P_1(\mu,X)$ formed by the simple functions.

(i) $K(\mu,X)$ is a Banach space.

(ii) $T : P_1(\mu,X) \longrightarrow ca(\Sigma,X)$ is a (linear) isometry into its image.

(iii) $T(P_1(\mu,X)) \subseteq K(\mu,X)$. *[Hint: Notice that* $Th \in K(\mu,X)$ *for each* h *of* $S(\mu,X)$ *and use Lemma 10.4.3.]*

3. *The mappings* E_π. We use the same notation as in Exercise 2.

(i) For each $\pi \in \Pi(\Omega)$, define $E_\pi : L^1(\mu) \longrightarrow L^1(\mu)$ by

$$E_\pi = \sum_{A\in\pi} \frac{\int_A f \, d\mu}{\mu(A)} e(A)$$

with the convention $0/0 = 0$. Then E_π is a bounded linear operator such that $E_\pi(L^\infty(\mu)) \subseteq L^\infty(\mu)$ and $\lim_\pi \|E_\pi f - f\| = 0$.

(ii) If $E_\pi{}^*$ denotes the adjoint of E_π, then $E_\pi{}^* = E_\pi|_{L^\infty(\mu)}$.

(iii) Given some fixed $G \in K(\mu,X)$, define $K : \ell^\infty(\Sigma) \longrightarrow X$ by $K(e(E)) = G(E)$ for each $E \in \Sigma$. Then K is a compact linear operator.

(iv) Define $S : L^\infty(\mu) \longrightarrow X$ such that $Sf = Kh$ whenever f is simple and h is any element belonging to the class defined by f. Then S is a compact linear operator and consequently its adjoint S^* is compact.

(v) One has that $S^*(X^*) \subseteq L^1(\mu)$.

(vi) The compactness of S^* implies that $E_\pi S^* \longrightarrow S^*$ in $\mathcal{L}(X^*,L^1(\mu))$ under the uniform convergence topology.

(vii) $SE_\pi \longrightarrow S$ in $\mathcal{L}(L^\infty(\mu),X)$ under the uniform convergence topology.

(viii) For each finite partition π of Ω by members of Σ, define $G_\pi(E) = \sum_{A\in\pi}(G(A)/\mu(A))\cdot\mu(E \cap A)$, $E \in \Sigma$ (agreeing that $0/0 = 0$). Then, we have that $G_\pi \longrightarrow G$ in $ca(\Sigma,X)$. *[Hint: Show that* $\|G_\pi - G\| \le 4\|SE_\pi - S\|$*.]*

4. *The space* $K(\mu,X)$ *is (isometric to) the completion of* $P_1(\mu,X)$. We use the same notation and definitions as in the previous exercises.

Prove that T is of dense range in order to conclude that T is an isometry from the completion of $P_1(\mu,X)$ onto $K(\mu,X)$. *[Hint: Assume that* $G \in K(\mu,X)$ *and define* G_π *as before. Mazur's theorem ensures that* $G_\pi(\Sigma)$ *is relatively compact, and consequently* $G_\pi \in K(\mu,X)$*; hence according to part (viii) of the previous exercise the set* $Q = \{G_\pi : G \in K(\mu,X), \pi \in \Pi(\Omega)\}$ *is dense in* $K(\mu,X)$. *Thus, defining* $f_\pi = \sum_{A\in\pi}(G(A)/\mu(A))\cdot e(A)$, *we* *have that* $Tf_\pi = G_\pi$ *for each* $\pi \in \Pi(\Omega)$. *As each* f_π *is a simple function, it follows that* $Q \subseteq T(P_1(\mu,X))$.]

5. *Uncompleteness of the space* $P_1(\mu,X)$. Let $K_\infty(\mu,X)$ denote the closure in $L^\infty(\mu,X)$ of the subspace $S(\mu,X)$ of the simple functions.

(i) Given $g \in L^\infty(\mu,X)$, $g \in K_\infty(\mu,X)$ if and only if g has an essentially relatively compact range.

(ii) If $g \in K_\infty(\mu,X)$ then the linear operator $T : L^1(\mu) \longrightarrow X$ defined by $Tf = (B)\int_\Omega f\cdot g\, d\mu$ for each $f \in L^1(\mu)$ is compact.

(iii) Let $f : \Omega \longrightarrow X$ be μ-measurable. If f is Dunford integrable and $F(E) := (D)\int_E f\, d\mu$ for each $E \in \Sigma$, then for each $\varepsilon > 0$ there exists some $E_\varepsilon \in \Sigma$ with $\mu(\Omega \setminus E_\varepsilon) < \varepsilon$ such that the set $\{F(E)/\mu(E) : E \in \Sigma,\ E \subseteq E_\varepsilon,\ \mu(E) > 0\}$ is relatively compact in norm. *[Hint: Use (ii) and Egoroff's theorem for* μ*-measurable vector-valued functions.]*

(iv) Let $\Omega = [0, 1]$, μ the Lebesgue measure on Ω and Σ the σ-algebra of all Lebesgue measurable subsets of the interval [0, 1]. Define $F : \Sigma \longrightarrow C[0,1]$ by $F(E)(t) = \mu(E \cap [0,t))$, where C[0,1] denotes the Banach space of real continuous functions defined in [0,1] endowed with the supremum norm. Use Ascoli's lemma to show that $F(\Sigma)$ is a relatively compact set of C[0,1].

(v) Assume there is an $f \in P_1(\mu,C[0,1])$ such that $F(E) = (P)\int_E f\, d\mu$ and conclude that there exists some $E_0 \in \Sigma$ such that $\mu(\Omega \setminus E_0) < 1/2$ and the set $\{F(A)/\mu(A) : A \in \Sigma,\ A \subseteq E_0,\ \mu(A) > 0\}$ is relatively compact in C[0,1].

(vi) Since $\mu(E_0) > 0$ and μ is atomless choose a partition $0 = t_1 < t_1 < \ldots < t_m = 1$ of [0,1] such that $\mu\left([t_{n-1}, t_n) \cap E_0\right) = \mu(E_0)/m$ for $1 \le n \le m$. Setting $E_i := [t_{i-1}, t_i) \cap E_0$, $1 \le i \le m$, show that if $1 \le i < j \le m$, then $\|F(E_i)/\mu(E_i) - F(E_j)/\mu(E_j)\|_\infty \ge 1$, contradicting (iv).

(vii) If $P_1(\mu,C[0,1])$ were complete, according to the previous exercise T would be an isometry from $P_1(\mu,C[0,1])$ onto $K(\mu,C[0,1])$. Thus we must conclude that $P_1(\mu,C[0,1])$ is not a Banach space.

6. *A result of Pelczynski.* Let X be a Banach space. A sequence (x_n) is called weak unconditionally Cauchy if $\sum_n |x^*x_n| < \infty$ for each $x^* \in X^*$. By $\ell^1_{weak}(X)$ we denote the Banach space of all weak unconditionally Cauchy sequences of X, equipped with the norm

$$\|(x_n)\|_1 = \sup \{\sum_n |x^*x_n| \ : \ x^* \in X^*, \ \|x^*\| \le 1\}.$$

The closed subspace of $\ell^1_{weak}(X)$ consisting of all $(x_n) \subseteq X$ such that the series $\sum_n x_n$ is unconditionally convergent will be denoted by $\ell^1\langle X\rangle$.

(i) Let (Ω,Σ,μ) be a finite measure space. If μ is purely atomic, then $P_1(\mu,X)$ is linearly isometric either to X^n for some $n \in \mathbb{N}$ or to $\ell^1\langle X\rangle$, while $D_1(\mu,X)$ coincides isometrically with some X^n or with $\ell^1_{weak}(X)$.

(ii) If X does not contain a copy of c_0, then $\ell^1_{weak}(X) = \ell^1\langle X\rangle$.

(iii) If $\ell^1_{weak}(X) = \ell^1\langle X\rangle$, then X does not contain a copy of c_0.

7. *A theorem of Bartle-Dunford-Schwartz.* Assume Σ is a σ-algebra and X is a Banach space. If $F : \Sigma \longrightarrow X$ is a countably additive vector measure then F has a relatively weakly compact range.

10.8 NOTES AND REMARKS.

For sections 10.1-10.4 our main source has been Diestel and Uhl's book [DU], apart from Theorem 10.2.3, which has been taken from [Hu], and Theorem 10.4.5. For section 10.5 we have followed Drewnowski et al. [DrFP1], and section 10.6, apart from Theorem 10.6.3, is based upon sections 2 and 3 of Díaz et al. [DFFP2], though they work with a σ-finite measure space. Other results concerning some normed spaces of weakly measurable vector-valued functions with (Ω,Σ,μ)-Boolean algebras of projections, where (Ω,Σ,μ) is an arbitrary positive measure space without atoms of finite measure, have been studied in [DrFP3].

Exercises 1, 2, 3 and 4 are taken from [DU], while Exercise 5 is an adaptation of Example 8 in [DU, Ch.III]. Assuming μ is the Lebesgue measure in $[0,1]$, B.J. Pettis [Pe] proved that $\mathcal{P}(\mu,L^2(\mu))$ is not a Banach space. On the other hand, L. Janicka and N.J. Kalton [JK] have shown that if X is an infinite-dimensional Banach space, $\mathcal{P}(\mu,X)$ is not a Banach space whenever μ is atomless. Exercise 6 is a form of a well-known theorem of A. Pelczynski, which the reader can find in [D, p.45]. Exercise 7 is a classic result of vector measure theory which the reader can see for instance in [DU, p.14].

11. A short list of open problems

We conclude our journey through metrizable barrelled spaces by collecting in this concluding chapter a rather whimsical selection of problems, perhaps some of them known, and whose answer we would like to know.

Problem 11.1 *Assume that* X *is an infinite-dimensional normed barrelled space. Under what conditions, if there are any,* X *is ultrabornological whenever* X *does not contain a copy of* ℓ_0^∞*?*

Problem 11.2 *Characterize in familiar terms the largest cardinal* κ *such that any subspace of codimension* $< \kappa$ *in a barrelled space is barrelled.*

Problem 11.3 *Is there a* $\mathfrak{b}$*-dimensional normed barrelled space?*

Problem 11.4 *Is there a normed barrelled space which is not suprabarrelled? Is there a normed barrelled space of class* n *which is not of class* n + 1*. Is there a normed barrelled space which is not barrelled of class* $\aleph_0$*? Is there a normed barrelled space which is not baireled?*

Problem 11.5 *Is there a normed barrelled space of class* $\aleph_0$ *which is not baireled?*

Problem 11.6 *Does each infinite-dimensional Fréchet space contain a dense barrelled subspace of class* n *which is not a barrelled space of class* n + 1*?*

Problem 11.7 *Does each infinite-dimensional Fréchet space contain a dense barrelled subspace of class* $\aleph_0$ *which is not baireled?*

Problem 11.8 *Does each infinite-dimensional Fréchet space contain a dense barrelled subspace of class* $\aleph_0$ *which is not totally barrelled?*

Problem 11.9 *Does each infinite-dimensional Fréchet space contain a dense baireled subspace which is not totally barrelled?*

Problem 11.10 *Is* $\ell_0^\infty(\Sigma)$*,* Σ *being a* σ*-algebra of subsets of* Ω*, baireled?*

Problem 11.11 *Obtain a characterization of all Boolean algebras* $\mathcal{A}$ *for which* $\ell_0^\infty(\mathcal{A})$ *is barrelled, i.e., all Boolean algebras with property (N).*

Problem 11.12 *Is there an algebra of subsets* Σ *with property (N) such that the corresponding space* $\ell_0^\infty(\Sigma)$ *is not suprabarrelled? For* $n \geq 1$, *is there an algebra of subsets* Σ_n *such that* $\ell_0^\infty(\Sigma_n)$ *is barrelled of class* n *but not of class* $n + 1$*?*

Problem 11.13 *Does* $\ell_0^\infty(\Omega,\Sigma)$, Σ *being an algebra with property* (f), *enjoy stronger barrelledness properties than that of being suprabarrelled?*

Problem 11.14 *Assuming* (Ω,Σ,μ) *is an atomless finite measure space and* X *is an infinite-dimensional normed space, is* $L^p(\mu,X)$, *with* $1 \leq p \leq \infty$, *a baireled space? Is it a totally barrelled space? Is it an unordered Baire-like space?*

Problem 11.15 *Assume that* (Ω,Σ,μ) *is a finite measure space and* X *is an infinite-dimensional normed space. Is* $L^p(\mu,X)$, $1 \leq p \leq \infty$, *baireled (totally barrelled) whenever* X *is baireled (totally barrelled)?*

Problem 11.16 *Assuming that* Σ *is a* σ*-algebra of subsets of a set* Ω *and* X *is an infinite-dimensional normed space, is* $B(\Sigma,X)$ *unordered Baire like (totally barrelled) whenever* X *is unordered Baire-like (totally barrelled)?*

Problem 11.17 *Assuming* Σ *is a* σ*-algebra of subsets of a set and* X *is an infinite-dimensional normed space, if it were possible, find conditions on* Σ *and* X *in order that bvca*(Σ,X) *be ultrabornological (barrelled of class* $n \geq 1$*) whenever* X *is ultrabornological (barrelled of class* n*).*

Problem 11.18 *Assuming* Σ *is an algebra of sets and* X *is an infinite-dimensional normed space, is* $B(\Sigma,X)$ *barrelled of class* $n \geq 1$ *(barrelled of class* $\aleph_0$*) if and only* X *is barrelled of class* $n \geq 1$ *(barrelled of class* $\aleph_0$*)?*

Problem 11.19 *Assuming* (Ω,Σ,μ) *is a finite measure space and* X *is a Banach space, is* $P_1(\mu,X)$ *a barrelled space of class* $\aleph_0$*? Is it baireled?*

Problem 11.20. *Assuming* (Ω,Σ,μ) *is a finite measure space and* X *is a Banach space, is* $P_1(\mu,X)$ *totally barrelled?*

Remark 11.21 Let us mention that Problem 11.10 has just been solved recently in the affirmative by M. López Pellicer [Lo].

References

[AEK] Adasch, N., Ernst, B. and Keim, D., *Topological vector spaces. The theory without convexity conditions.* Lect. Notes in Math. 639. Springer-Verlag 1978.

[A] Amemiya, I., *Some examples of (F) and (DF) spaces.* Proc. Japan. Acad. 33, 169-171 (1957).

[AK] Amemiya, I. and Kōmura, Y., *Über nicht vollständige Montelräume* Math. Ann. 177, 273-277 (1968).

[Ar1] Arias de Reyna, J., *Dense hyperplanes of first category.* Math. Ann. 249, 111-114 (1980).

[Ar2] Arias de Reyna, J., $\ell_0^\infty(\Sigma)$ *no es totalmente tonelado.* Rev. Real Acad. Ciencias Madrid 79, 77-78 (1980).

[BDV] Batt, J., Dierolf, P. and Voigt, J., *Summable sequences and topological properties of* $m_0(I)$. Arch. Math. 28, 86-90 (1977).

[BK1] Bennet, G. and Kalton, N., (FK)-*spaces containing* c_0. Duke Math. J. 39, 561-582 (1972).

[BK2] Bennet, G. and Kalton, N., *Addendum to* '(FK)-*spaces containing* c_0'. Duke Math. J. 39, 819-821 (1972).

[BK3] Bennet, G. and Kalton, N., *Inclusion theorems for k-spaces.* Canad. J. Math. 25, 511-524 (1973).

[BM] Bierstedt, K.D. and Meise, R., *Distinguished echelon spaces and the projective description of weighted inductive limits of type* $\mathcal{V}$C(X). Aspects of Mathematics and its Applications, 169-226. North Holland Math. Library. North Holland 1986.

[BB] Bierstedt, K.D. and Bonet, J., *Stefan Heinrich's density condition for Fréchet spaces and the characterization of the distinguished Köthe echelon spaces.* Mat. Nach. 135, 149-180 (1988).

[BP1] Bonet, J. and Pérez Carreras, P., *On the three-space-problem for certain classes of Baire-like spaces.* Bull. Soc. R. Sci. Liège 51, 381-385 (1982).

[BP2] Bonet, J. and Pérez Carreras, P., *Remarks on the stability of barrelled-type properties.* Bull. Soc. R. Sci. Liège 52, 313-318 (1983).

[B] Bourbaki, N., *Elements de mathématiques. Libre V.* Espaces vectoriels topologiques. Paris. Hermann 1964.

[BT] Burke, M.R. and Todorcevic, S., *Bounded sets in topological vector spaces.* Submitted.

[CT] Catalán, F.X. and Tweddle, I., *Countable enlargements of norm topologies and the quasidistinguished property.* Glasgow Math. J. 35, 235-238 (1993).

[DGR] De Guzmán, M. y Rubio, B., *Integración: Teoría y técnicas.* Madrid 1979.

[DW] De Wilde, M., *Closed graph theorems and webbed spaces.* Research Notes in Math. 19. Pitman, London 1978.

[DWH] De Wilde, M. and Houet, C., *On increasing sequences of absolutely convex sets in locally convex spaces.* Math. Ann. 192, 257-261 (1971).

[DF] Defant, A. and Floret, K., *Tensor norms and operator ideals.* North Holland Math. Studies 176. North Holland 1993.

[DFP] Díaz, J.C., Florencio, M. and Paul, P.J., *A uniform boundedness theorem for* $L^{\infty}(\mu,E)$. Arch. Math. 60, 73-78 (1993).

[DLR] Díaz, J.C., López Molina, J.A. and Rivera, M.J., *Distinguished injective tensor products of Fréchet spaces.* Proc. R. Ir. Acad. 92A, 2, 165-174 (1992).

[DM] Díaz, J.C. and Miñarro, M.A., *Distinguished Fréchet spaces and projective tensor products.* Doga Tr. J. Math. 14, 191-208 (1990).

[DDFFP] Díaz, S., Drewnowski, L., Fernández, A., Florencio, M. and Paúl, P., *Barrelledness and bornological conditions on spaces of vector-valued* μ*-simple functions.* Resultate der Math. 21, 289-298 (1992).

[DFFP1] Díaz, S., Fernández, A., Florencio, M. and Paúl, P.J., *An abstract Banach-Steinhaus theorem and applications to function spaces.* Resultate der Math., to appear.

[DFFP2] Díaz, S., Fernández, A., Florencio, M. and Paúl, P.J., *A wide class of ultrabornological spaces of measurables functions.* J. Math. Anal. Appl., to appear.

[De] Dierolf, S., *On spaces of continuous linear mappings between locally convex spaces.* Note di Mat. 5, 147-255 (1985).

[DDD] Dierolf, P., Dierolf, S. and Drewnowski, L., *Remarks and examples concerning unordered Baire-like and ultrabarrelled spaces.* Colloq. Math. 39, 1, 109-116 (1978).

[DK] Dierolf, S. and Kakol, J., *On S-barrelled spaces.* Result. in Math., to appear.

[D] Diestel, J., *Sequences and series in Banach spaces.* Springer-Verlag 1984.

[DF] Diestel, J. and Faires, B., *On vector measures.* Trans. Amer. Math. Soc. 198, 253-271 (1974).

[DU] Diestel, J. and Uhl, J.J., *Vector measures.* Math. Surveys 15. Amer. Math. Soc. Providence 1977.

[Di] Dieudonné, J., *Bounded sets in* (F) *spaces.* Proc. Amer. Math. Soc. 6, 729-731 (1955).

[DiS] Dieudonné, J. and Schwartz, L., *La dualité dans les espaces* (F) *et* (DF). Ann. Inst. Fourier 1, 61-101 (1949).

[Dr1] Drewnowski, L., *Un Théorème sur les opérateurs de* $\ell_{\infty}(\Gamma)$. C. R. Acad. Sc. Paris 281 (Serie A), 967-969 (1975).

[Dr2] Drewnowski, L., *An extension of a theorem of Rosenthal on operators acting from* $\ell_{\infty}(\Gamma)$. Studia Math. 57, 209-215 (1976).

[DrFP1] Drewnowski, L., Florencio, M. and Paúl, P.J., *The space of Pettis integrable functions is barrelled.* Proc. Amer. Math. Soc. 114, 687-694 (1992).

[DrFP2] Drewnowski, L., Florencio, M. and Paúl, P.J., *Barrelled function spaces.* Progress in Functional Analysis. K.D. Bierstedt, J. Bonet, J. Horváth & M. Maestre (Eds.), Notas de Mat. 170, North Holland, Amsterdam, London, New York, Tokyo 1992, 191-199.

[DrFP3] Drewnowski, L., Florencio, M. and Paúl, P.J., *Uniform boundedness of operators and barrelledness in spaces with Boolean algebras of projections.* Atti. Sem. Mat. Fis. Univ. Modena 41, 317-329 (1993).

[DrFP4] Drewnowski, L., Florencio, M. and Paúl, P.J., *Barrelled subspaces of spaces with subseries decompositions or Boolean rings of projections.* Glasgow Math. J. 36, 57-69 (1994).

[DrFP5] Drewnowski, L., Florencio, M. and Paúl, P.J., *On the barrelledness of spaces of bounded vector functions.* Arch. Math., to appear.

[DS] Dunford, N. and Schwartz, J.T., *Linear Operators*, Part I. Interscience 1958.

[E] Eidelheit, M., *Zur Theorie der Systeme linearer Gleichungen.* Studia Math. 6, 139-148 (1936).

[F] Faires, B.T., *On Vitali-Hahn-Saks-Nikodym type theorems.* Ann. Inst. Fourier 21, 3-13 (1971).

[Fe1] Ferrando, J.C., *Subespacios totalmente tonelados en espacios de Banach con una base.* Rev. Real Acad. Cien. Madrid 86, 211-216 (1992).

[Fe2] Ferrando, J.C., *A projective description of the simple scalar function space.* Rev. Real Acad. Cien. Madrid 86, 231-236 (1992).

[Fe3] Ferrando, J.C., *On the barrelledness of the space* $L_\infty(\mu,X)$. Bull. Soc. Math. Belg. 44 (Serie B), 165-169 (1992).

[Fe4] Ferrando, J.C., *Ultrabornological Bochner integrable function spaces.* Bull. Austral. Math. Soc. 47, 119-126 (1993).

[Fe5] Ferrando, J.C., *On the barrelledness of the vector-valued bounded function space.* J. Math. Anal. Appl. 184, 437-440 (1994).

[Fe6] Ferrando, J.C., *Strong barrelledness properties in certain* $\ell_0^\infty(\mathcal{A})$ *spaces.* J. Math. Anal. Appl. 190, 194-202 (1995).

[Fe7] Ferrando, J.C., *Unordered Baire-like vector-valued function spaces.* Bull. Soc.Math. Belg, to appear.

[Fe8] Ferrando, J.C., *Barrelledness properties of the vector-valued bounded function space.* In preparation.

[FF] Ferrando, J.C. and Ferrer, J., *On certain non-barrelled normed spaces.* Math. Japonica 38, 161-164 (1993).

[FFL] Ferrando, J.C., Ferrer, J. and López-Pellicer, M., *Strong barrelledness properties in* $L_p(\mu,X)$. Bull. Belg. Math. Soc. 1, 73-78 (1994).

[FL1] Ferrando, J.C. and López Pellicer, M., *On ordered suprabarrelled spaces.* Arch. Math. 53, 405-410 (1989).

[FL2] Ferrando, J.C. and López Pellicer, M., *Quasi-suprabarrelled spaces.* J. Austral. Math. Soc. (Series A) 46, 137-145 (1989).

[FL3] Ferrando, J.C. and López Pellicer, M., *Strong barrelledness properties in* $\ell_0^\infty(X,\mathcal{A})$ *and bounded finite additive measures.* Math. Ann. 287, 727-736 (1990).

[FL4] Ferrando, J.C. and López Pellicer, M., *Una propiedad de M. Valdivia, espacios de tipo* (L) *y condiciones débiles de tonelación.* Rev. Real Acad. Cien. Madrid 83, 25-34 (1990).

[FL5] Ferrando, J.C. and López Pellicer, M., *An ordered suprabarrelled space.* J. Austral. Math. Soc. (Series A) 51, 106-117 (1991).

[FL6] Ferrando, J.C. and López Pellicer, M., *Barrelled spaces of class n and of class* $\aleph_0$. Disert. Sem. Mat. Fund. UNED, 4 (1992).

[FL7] Ferrando, J.C. and López Pellicer, M., *Vector valued function spaces which are barrelled of class* $\aleph_0$ *but not totally barrelled.* Math. Japonica 37, 117-121 (1992).

[FL8] Ferrando, J.C. and López Pellicer, M., *A note on a theorem of J. Diestel and B. Faires.* Proc. Amer. Math. Soc. 115, 1077-1081 (1992).

[FL9] Ferrando, J.C. and López Pellicer, M., *Barrelled spaces of class* $\aleph_0$ *which are not totally barrelled.* Math. Japonica 37, 465-471 (1992). Correction, 38, 1197-1199 (1993).

[FM] Ferrando, J.C. and Más, J., *Quasi-totally barrelled spaces.* Portugaliae Math. 47, 2, 205-213 (1990).

[FS1] Ferrando, J.C. and Sánchez Ruiz, L.M., *On sequential barrelledness.* Arch. Math. 57, 597-605 (1991).

[FS2] Ferrando, J.C. and Sánchez Ruiz, L.M., *A maximal class of spaces with strong barrelledness conditions.* Proc. R. Ir. Acad. 92A, No. 1, 69-75 (1992).

[FS3] Ferrando, J.C. and Sánchez Ruiz, L.M., *On a barrelled space of class* $\aleph_0$ *and measure theory.* Math. Scand. 73, 41-48 (1993).

[FS4] Ferrando, J.C. and Sánchez Ruiz, L.M., *Countably quasi-suprabarrelled spaces.* Bull. Austral. Math. Soc. 48, 1-6 (1993).

[FS5] Ferrando, J.C. and Sánchez Ruiz, L.M., *Strong barrelledness properties in* $L_\infty(\mu,X)$. Math. Scand. 73, 41-48 (1993).

[FS6] Ferrando, J.C. and Sánchez Ruiz, L.M., *A note on vector-valued function spaces.* Topological Vector Spaces, Algebras & Related Areas (Eds., A. Lau and I. Tweddle), 20-23. Pitman Research in Math. Series 316. Longman 1994.

[FS7] Ferrando, J.C. and Sánchez Ruiz, L.M., *Strong barrelledness properties in* B(Σ,X). Bull. Austral. Math. Soc., to appear.

[FL] Ferrer, J. and López Pellicer, M. *Algebras of sets generating non-barrelled spaces.* Periodica Math. Ungarica 29(3), 207-211 (1994).

[FLS1] Ferrer, J.R., López Pellicer, M. and Sánchez Ruiz, L.M., *A strongly barrelled space.* Math. Japonica 39, 1, 89-94 (1993).

[FLS2] Ferrer, J.R., López Pellicer, M. and Sánchez Ruiz, L.M., *Some* $\mathcal{L}$*-barrelled subspaces in Banach spaces.* Math. Japonica 40, 1, 93-97 (1994).

[FP] Florencio, M. and Paúl, P.J., *Barrelledness conditions on certain vector valued sequence spaces.* Arch. Math. 48, 153-164 (1987).

[Fr1] Freniche, F.J., *Barrelledness conditions of the space of vector valued and simple functions.* Math. Ann. 267, 479-489 (1984).

[Fr2] Freniche, F.J., *The Vitali-Hahn-Saks theorem for Boolean algebras with the subsequential interpolation property.* Proc. Amer. Math. Soc. 92, 362-366 (1984).

[Ga] Gassó, M.T., *Algebras with the local interpolation property.* Rocky J. Math. 22, 181-196 (1992).

[GJ] Gillman, L. and Jerison, M., *Rings of continuous functions.* Van Nostrand-Reinhold Comp. 1960.

[GW] Graves, W.H. and Weeler, R.F., *The Grothendieck and Nikodym properties.* Rocky J. Math. 13, 333-353 (1983).

[G1] Grothendieck, A., *Sur les applications lineaires faiblement compactes d'espaces du type* C(K). Canad. J. Math. 5, 129-173 (1953).

[G2] Grothendiek, A., *Sur les espaces* (F) *et* (DF). Summa Brasil Math. 3, 57-123 (1954).

[G3] Grothendieck, A., *Produits tensoriels topologiques et espaces nucléaires.* Mem. Amer. Math. Soc. 16 (1955).

[H] Horváth, J., *Topological vector spaces and distributions* I. Addison-Wesley. Reading, Massachusetts 1966.

[Hu] Huff, R., *Remarks on Pettis integrability.* Proc. Amer. Math. Soc. 96, 402-404 (1986).

[Hus] Husain, T. *Two new classes of locally convex spaces.* Math. Ann. 166, 289-299 (1966).

[JK] Janicka, L. and Kalton, N.J., *Vector measures of infinite variation.* Bull. Acad. Polon. Sci. Sér. Sci. Math. Astronom. Phys. 25, 239-241 (1977).

[J] Jarchow, H., *Locally convex spaces.* B.G. Teubner Stuttgart 1981.

[Ka1] Kakol, J., *Topological linear spaces with some Baire-like properties.* Functiones et Approx. 13, 109-116 (1982).

[Ka2] Kakol, J., *Unordered Baire-like spaces without local convexity.* Collect. Math. 43, 43-53 (1992).

[KR1] Kakol, J., Roelcke, W., *Topological vector spaces with some Baire-type properties.* Note di Mat. 12, 77-92 (1992).

[KR2] Kakol, J., Roelcke, W., *On the barrelledness of* ℓ_p*-direct sums of seminormed spaces for* $1 \le p \le \infty$. Arch. Math. (Basel) 62, 331-334 (1994).

[KSW] Kakol, J., Sliwa, W. and Wójtowicz, M., CS-*barrelled spaces.* Preprint.

[Ka] Katekov, M., *Remarks on characters and pseudocharacters.* Comm. Mat. Univ. Car. 1, 20-25 (1960).

[KW] Klee, V. and Wilansky, A., *Research problems no. 13.* Bull. Amer. Math. Soc. 72, 656-656 (1966).

[K1] Köthe, G., *Topological vector spaces* I. Springer-Verlag 1969.

[K2] Köthe, G., *Topological vector spaces* II. Springer-Verlag 1979

[Ku] Kunzinger, M., *Barrelledness, Baire-like and* (LF)*-spaces.* Pitman Research Notes in Math. Series 298. Longman, 1993.

[La] Labuda, I., *Exhaustive measures in arbitrary topological vector spaces.* Studia Math. 58, 239-248 (1976).

[L] Lacey, E., *Separable quotients of Banach spaces.* An. Acad. Brasil. Cienc. 44, 185-189 (1972).

[LS] Levin, M. and Saxon, S.A., *A note in the inheritance of properties of locally convex spaces by subspaces of countable codimension.* Proc. Amer. Math. Soc. 29, 1, 97-102 (1971).

[LR] López Molina, J.A. and Rivera, M.J., *Distinguished* α_{pq}*-tensor products of Fréchet spaces.* Rend. Circ. Mat.Palermo 33, 363-374 (1993).

[Lo] López Pellicer, M. *Webs and finite additive measures.* Preprint.

[MS] Más Marí, J. and Sánchez Ruiz, L.M., *On* $\mathcal{L}$-(DF) *spaces.* Bull. Soc. Math. Belg. 43 (Serie B), 31-36 (1991).

[Mc] McArthur, C.W., *On a theorem of Orlicz and Pettis.* Pacific J. Math. 22, 297-302 (1967).

[Me1] Mendoza, J., *Barrelledness conditions on* S(Σ,E) *and* B(Σ,E). Math. Ann. 261, 11-22 (1982).

[Me2] Mendoza, J., *A barrelledness criterion for* $c_0(E)$. Arch. Math. 40, 156-158 (1983).

[Mo1] Moltó, A., *On the Vitali-Hahn-Saks theorem.* Proc. Roy. Soc. Edinburgh 90A, 163-173 (1981).

[Mo2] Moltó, A., *On uniform boundedness properties in exhausting additive set function spaces.* Proc. Roy. Soc. Edinburgh 90A, 175-184 (1981).

[NS] Narayanaswami, P.P and Saxon, S.A., (LF)*-spaces, quasi-Baire spaces and the strongest locally convex topology.* Math. Ann. 274, 627-641 (1986).

[NoS] Noll, D. and Stadler, W., *Abstract sliding hump technique and characterization of barrelled spaces.* Studia Math. 94, 103-120 (1989).

[P] Pérez Carreras, P., *Sobre ciertas clases de espacios vectoriales topológicos.* Rev. Re. Ac. Ci. Madrid 76, 585-594 (1982).

[PB1] Pérez Carreras, P. and Bonet, J., *Remarks and examples concerning suprabarrelled and totally barrelled spaces.* Arch. Math. 34, 340-347 (1982).

[PB2] Pérez Carreras, P. and Bonet, J., *Barrelled locally convex spaces.* Math. Studies 131. North-Holland 1987.

[Pe] Pettis, B.J., *On integration in vector spaces.* Trans. Amer. Math. Soc. 44, 277-304 (1938).

[Pi] Pietsch, A., *Nuclear locally convex spaces.* Ergebnisse der Mathematik und ihrer Grenzgebiete 66. Springer 1972.

[PT] Popoola, J.O. and Tweddle, I., *On the dimension of a complete metrizable topological vector space.* Canad. Math. Bull. 20, 271-272 (1977).

[Pt] Pták, V., *Completeness and the open mapping theorem.* Bull. Soc. Math. France 86, 41-74 (1958).

[RR] Robertson, A.P. and Robertson, W.J., *Topological vector spaces.* Cambridge University Press 1973.

[R1] Robertson, W., *Completions of topological vector spaces.* Proc. London Math. Soc. 8, 242-257 (1958).

[R2] Robertson, W.J., *On properly separable quotients of strict* (LF) *spaces.* J. Austral. Math. Soc. (Series A) 47, 307-312 (1989).

[RSR] Robertson, W.J., Saxon, S.A. and Robertson, A.P., *Barrelled spaces and dense vector subspaces.* Bull. Austral. Math. Soc. 37, 383-388 (1988).

[RTY] Robertson, W.J., Tweddle, I. and Yeomans, F.E., *On the stability of barrelled topologies* III. Bull. Austral. Math. Soc. 22, 99-112 (1980).

[Ro1] Rodríguez Salinas, B., *Sobre el teorema de la gráfica cerrada. Aplicaciones lineales subcontinuas.* Rev. Real Acad Cien. Madrid 74, 5, 811-825 (1980).

[Ro2] Rodríguez Salinas, B., *Sobre la clase del espacio tonelado* $\ell_0^\infty(\Sigma)$. Rev. Real Acad. Cien. Madrid 74, 5, 827-829 (1980).

[RD] Roelcke, W. and Dierolf, S., *On the three-space-problem for topological vector spaces.* Collect. Math. 32, 1, 3-25 (1981).

[Rs1] Rosenthal, H.P., *Quasicomplemented subspaces of Banach spaces.* J. Funct. Anal. 4, 176-214 (1969).

[Rs2] Rosenthal, H.P., *On relatively disjoint families of measures, with some applications to Banach space theory.* Studia Math. 37, 13-36 (1970).

[Rt] Rothberger, F., *Sur un ensembre toujours de première categorie qui est depourvu de la propriété λ.* Fund. Mat. 32, 294-300 (1939).

[Ru] Ruess, W., *Generalized inductive limit topologies and barrelledness properties.* Pacific J. Math. 63, 2, 499-516 (1976).

[Sa1] Sánchez Ruiz, L.M., *Some examples on* $\mathcal{L}$*-barrelled spaces.* Bull. Soc. Math. Belg. 43 (Serie B), 211-215 (1991).

[Sa2] Sánchez Ruiz, L.M., *On the Banach-Steinhaus theorem between topological vector spaces and locally convex spaces.* Math. Japonica 36, 143-145 (1991).

[Sa3] Sánchez Ruiz, L.M., *On the closed graph theorem between topological vector spaces and Fréchet spaces.* Math. Japonica 36, 271-275 (1991).

[Sa4] Sánchez Ruiz, L.M., *On sequential barrelledness without locally convex conditions.* Bol. Soc. Mat. Mexicana 36, 29-33 (1991).

[Sa5] Sánchez Ruiz, L.M., *Two maximal classes of topological vector spaces.* Math. Japonica 36, 643-646 (1991).

[Sa6] Sánchez Ruiz, L.M., *On completeness and the closed graph theorem.* Math. Japonica 36, 891-894 (1991).

[Sa7] Sánchez Ruiz, L.M., *The closed graph theorem for locally bounded linear mappings between topological vector spaces and Fréchet spaces.* Math. Japonica 37, 877-881 (1992).

[Sa8] Sánchez Ruiz, L.M., *A countable number of $\mathcal{L}$-barrelled spaces.* Bull. Belg. Math. Soc. 45 Ser. B, 1, 59-67 (1993).

[S1] Saxon, S.A., *Nuclear and product spaces, Baire-like spaces and the strongest locally convex topology.* Math. Ann. 197, 87-106 (1972).

[S2] Saxon, S.A., *Two characterizations of linear Baire spaces.* Proc. Amer. Math. Soc. 45, 2, 204-208 (1974).

[S3] Saxon, S.A., *Some normed barrelled spaces which are not Baire.* Math. Ann. 209, 153-160 (1974).

[S4] Saxon, S.A., *Review of the book 'Barrelled locally convex spaces'.* Bull. Amer. Math. Soc. (New Series) 24, 2, 424-435 (1991).

[S5] Saxon, S.A., *Non-Baire hyperplanes in non-separable Baire spaces.* J. Math. Anal. Appl. 168, 2, 460-468 (1992).

[S6] Saxon, S.A., *The codensity character of topological vector spaces.* Topological Vector Spaces, Algebras & Related Areas (Eds.: A. Lau and I. Tweddle), 24-36. Pitman Research in Math. Series 316. Longman 1994.

[S7] Saxon, S.A., *Metrizable barrelled countable enlargements.* Preprint.

[SL] Saxon, S.A. and Levin M., *Every countable-codimensional subspace of a barrelled space is barrelled.* Proc. Amer. Math. Soc. 29, 91-96 (1971).

[SN1] Saxon, S.A. and Narayanaswami, P.P., *Metrizable (LF)-spaces, (db)-spaces and the separable quotient problem.* Bull. Austral. Math. Soc. 23, 65-80 (1981).

[SN2] Saxon, S.A. and Narayanaswami, P.P., *Metrizable [normable] (LF)-spaces and two classical problems in Fréchet [Banach] spaces.* Studia Math. 93, 1-16 (1989).

[SS1] Saxon, S.A. and Sánchez Ruiz, L.M., *Optimal cardinals for metrizable barrelled spaces.* J. London Math. Soc. 51, 137-147 (1995).

[SS2] Saxon, S.A. and Sánchez Ruiz, L.M., *Barrelled countable enlargements and the bounding cardinal.* J. London Math. Soc., to appear.

[SS3] Saxon, S.A. and Sánchez Ruiz, L.M., *Barrelled countable enlargements and the dominating cardinal.* Preprint.

[SS4] Saxon, S.A. and Sánchez Ruiz, L.M., *Quasidistinguished countable enlargements of normed spaces.* Glasgow Math. J., to appear.

[SS5] Saxon, S.A. and Sánchez Ruiz, L.M., *Dual locally complete spaces and spaces with property* (L). In preparation.

[SS6] Saxon, S.A. and Sánchez Ruiz, L.M., *Reinventing weak barrelledness*. In preparation.

[ST] Saxon, S.A. and Tweddle, I., *The fit and flat components of barrelled spaces.* Bull. Austral. Math. Soc., to appear.

[SW] Saxon, S.A. and Wilansky, A., *The equivalence of some Banach spaces.* Colloq. Math. 37, 217-226 (1977).

[Sc] Schachermayer, W., *On some classical measure-theoretic theorems for non σ-complete Boolean algebras.* Dissertations Math. Rozprawy Mat. 214, 1-36 (1982).

[Sh] Schaefer, H.H., *Topological vector spaces.* Springer Varlag 1986.

[Sm] Schmets, J., *Spaces of vector-valued continuous functions.* Lect. Notes in Math. 1003. Springer-Verlag 1983.

[Se] Seever, G.L., *Measures on F-spaces.* Trans. Amer. Math. Soc. 133, 267-280 (1968).

[Sd] Semadeni, Z., *Banach spaces of continuous functions* I. Polish Scientific Publishers, Warszawa 1971.

[Si] Sikorski, R., *Boolean Algebras.* Springer, Berlin 1964.

[TS] Todd, A.R. and Saxon, S., *A property of locally convex Baire spaces.* Math. Ann. 206, 23-34 (1973).

[Ts] Tsirulnikov, B., *On the locally convex noncomplete quasi-distinguished spaces.* Bull. Soc. Roy. Sci. Liège 47, 147-152 (1978).

[Tw] Tweddle, I., *Barrelled spaces whose bounded sets have at most countable dimension.* J. London Math. Soc. (2) 29, 276-282 (1984).

[TC] Tweddle, I. and Catalán, F.X., *Countable enlargements and weak barrelledness conditions.* Topological Vector Spaces, Algebras & Related Areas (Eds., A. Lau and I. Tweddle), 56-65. Pitman Research in Math. Series 316. Longman 1994.

[TSS] Tweddle, S.A., Saxon, S.A. and Sánchez Ruiz, L.M., *Barrelled countable enlargements.* Topological Vector Spaces, Algebras & Related Areas (Eds., A. Lau and I. Tweddle), 3-15. Pitman Research in Math. Series 316. Longman 1994.

[TY] Tweddle, S.A. and Yeomans, F.E., *On the stability of barrelled topologies* II. Glasgow Math. J. 21, 91-95 (1980).

[V1] Valdivia, M., *Absolutely convex sets in barrelled spaces.* Ann. Inst. Fourier 21, 3-13 (1971).

[V2] Valdivia, M., *A class of bornological barrelled spaces which are not ultrabornological.* Math. Ann. 194, 43-51 (1971).

[V3] Valdivia, M., *Sobre el teorema de la gráfica cerrada.* Collect. Math. 22, 51-72 (1971).

[V4] Valdivia, M., *Mackey convergence and the closed graph theorem.* Arch. Math. 25, 649-656 (1974).

[V5] Valdivia, M., *On certain barrelled normed spaces.* Ann. Inst. Fourier 29, 39-56 (1979).

[V6] Valdivia, M., *On Baire-hyperplane spaces.* Proc. Edinburgh Math. Soc. 22, 247-255 (1979).

[V7] Valdivia, M., *On suprabarrelled spaces.* Lect. Not. in Math. 843. Func. Anal. Hol. and Approx. Theory, 572-580. Springer 1981.

[V8] Valdivia, M., *Classes of barrelled spaces related with the closed graph theoerem.* Portugaliae Math. 40, 3, 345-366 (1981).

[V9] Valdivia, M., *Topics in locally convex spaces.* Math. Studies 67. North-Holland 1982.

[V10] Valdivia, M., *A class of locally convex spaces without C-webs.* Ann. Inst. Fourier 32, 2, 261-269 (1982).

[V11] Valdivia, M., *A property of Fréchet spaces.* Func. Anal. Hol. and Approx. Theory II (ed. G.I. Zapata), 469-477. North Holland 1984. Func. Anal. Hol. and Approx. Theory, 572-580. Springer 1981.

[V12] Valdivia, M., *Products of Baire topological vector spaces.* Fund. Math. 125, 61-80 (1985).

[V13] Valdivia, M., *Three-space problems and projective tensor products of locally convex spaces.* Proc. R. Ir. Acad., 86A, 2, 127-141 (1986).

[VP1] Valdivia, M. and Pérez Carreras, P., *On totally barrelled spaces.* Math. Z. 178, 263-269 (1981).

[VP2] Valdivia, M. and Pérez Carreras, P., *Sobre espacios* (LF) *metrizables.* Collect. Math. 33, 297-303 (1982).

[vD] van Dowen, E.K., *The integers and topology, Handbook of set-theoretic topology* (Eds. K. Kunen and J.E. Vaughan), 111-168. North Holland 1984.

[W] Webb, J.H., *Sequential convergence in locally convex spaces.* Proc. Cambridge Phil. Soc. 64, 341-364 (1986).